BASIC MATHEMATICAL CONCEPTS

Second Edition

BASIC MATHEMATICAL CONCEPTS

F. Lynwood Wren

Professor of Mathematics
California State University, Northridge

McGRAW-HILL BOOK COMPANY

New York   St. Louis   San Francisco   Düsseldorf   Johannesburg
Kuala Lumpur   London   Mexico   Montreal   New Delhi
Panama   Rio de Janeiro   Singapore   Sydney   Toronto

BASIC MATHEMATICAL CONCEPTS

1 2 3 4 5 6 7 8 9 0   KPKP   7 9 8 7 6 5 4 3

Library of Congress Cataloging in Publication Data

Wren, Frank Lynwood, 1894–
    Basic mathematical concepts.

    Bibliography: p.
    1. Arithmetic—Foundations.   2. Geometry—
Foundations.   I. Title.
QA248.3.W7   1973   513′.133   72-10154
ISBN 0-07-071907-1

This book was set in Vega Light by York Graphic Services, Inc.
The editors were Jack L. Farnsworth and Laura Warner;
the designer was J. E. O'Connor;
and the production supervisor was John A. Sabella.
The drawings were done by York Graphic Services, Inc.
The printer and binder was Kingsport Press, Inc.

TO ALLEYNE

CONTENTS

vii

PREFACE TO THE SECOND EDITION

The cordial reception of the first edition presented a challenge to prepare the second edition of this fundamental text for teachers of mathematics in the elementary school. Although the general theme and basic framework of the book have been maintained, there have been major revisions in the presentation.

The first two chapters have been interchanged in order to give a brief historical orientation for the presentation of the content as well as to provide for its smoother sequential development. The chapter on modular arithmetic has been moved forward to its more natural position immediately following the discussion of integers. The first presentations of the division algorithm, the euclidean algorithm, and prime numbers are now made in connection with natural numbers, to provide for more extended and flexible use of these numbers.

Throughout the book there appears starred material, usually of greater difficulty and stronger challenge to the student. Omission of such material will not interfere with the sequential development of the essential content of the book.

There are additional illustrative examples and figures in this edition. The exercises have been increased very materially in number and revised in content.

The treatment of the nature of proof has been completely modified. Each presentation of a proof is made in a parallel-column format. The left column carries through the argument in a specific case, while the general case is developed in the right column; each matches the other step by step, with the supporting reasons in the middle column. Such presentation provides the student not only with a contrast between the two arguments but also with a ready reference to a concrete example to aid in the interpretation of an abstract idea.

The present edition is supplemented by a Practice Book which offers slightly different (but briefer) developmental discussion, additional examples, and many more exercises. In the Practice Book answers are given for all exercises, to provide students with immediate checks on the success of their efforts.

The content of the text can be adapted very readily to a one-quarter (four-unit) course, a one-semester (three-unit) course, or a two-semester (six-unit) course. These suggestions are offered as aids in making such adaptations:

One quarter: Chapters 2 to 4 and 6, omitting starred sections and exercises
One semester: Chapters 1 to 4 and 6, with starred sections and exercises optional

Two semesters: First semester: Chapters 1 to 5; second semester: Chapters 6 to 10. The additional time and credit offer distinct opportunities for advantageous use of the Invitations to Extended Study sections.

The author is grateful to students and colleagues for their suggestions and criticism, which have been most helpful in this revision. He is also indebted to his wife, Alleyne T. Wren, for her aid in reading and typing the manuscript.

F. Lynwood Wren

PREFACE TO THE FIRST EDITION

This book has been written with the hope that it will provide for the person with slight background in mathematics an opportunity to acquire an understanding and appreciation of the basic structure of elementary mathematics. The nature of the number system we use is traced carefully from its beginnings in the natural number system to its full structure in the field of complex numbers. A very significant theme of this development is the important contribution which *place value* makes to the simplicity of our system of numeration and to the efficiency of our computational techniques. Other basic concepts and principles which contribute to the subject-matter content of the book are position, shape, and size; measurement, both direct and indirect; relation and function; and problem solving.

No discussion of the structure of number systems could have any semblance of completeness without some attention to the nature and importance of deduction. The limitations imposed here by the context of minimum prerequisites restrict this development to a brief consideration of the nature of implication. Some of the simpler theorems are proved, while others are left as exercises for the reader to develop. Both the formality of deduction and the informality of intuition play very important roles in shaping the content of this presentation.

A prefatory section, Guidelines for Careful Study, appears at the beginning of each chapter. These guidelines include questions designed to assist the reader in getting the most from the chapter. In the main, they point out key concepts and principles which the discussion of the chapter develops. Some of the questions serve as refresher ties with previous chapters. Each set of guideline questions thus can serve in a dual capacity for the reader: (1) as a partial review of previous chapters and (2) as a preview of the new chapter.

At the end of each chapter there is a section entitled Invitations to Extended Study. Here the reader will find questions and suggestions for pursuing the development of the preceding material in further detail and also, in some cases, to more advanced levels. It is hoped that at least some readers will be challenged to accept some, if not all, such invitations.

Finally, a fairly extensive bibliography of pertinent references follows the final chapter. In these references the reader will find more elaborate treatment of some subjects and, in general, effective support of all topics discussed.

It is the sincere hope of the author that this book will answer, both authentically and satisfactorily, the question of what is meant by the word "structure" when applied to mathematics. Also, it is hoped that at least some readers may be challenged to further study of the basic concepts and principles of elementary mathematics. The principal motivation which has inspired the author from the beginning, however, has been a desire to provide, in language not too highly technical in nature, a discussion which would present mathematics—even arithmetic—as something more than a mere composite of figures and formulas to be used as tools for more efficient puzzle solving, bookkeeping, scorekeeping, and tax computation.

The content of the book, in its present organization, has evolved out of lecture notes gathered over several years of attempting to provide a stronger mathematics background for those aspiring to become teachers in the elementary school. It has been checked carefully against the "Recommendations for Level I" of the Committee on the Undergraduate Program in Mathematics (CUPM). There is substantial treatment of the topics recommended in their preliminary report, *Course Guides for the Training of Teachers of Elementary School Mathematics.*

The author is deeply indebted and sincerely grateful to many individuals for their suggestions and criticisms in the preparation of the manuscript, namely, (1) his students, who have served both as a source of inspiration and as a leaven of guidance; (2) Professors J. Houston Banks, Irving H. Brune, John R. Hatcher, Johnathan W. Lindsay, John W. McGhee, Jr., Maria A. Steinberg, James C. Smith, Warren C. Willig, and Mr. Sidney Sharron, who gave him the benefit of their comments and criticisms after either reading the manuscript or using the preliminary edition as a text in their classes; and (3) certain unknown reviewers for their evaluation and criticism. Finally, the author owes no greater debt of appreciation and gratitude than that due his wife, Alleyne T. Wren, who not only typed the manuscript but also criticized, counseled, and encouraged through the months of writing and rewriting necessary to get it in its final form.

F. Lynwood Wren

SYSTEMS OF NUMERATION

The fundamental theme of this book is the study of number systems and their associated systems of numeration along with some of their uses in the analysis of quantitative concepts and situations. To aid the student in acquiring a better appreciation of our own system of numeration, it will be of value to review briefly a few of those numeration systems which have been especially important in the evolution of our system.

In this chapter brief analyses are made of some of the more significant of the primitive systems of numeration. A careful study of these patterns of notation will reveal that there are certain problems basic to the structure of any truly functional numeration system. In the first place, number names and symbols are necessary before one can operate with numbers or record the results of such operations. To allow for greater flexibility in dealing with numerical concepts there developed the need for a number (*base*) to be used in grouping, a pattern for indicating positional value, and a principle for combining different values.

The decimal system, in which ten is used as a base, seems to have evolved from the convenience of the physiological structure of our hands. It is informative to analyze the struggle to give its use the greatest possible efficiency. Also, the study of numeration systems using bases other than ten provides a context for significant comprehension of the advantages and limitations of the decimal system.

The questions which follow are designed to serve as guidelines for careful study of this chapter. They can be used (1) as a preview of the chapter, (2) as an outline for study, and (3) upon completion of the chapter, as an aid to helpful review.

1 Observe carefully the nature of the number symbolism characteristic of each culture discussed. What base was used? Did it make use of any form of place value? Did it make use of a zero symbol?

2 What are the additive and multiplicative principles which are used in different systems as aids in forming numerals?

3 How readily did each of the numeration systems adapt to computation and other quantitative uses?

4 Did the absence of a concept of place value seem to affect the efficiency of any of the systems?

5 Did the absence of simplicity of form seem to affect the efficiency of any of the systems?

6 What are the essential characteristics of a decimal system of numeration?

7 What are the basic characteristics of our numeration system which contribute to its effectiveness?

8 What is the historical origin of our system of numeration?

9 Does our decimal system of numeration make use of either the additive or the multiplicative principle in its numeral structure?

3

10   What do you think might be some of the reasons why the writer Tobias Dantzig attributed such great significance to the discovery of "the symbol for an empty column"?

11   In what sense does a numeral such as 5,204 represent a pattern of grouping?

12   What is the contrast between the digit value and the place value of each digital symbol in the symbol for any number larger than the base of a numeration system?

13   Does each of the digital symbols 0, 1, 2, 3, 4, 5, 6, 7, 8, 9 have significance both as a number symbol and as a placeholder in symbols for large numbers in a decimal system?

14   Why do symbols such as 10 and 20,456 have no meaning until they are given orientation within the context of a specified base?

15   Under what conditions does 10 not mean ten?

16   What are the basic laws of exponents?

INTRODUCTION

What is the one characteristic of the Hindu-Arabic numeration system in use today that has caused it to supersede other numeration systems? Any informed person would most likely answer "the use of place value in the forming of our number symbols." A study of the numeration systems of some of the more significant of ancient cultures clearly reveals the difficulty in operating with these systems due to the absence of the simplifying contribution of *place value* in the structure of the numerals used.

Place value is of great significance

1-1   THE EGYPTIAN SYSTEM

The approximate date 3500 B.C. is accepted as the beginning of the use of numerals in a definite number system, with possibly the earliest known type being a *simple grouping system*. Such a pattern made use of a selected number as a *base* with repetition and addition providing the means for building each respective numeral. Two of the earliest known such systems were those used by the Babylonians and the Egyptians. The Babylonian system was sexagesimal (base sixty) and used a pattern of wedge-shaped (cuneiform) characters in the formation of numerals, while the Egyptian system was decimal (base ten) and used a pattern of hieroglyphic symbols. Neither system had a zero symbol or made any use of the concept of positional, or place, value.

Simple grouping system

In the Egyptian system distinct symbols represented the different powers of ten (10; 100; 1,000; 10,000; 100,000; and 1,000,000). Figure 1-1 reveals the symbolism used for each of the basic denominations. Multiples of each specific power of ten were symbolized by repetition of the appropriate

Egyptian numerals

OUR NUMBER	HIEROGLYPHIC SYMBOL	OBJECT REPRESENTED
1			vertical staff
10	∩	heel bone
100	៙	scroll
1,000	⚇	lotus flower
10,000	⌒	pointing finger
100,000	⌇	burbot fish
1,000,000	⚌	man in astonishment

FIGURE 1-1 The Egyptian numerals

symbol. The symbol for 40, for example, was 4 of the symbols for ten, the symbol for 500 was 5 of the symbols for one hundred, the symbol for 9,000 was 9 of the symbols for one thousand, and so on.

In composing large numerals the Egyptians used only the principles of repetition and addition. Thus ៙៙∩∩∩ ∩∩ ||| ||| would represent†

100 + 100 + 10 + 10 + 10 + 10 + 10 + 1 + 1 + 1 + 1 + 1 + 1

or two hundreds plus five tens plus six ones; we write this number as 256.

Rhind papyrus The Rhind, or Ahmes, papyrus,‡ a rather comprehensive Egyptian arithmetic written about 1659 B.C. and representing the highest arithmetical attainments of the Egyptians, gives Egypt the distinction of having one of the most ancient of the known decimal systems. Although their knowledge was probably not so profound as we might expect, it shows marked proficiency in mathematics at the time when Abraham visited Egypt and brought the art of numbers from Chaldea to Egypt.

1-2 THE GREEK AND ROMAN SYSTEMS

Greek numerals In writing their numerals the Greeks used the 24 letters of their own alphabet and three special symbols.§ Their method was to assign the numbers from one to nine to the first nine letters of the alphabet; then 10, 20, 30, . . . ,

† The symbol has been written here to read from left to right, although it was the practice among the Egyptians to read from right to left.

‡ A. B. Chace, L. S. Bull, H. P. Manning, and R. C. Archibald (eds.), "The Rhind Mathematical Papyrus," vols. I and II, Mathematical Association of America, Inc., Buffalo, N.Y., 1927 and 1929.

§ Howard Eves, "An Introduction to the History of Mathematics," 3d ed., p. 14, Holt, Rinehart and Winston, Inc., New York, 1969.

1	α	alpha	10	ι	iota	100	ρ	rho	1,000	α'
2	β	beta	20	κ	kappa	200	σ	sigma	2,000	β'
3	γ	gamma	30	λ	lambda	300	τ	tau	3,000	γ'
4	δ	delta	40	μ	mu	400	υ	upsilon	4,000	δ'
5	ϵ	epsilon	50	ν	nu	500	ϕ	phi	5,000	ϵ'
6	F	digamma*	60	ξ	xi	600	χ	chi	6,000	F'
7	ζ	zeta	70	o	omicron	700	ψ	psi	7,000	ζ'
8	η	eta	80	π	pi	800	ω	omega	8,000	η'
9	θ	theta	90	φ	koppa*	900	$\lambdabar$	sampi*	9,000	θ'

* Obsolete symbols

FIGURE 1-2 The Greek numerals

90, to nine more; and 100, 200, . . . , 900 to nine more. There being only 24 letters in the Greek alphabet, three more became necessary. Thus they made use of three obsolete letters. An examination of Fig. 1-2 reveals that the obsolete letters were used as symbols for 6, 90, and 900.

The additive principle only was used in numbers smaller than 10,000. For example, $\sigma\mu\beta$ was interpreted to mean 200 + 40 + 2 or, as we should write it, 242. An accented letter was used to indicate multiplication by 1,000. For example, δ' became the symbol for 4,000, and $\delta'\rho\lambda\delta$ the symbol for 4,134. For still larger numerals an additional symbol, M, was introduced to indicate multiplication by 10,000. There seems to be some ambiguity as to where the number to be multiplied would be written. It occurs in print sometimes above, sometimes before, and sometimes after the M. Thus, $\overset{\sigma}{M}$, σM, and Mσ might be found as the symbol for 200 $\times$ 10,000 = 2,000,000. We shall use σM. In writing larger numerals both the multiplicative principle and the additive principle were used. For example, ψM$\gamma'\psi\pi\eta$ would be interpreted as

$$7,000,000 + 3,000 + 700 + 80 + 8$$

or, as we should write it, 7,003,788.

Although the numeral system used by the Greeks was decimal, it was so clumsy that they made very little progress in calculations. It seems from the accounts of most historians that when the fingers were not adequate, some kind of abacus was used. According to tradition, Pythagoras first introduced this instrument into Greece after his travels in Egypt.† As to the kind of abacus, we know very little. Herodotus is given credit for the statement that "the Egyptians calculate with pebbles moving the hand from right to left." This would seem to indicate some notion of the principle of positional value. In Fig. 1-3 the colored disks show how the number 7,510,269 would be recorded on a modern abacus.

† F. Cajori, "A History of Mathematics," rev. ed., p. 52, The Macmillan Company, New York, 1919.

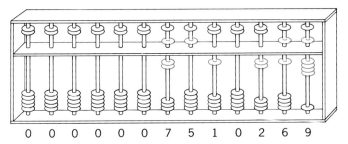

0 0 0 0 0 0 7 5 1 0 2 6 9

FIGURE 1-3   Abacus showing 7,510,269

Roman numerals   The system of notation used by the Romans was much better suited to computation than that of either the Greeks or the Egyptians. It made use not only of the addition principle but also of the principle of subtraction, which was represented occasionally among the cuneiform characters of the Babylonians. The use of these two principles is seen in the three numbers X (ten), IX (10 − 1, or nine), and XI (10 + 1, or eleven). The subtraction principle was restricted to the formation of only six numerals: IX, IV (four), XL (forty), XC (ninety), CD (four hundred), and CM (nine hundred). This notational pattern was a decided improvement over that used by the Greeks, since it was less cumbersome and did not burden the memory with numerous symbols, only seven being necessary, I, V, X, L, C, D, M. In large numbers, an overscoring bar was used to indicate multiplication by 1,000; for example, $\overline{X} = 10 \times 1,000 = 10,000$. The number 2,070,849 would be written as the Roman numeral $\overline{\text{MMLXX}}\text{DCCCXLIX}$.

For several centuries the Greek and Roman systems contended for popular favor among mathematicians, the Roman system having been finally adopted for reckoning in medieval Europe. It has been said that the Roman system of calculating on the reckoning board was one of the most important steps in the development of the decimal system.

*1-3   THE CHINESE-JAPANESE SYSTEM

The decimal numeration systems discussed in the previous sections were very limited in computational effectiveness due to the absence of a zero symbol and of any concept of positional value. In forming large numerals they appealed to differing forms of repetition, addition, subtraction, and *Chinese-* multiplication. The Chinese-Japanese system is also decimal, has no zero *Japanese* symbol, and makes no use of positional value.† This system is of interest *numerals* and significance in the present discussion because of its close approach to the concept of place value, accomplished through the use of special symbols for the respective powers of ten. As in any truly decimal system there is an individual symbol for each digit 1 to 9. The multiplicative principle

† Howard Eves, *op. cit.*, p. 13.

1 一	10 十	8,429
2 二	100 百	
3 三	1,000 千	
4 四		
5 五		
6 六		
7 七		
8 八		
9 九		

八 千 } 8 × 1,000

四 百 } 4 × 100

二 十 } 2 × 10

九

$8 \times 1{,}000 + 4 \times 100 + 2 \times 10 + 9 = 8{,}429$

FIGURE 1-4 The Chinese-Japanese numerals

is used to interpret the use of a digit symbol followed by some power of ten. The additive principle is then used to combine these values to obtain the number represented by the numeral. Figure 1-4 exhibits the digit symbols along with those for 10, 100, and 1,000. Also in vertical array, in accordance with custom, the numeral is written to represent the quantity which we symbolize by 8,429.

EXERCISES

1 Which of the characteristics listed can be used to describe the different numeration systems we have studied?

Numeration systems   *Characteristics*
Babylonian   (a) Absence of positional value
Chinese-Japanese   (b) Absence of a zero symbol
Egyptian   (c) Use of principle of addition
Greek   (d) Use of principle of multiplication
Roman   (e) Use of principle of subtraction
                               (f) Use of principle of repetition
                               (g) Use of ten as a base for grouping

*2 Contrast the numbers represented by each pair of numerals. What common characteristic of the Egyptian, Roman, and Greek systems is illustrated?

(a) [Egyptian, Roman, and Greek numeral pairs]

(b) DMCL   CDLM
(c) χοδδ′   δοχδ′

3 Can you use other forms in the same language to represent the numbers indicated in each portion of Exercise 2?

*4 What contrast in principle of forming numerals is illustrated by the Egyptian and Chinese-Japanese symbols used to represent the number which we represent by 2,357?

In Exercises 5 to 8 construct examples and write explanations of how each indicated principle was used in the different numeration systems discussed previously.

5 The principle of addition
6 The principle of multiplication
7 The principle of subtraction
8 The principle of repetition
9 Discuss the effect the absence of a zero symbol had on each of the numeration systems.

*1-4 THE MAYAN SYSTEM

Mayan numerals

To the Maya of Central America seems to belong the distinction of first having the principle of place value and the use of a zero symbol in a fully developed numeral system. Research has revealed numerous records of their calendars and chronology which have been successfully deciphered, disclosing the fact that the Maya possessed this system five or six centuries before any Asiatic country.† Their system, as found in their calendars, rituals, and astronomical observations, was not decimal but vigesimal (base twenty) in nature, except for one position. "Twenty units of the lowest order (kins) make one unit of the next higher order (uinal or 20 kins), 18 uinals make one unit of the third order (tun or 360 kins), 20 tuns make one unit of the fourth order (katun or 7,200 kins), 20 katuns make one cycle or 144,000 kins, and finally 20 cycles make one great cycle, or 2,880,000 kins." ‡ It would be very easy to build this scale as high as one might wish. The number of kins in each new higher order could be obtained by multiplying the positional value last obtained by 20.

Since this system has a base of 20 it requires digit symbols for values from 1 to 19. These are shown in Fig. 1-5. They are represented by dots and bars, each dot standing for 1 unit and each bar for 5. The additive principle is used to determine the cumulative value of each symbol. In writing the numeral for 20 the principles of place, or positional, value and of multiplication enter for the first time. Figure 1-5 also shows symbols for 20, 38, 300, and 364. Equivalent values in our numerals are used to illustrate how these symbols are to be interpreted. Note that the numerals are written in vertical array

† F. Cajori, The Zero and Principle of Local Value Used by the Maya of Central America, *Science*, **44:**715 (1916).
‡ *Ibid.*, p. 716.

0 ⬮	5 —	10 =	15 ≡	20 ⬮
1 .	6 .	11 ..	16 ..	38 ...
2 ..	7 ..	12 ...	17 ...	300 ⬮
3 ...	8	13	18	
4	9	14	19	364 ⬮

FIGURE 1-5 The Maya numerals

$$⬮ = 1 \times 20 + 0 = 20 + 0 = 20$$

$$⬮ = 1 \times 20 + 18 = 20 + 18 = 38$$

$$⬮ = 15 \times 20 + 0 = 300 + 0 = 300$$

$$⬮ = (1 \times 18 \times 20) + 0 + 4 = 360 + 4 = 364$$

There is present in this system some ambiguity in notation. For example, one might well question whether ⫶ is the symbol for 12 or the symbol for $2 \times 20 + 10 = 50$. In this discussion we attempt to take care of this through the use of spacing. The above symbol would be interpreted as 12, while the symbol for 50 would have more space between the dots and bars, and occur as ⫶.

Example Write the Mayan numeral 3 great cycles, 0 cycles, 10 katuns, 18 tuns, 2 uinals, 11 kins in their vertical symbolic form. Convert to an equivalent numeral in our system.

The *kin* of the Mayan system has the same value as the *one* of our decimal system. Therefore, any numeral of the Mayan system may be converted into its equivalent numeral in our system by changing each position symbol into its equivalent number of kins. Note the use of the multiplicative and additive principles.

⋯ 3 great cycles	=	$3 \times 2{,}880{,}000$ kins =	8,640,000 kins
⬮ 0 cycles	=	$0 \times 144{,}000$ kins =	0 kins
= 10 katuns	=	$10 \times 7{,}200$ kins =	72,000 kins
≣ 18 tuns	=	18×360 kins =	6,480 kins
⋯ 2 uinals	=	2×20 kins =	40 kins
= 11 kins	=	11 kins =	11 kins
			8,718,531 kins

The equivalent numeral in our decimal system is 8,718,531.

Example Convert 6,192,204 into an equivalent Mayan numeral.

Since 1 great cycle = 2,880,000 kins and 1 kin = 1 one, we can determine the number of great cycles in the given number by division.

6,192,204 ÷ 2,880,000 gives a quotient of 2	••	great cycles
with a remainder of 432,204.		
432,204 ÷ 144,000 = 3 with a remainder	•••	cycles
of 204. In sequence we then have		
204 ÷ 7,200 = 0 with 204 remainder	⬭	katuns
204 ÷ 360 = 0 with 204 remainder	⬭	tuns
204 ÷ 20 = 10 with 4 remainder	══	uinals
	••••	kins

Thus the Mayan numeral is 2 great cycles, 3 cycles, 0 katuns, 0 tuns, 10 uinals, and 4 kins.

Conant says: "In the Maya scale we have one of the best and most perfect examples of vigesimal numeration ever developed by any race."† There is no evidence of digital numeration in the first 10 units, but, judging from the almost universal practice of the Indian tribes of both North and South America, this may have been the origin of the Maya counting system. Can you think of any illustration in our use of number that implies a scale of twenty?

EXERCISES

1 Translate each of these symbols from the Egyptian numerals to our numerals:

(a) ∩∩∩ ||| / ∩∩ ||| (b) 99 9 / 9 | (c) 9 9 ∩∩∩ ||| / 99 ∩∩∩ ||| / 99 ∩∩∩ ||| (d) 𝒸𝒸 ∩∩∩ || / ∩∩∩ ||

(e) 𝕏𝕏 999 ||| / 𝕏𝕏𝕏 999 ||| / 999 ||| (f) 𝕏 𝒸𝒸𝒸 ∩∩ ||| / 𝒸𝒸 ∩∩ ||

2 Translate each of these symbols from the Greek numerals to our numerals:

(a) ψπη   (b) ε′ε   (c) λ′μ   (d) δMρα   (e) η′MχMγ′ωλδ
(f) δ′ψπθ   (g) θ′Ϙθ   (h) ρMε′   (i) γ′Mγ′γ   (j) Ϝ′χμϜ

3 Translate each of these symbols from the Roman numerals to our numerals:

(a) CCXLVII   (b) MI   (c) MCMLXII
(d) MCDXCII   (e) M̄MMIV   (f) M̄DCXCCX

†L. L. Conant, "Number Concept," p. 200, The Macmillan Company, New York, 1931.

*4 Translate each of these symbols from the Mayan numerals to our numerals:

(a) (b) (c) (d) (e)

5 What major contribution did the abacus make to those primitive numeration systems with which it was used?

*6 Translate each of these symbols to equivalent numerals in the Chinese-Japanese system:

(a) 25   (b) 341   (c) 3,041   (d) 3,401   (e) 6,524
(f) 4,002   (g) 89   (h) 101   (i) 5,555   (j) 7,800

*7 Translate each of these symbols from the Chinese-Japanese numerals to our numerals:

(a) (b) (c) (d) (e)

8 Arrange the numerals in each of the given sets so that they will be in order from left to right with the numeral representing the smallest number on the left.

(a)

(b)

(c) $\{\tau\mu\delta, \tau\mu\eta, \tau\mu, \tau\nu, \tau\gamma\}$
(d) $\{\beta'\sigma\kappa, \beta'\phi, \beta'\mu\theta, \beta'\phi\alpha\}$
(e) {XIX, XXI, XIV, XVI}
(f) {MCM, MMC, MD, MCCCC}

*(g)

*(h)

9 Translate each of these symbols first into Egyptian numerals, then into Greek numerals, then into Roman numerals, and then into Mayan numerals:

(a) 467   (b) 2,508   (c) 40,720   (d) 1,492
(e) 3,004   (f) 400,500   (g) 1,610,430   (h) 4,201,897

1-5 THE DECIMAL SYSTEM

Ten as the base of a numeral system

The numeral system we use is called a decimal system, since it uses ten as its base, or radix. Any such system, which also uses the principle of positional value, must have ten primary digital symbols.† We use the nine digit numerals 1, 2, 3, 4, 5, 6, 7, 8, and 9 to represent the presence of value and the numeral 0 to represent the absence of value. These symbols,

Place (positional) value

along with the use of the concept of positional value, or place value, and the multiplicative and additive principles, enable us to write any numeral we wish, no matter how large or how small. Positional value simply means the value of any one of the above symbols as determined by its relative position in the numeral. The additive principle tells us that the value of any

Additive principle

numeral is found by getting the sum of all the respective implied products of digit and place values. For a clearer comprehension of the intrinsic part the ten digital symbols play in the forming of numerals in our system of numeration, let us consider a symbol such as 1,203. For this particular illustration let us think of this numeral in parallel with the word "teeter" ($t_{ee}t_e{}^r$). Because of the place the symbol 1 holds in 1,203, it has a place value of "one thousand." How many "thousands" are in 1,203? We say "one," but what is "one"? In *this* illustration, it is the number that is associated with the concepts of "thousands in 1,203" and "the number of times the letter 'r' is used in the word 'teeter'." Thus, in addition to its *place* value of thousand, the digit symbol 1 in the number 1,203 has a *digit* value we call "one." Similarly, in this illustration, the symbol 2 has a place value of "one hundred" and a digit value of the number that is associated with the concepts of "hundreds in 1,203" and "the number of times the letter 't' is used in the word 'teeter'." In like manner, the symbol 0 has a place value of "ten" in the number 1,203 and a digit value of the number to be associated with the concepts of "tens in 1,203" and "the number of times the letter 'a' is used in the word 'teeter'." The symbol 3 has a place value of "one" and a digit value associated with "ones in 1,203" and "the number of times the letter 'e' is used in the word 'teeter'." Thus,

Digit value and place value in numerals

in the structure of any numeral, the digital symbols 0, 1, 2, 3, 4, 5, 6, 7, 8, and 9 have, each in its own right, both a *place* value and a *digit* value. They are used *both* as placeholders *and* to represent numbers.

---

† The word "digit" is derived from a Latin word which means "finger" or "toe." This association has led historians to surmise that our decimal system of numeration found its origin in the fact that we have ten fingers.

Relative place values in numerals

The multiplicative and additive principles then give the numerical significance of the numeral 1,203 as being $1(1,000) + 2(100) + 0(10) + 3(1)$, or $1,000 + 200 + 3$. It is clearly understood that in a decimal system of numeration, the value of each position in any given numeral is ten times the value of the next position to its right and one-tenth the value of the next position to its left. This fact is more explicitly displayed when 1,203 is written as $1(10^3) + 2(10^2) + 0(10^1) + 3(1)$. Counting numbers used as exponents

Use of exponents in numeral structure

simply tell how many times a given number is used as a factor. In other words, $10^3 = 10 \times 10 \times 10 = 1,000$; $10^2 = 10 \times 10 = 100$; and $10^1 = 10$.† With this notation, increasing the size of the exponent of 10 is all that is needed to indicate positional values of greater value: $10^4 = 10 \times 10 \times 10 \times 10 = 10,000$; $10^6 = 10 \times 10 \times 10 \times 10 \times 10 \times 10 = 1,000,000$; and so on. Decreasing the size of the exponent of 10 indicates positional values of smaller value: $10^1 = 10$; $10^0 = 1$. Also, we have $10^{-1} = \frac{1}{10}$; $10^{-2} = \frac{1}{10^2} = \frac{1}{10 \times 10} = \frac{1}{100}$; and so on. This use of 0 and negative numbers as exponents is in accordance with the definition:

Definition For any nonzero number a, $a^0 = 1$ and $a^{-n} = \frac{1}{a^n}$ where n is any *counting* number.

Thus $10^0 = 1$ and $10^{-2} = \frac{1}{10^2}$.

Significance of place value

The real significance of such a numeration system lies in its compactness and adaptability in form and in the great simplification it contributes to computation. The principle of place value is the most important single attribute of our system of numeration and has given to it the power and usefulness that has resulted in its almost universal acceptance by the nations of the world.

1-6 EXPONENTS

Analyzing a numeral

In the previous section the numeral 1,203 was written in the expanded form $1(1,000) + 2(100) + 0(10) + 3$, or $1(10^3) + 2(10^2) + 0(10) + 3$. Henceforth, we shall indicate the use of this procedure by saying that we shall *analyze the numeral*. The expanded form vividly portrays the use of both the multiplicative principle and the additive principle in the formation of our decimal numerals.

Since this practice will call for a limited use of exponents, a brief review

---

† As a consequence of this equality it is the custom, generally, not to use the explicit exponential form when the exponent is 1. In other words, the general practice is to write 10 rather than 10^1; similarly, we write $\frac{1}{10}$ rather than $\frac{1}{10^1}$.

of the pertinent laws of exponents is desirable at this point. No effort will be made to review the proofs of these principles, but general illustrations will be given to provide a basis for the intuitive acceptance of each law.

If m is a counting number, as in the previous section, then $a^m = a \times a \times a \times \cdots \times a$, where a occurs m times as a factor. In such usage a is called the *base;*† n, the *exponent;* and a^m, the mth *power* of a.

Illustration 1 $a^m \times a^n = a^{m+n}$

$$a^3 = a \times a \times a$$
$$a^4 = a \times a \times a \times a$$
$$a^3 \times a^4 = (a \times a \times a) \times (a \times a \times a \times a)$$
$$= a^7 = a^{3+4}$$

Illustration 2 $(a^m)^n = a^{mn}$

$$(a^3)^2 = a^3 \times a^3$$
$$= (a \times a \times a) \times (a \times a \times a)$$
$$= a^6 = a^{3 \times 2}$$

Illustration 3 $(a \times b)^m = a^m \times b^m$

(i)
$$(4 \times 5)^2 = (4 \times 5) \times (4 \times 5)$$
$$= (4 \times 4) \times (5 \times 5)$$
$$= 4^2 \times 5^2$$

(ii)
$$(2a)^3 = (2 \times a)^3$$
$$= 2^3 \times a^3 = 8a^3$$

Although the laws stated in Illustrations 4 and 5 do not apply until Chap. 6, they are stated here for completeness. In each case a is not equal to zero. This condition may be written $a \neq 0$.

Illustration 4 $\dfrac{a^m}{a^n} = a^{m-n}$ if m is larger than n

$$= \frac{1}{a^{n-m}} \text{ if } n \text{ is larger than } m$$

(i)
$$\frac{3^4}{3^2} = \frac{3 \times 3 \times 3 \times 3}{3 \times 3} = 3 \times 3$$
$$= 3^2 = 3^{4-2}$$

(ii)
$$\frac{3^2}{3^4} = \frac{3 \times 3}{3 \times 3 \times 3 \times 3} = \frac{1}{3 \times 3}$$
$$= \frac{1}{3^2} = \frac{1}{3^{4-2}}$$

† Note that the concept of "base" is used here in connection with exponents. When it is so used it carries an entirely different connotation from that which it has in the interpretation of numerals.

Illustration 5 $\left(\dfrac{a}{b}\right)^n = \dfrac{a^n}{b^n}$ for $b \neq 0$

$$\left(\frac{3}{2}\right)^4 = \frac{3}{2} \times \frac{3}{2} \times \frac{3}{2} \times \frac{3}{2}$$

$$= \frac{3 \times 3 \times 3 \times 3}{2 \times 2 \times 2 \times 2} = \frac{3^4}{2^4}$$

Example Simplify each of these expressions:

(*a*) $10^3 \times 10^8 = 10^{3+8} = 10^{11}$

(*b*) $\dfrac{10^5}{10} = 10^{5-1} = 10^4$; $\dfrac{10^2}{10^7} = \dfrac{1}{10^{7-2}} = \dfrac{1}{10^5}$

(*c*) $\left(\dfrac{2}{3}\right)^2 = \left(2 \times \dfrac{1}{3}\right)^2$

$\qquad = 2^2 \times \left(\dfrac{1}{3}\right)^2$

$\qquad = 4 \times \dfrac{1}{3^2} = \dfrac{4}{9}$

or $\left(\dfrac{2}{3}\right)^2 = \dfrac{2^2}{3^2} = \dfrac{4}{9}$

(*d*) $4^0 \times 4^3 = 4^{0+3} = 4^3 = 64$

or $4^0 \times 4^3 = 1 \times 4^3 = 4^3 = 64$

This illustration offers some support for the definition $a^0 = 1$ for $a \neq 0$ (Definition 1-1).

EXERCISES

In each of the expressions given first use the proper law of exponents and then expand. Example: $2^2 \times 2^5 = 2^{2+5} = 2^7 = 128$.

1 (*a*) $3^2 \times 3^4$ (*b*) $10^3 \times 10^5$ (*c*) $2^4 \times 2^5$
    (*d*) $10^1 \times 10^2 \times 10^3$ (*e*) $5^4 \times 5^0 \times 5^3$ (*f*) $3^7 \times 3 \times 3^4$

2 (*a*) $5^4 \times 5^3$ (*b*) $10^3 \times 10$ (*c*) $10^1 \times 10$
    (*d*) $2^6 \times 2^3 \times 2^3$ (*e*) $10^2 \times 10^4$ (*f*) $10^3 \times 10^2$

3 (*a*) $(2^3)^2$ (*b*) $(10^4)^3$ (*c*) $(10^2)^3$ (*d*) $(10^4)^2$

4 (*a*) $(3 \times 7)^2$ (*b*) $(2 \times 10)^3$ (*c*) $(3 \times 2)^4$
    (*d*) $(5 \times 6)^2$ (*e*) $(2a)^4$ (*f*) $(10a)^2$
    (*g*) $(3a)^3$ (*h*) $(4m)^3$

5 (*a*) $\dfrac{2^4}{2}$ (*b*) $\dfrac{10^5}{10^2}$ (*c*) $\dfrac{3^5}{3^2}$ (*d*) $\dfrac{5^7}{5^4}$

   (*e*) $\dfrac{10^3}{10^4}$ (*f*) $\dfrac{8^5}{8^7}$ (*g*) $\dfrac{10}{10^3}$ (*h*) $\dfrac{2^7}{2^{10}}$

6   (a) $\left(\dfrac{2}{3}\right)^5$ (b) $\left(\dfrac{3}{4}\right)^2$ (c) $\left(\dfrac{1}{2}\right)^3$ (d) $\left(\dfrac{5}{2}\right)^2$

7   (a) $2^0 \times 2^4 \times 2^3$ (b) $10^2 \times 10^3 \times 10^5$
   (c) $10 \times 10^0 \times 10^2$ (d) $10^4 \times 10^3 \times 10^1$
   (e) $4^5 \times 4 \times 4^2$ (f) $3^0 \times 3^2 \times 3^4$

8   (a) $\left(\dfrac{1}{2}\right)^0$ (b) $\left(\dfrac{2}{3}\right)^0$ (c) $\left(\dfrac{3 \times 2}{5 \times 4}\right)^0$ (d) $(6,724)^0$

9   Analyze these numerals. Write the results in two forms.

   (a) 3,421   (b) 202   (c) 2,624,321
   (d) 100,200   (e) 780   (f) 5,006,703

10   Use the laws of exponents to find these products. Check your results by first expanding and then finding the product. Examples:

$$10^4 \times 10^2 = 10^{4+2} = 10^6$$
$$10^4 \times 10^2 = 10{,}000 \times 100 = 1{,}000{,}000 = 10^6$$

   (a) $10^2 \times 10^3$ (b) $10^7 \times 10^2$ (c) $10^4 \times 10^0$
   (d) $10^2 \times 10^6$ (e) $10^0 \times 10^3$ (f) $10^5 \times 10^1$

1-7   ORIGIN AND DEVELOPMENT OF OUR NUMERALS

It is not known just who was the originator of the numeral symbols which we use, nor just what process of reasoning affected their forms. Many theories have been advanced as to their origin, but the one which seems to be the most generally accepted is that they are of Hindu origin. The authorities who adhere to this hypothesis seem also to be agreed upon the fact of Arabic influence in transmitting these symbols to other civilized nations and ultimately to the civilization of the present. For this reason our *Hindu-Arabic numerals* numeral system is usually called the Arabic, Hindu-Arabic, or Indo-Arabic numeral system, with Hindu-Arabic probably the most generally accepted name.

   The value of these numerals was not recognized at once, nor were they readily accepted. Even as late as the fifteenth and sixteenth centuries their use was very rare, since merchants still made use of the Roman system for keeping their accounts. In fact, the transition from the Roman numerals was incomplete even up to the time of the first Queen Elizabeth of England (1558–1603).

   The forms of the symbols 0, 1, 6, 8, and 9 have not changed much; those for 2, 3, and 7 have changed some; and the forms for 4 and 5 have been greatly altered. The changes which these forms underwent were brought about not so much from a desire for improvement as by the fact that all manuals and documents were written by hand. It was with the invention of printing that the numerals assumed some permanency of form. So it is that our symbols of today are practically the same in appearance as those of the fifteenth century.

0 as a symbol The date of the invention of 0 as a symbol in our numeral system is not known. Maximus Planudes, a Byzantine writer, thought that the Hindu numeral system made use of such a symbol and gave A.D. 738 as the first known instance of it.† There is additional evidence for the theory that the earliest Hindu notation neither contained such a symbol nor used the principle of place value. The rules used by Aryabhatta (A.D. 476) for extracting roots involved the principle of position and the zero, making it appear that they were both introduced about the fifth century. In an Indian inscription of A.D. 876, the numbers fifty and two hundred seventy are both written with zero. Dantzig attributes the discovery of zero to an attempt to record a permanent counting-board operation, for without some symbol to represent the empty columns "such an entry as $\equiv$ = might represent 32, 302, 320, 3002, or many other numbers." He further states:‡

> Conceived in all probability as the symbol for an empty column on a counting board, the Indian *sunya* was destined to become the turning-point in a development without which the progress of modern science, industry, or commerce is inconceivable. And the influence of this great discovery was by no means confined to arithmetic. By paving the way to a generalized number concept, it played just as fundamental a rôle in practically every branch of mathematics. In the history of culture the discovery of zero will always stand out as one of the greatest single achievements of the human race.

The ancient Babylonians used in their documents, but not in their calculations, a symbol to indicate that the units of a certain class were missing, and as early as the second century B.C. the Greek astronomers made use of their letter *omicron* for the same purpose. Because of this historical and notational emphasis given to 0 as a symbol for the "empty column," or "absence of value," incorrect interpretations of the full significance of the symbol are sometimes made. Some writers have been inclined to argue that zero is not a symbol for a number but is merely a placeholder. This is an entirely false argument since the symbol 0, like each of the remaining nine digital symbols of our numeral system, has distinct stature both as a symbol for a number and as a placeholder.

1-8 THE NUMERATION OF LARGE NUMBERS

After printing had, to a certain degree, stabilized the shape of the numerals used in counting and computation, there still remained the question of

† M. Cantor, *Vorlesungen über Geschichte der Mathematik*, vol. I, p. 603, B. G. Teubner Verlagsgesellschaft, mbH, Leipzig, 1922.
‡ Tobias Dantzig, "Number, the Language of Science," 3d ed., p. 35, The Macmillan Company, New York, 1945. Used by permission of the Macmillan Company.

notation and terminology to enable one to read and write the various combinations that might be made from these 10 digital symbols. Up to the seventeenth century many different methods were used in the numeration of large numbers.

Names for large numbers   A decided advance was made by the introduction of the word *millione*. It occurs first in the year 1484 in the "Arithmetic" of the Italian, Piero Borgi, and next in the "Suma" of Luca Pacioli. During the sixteenth century the use of *millione* to denote $(1,000)^2$ spread to other European countries and to England, where Tonstall in 1522 speaks of it as being commonly used but "rejects it as barbarous."

The words "billion," "trillion," etc., date back almost as far as "million," as they seem to have been used first by the Frenchman, Nicholas Chuquet, in 1520. He used the words *byllion, tryllion, quadrillion, quyllion, sixlion, septyllion, octyllion*, and *nonyllion* to denote the second, third, etc., powers of one million, or 10^{12}, 10^{18}, etc. Names with the same significance appeared in Germany in 1681 and in England in 1687, but did not become

Ambiguity of the symbol for one billion   used generally until the eighteenth century. When the people of France began grouping digits in groups of three, "billion" was assigned the new meaning of a thousand millions or 10^9, "trillion" a thousand billions or 10^{12}, etc. This difference in notation and terminology still prevails today, for in England, Germany, and certain Northern European countries "billion," "trillion," etc., are used for 10^{12}, 10^{18}, etc., while in the United States, France, and certain Southern European countries they designate 10^9, 10^{12}, etc. In other words, in the United States one billion is one thousand million or 1,000,000,000, while in Britain it is one million million or 1,000,000,-

Flexibility of our numeral system   000. Such ambiguity in interpretation loses its significance when contrasted with the effectiveness of a system of numeration which can express the estimated weight of one atom of hydrogen as 0.0000000000000000000-0000175 gram (175 hundred-septillionths gram) and at the same time tell us that a handful of uranium will produce as much useful power as about 50 trillion (50,000,000,000,000) pounds of coal. If these numbers do not impress the reader, there is the estimated mass of an electron, 0.000000-00000000000000000000000199 pound (199 hundred-nonillionths pound), the number of times a voice is amplified in a coast-to-coast telephone call† ($10^{3,000}$), Sir Arthur Eddington's estimate of the number (136×2^{256})

Googol and googolplex, and Skewes' number   of protons in the universe, the googol (10^{100}), the number sometimes mentioned as the total possible number of moves in a chess game ($10^{10^{50}}$), the googolplex‡ ($10^{10^{100}}$), or the really big number known as Skewes' number,

$$10^{10^{10^{34}}}$$

These are all finite numbers which theoretically can be written out by using the concept of place value, the nine digits, and zero. The googol would be fairly simple to write, as it is 1 followed by 100 zeros; but do not try

† *Telephone News*, August, 1963, Pacific Telephone, Los Angeles.
‡ Edward Kasner and James Newman, "Mathematics and the Imagination," pp. 18–35, Simon & Schuster, Inc., New York, 1940.

to write the googolplex, for it has been said that "there would not be enough room to write it if you went to the farthest star, touring all the nebulae and putting down zeros every inch of the way."†

EXERCISES

1 Why is our system of numerals referred to as the Hindu-Arabic system?
2 What is meant by the statement that our system of numeration is a decimal system?
3 Use the symbol 1,056.013 to illustrate what is meant by place value.
4 Contrast the Hindu-Arabic system of numerals with (a) the Mayan system, (b) the Babylonian system, (c) the Egyptian system, (d) the Roman system, (e) the Chinese-Japanese system. Point out the major similarities and differences.
5 What are some of the more important reasons that the Hindu-Arabic system has prevailed over each of the other systems of numeration named in Exercise 4?
6 If one digit is annexed to the right of the numeral 342, how is the value of each digit in the symbol affected?
7 If one zero is removed from the right of 26,500, how is each digit of the symbol affected?
8 Answer the question of Exercise 7 if two zeros are removed from the right of 26,500.
9 How could one-dollar bills (or pieces), dimes, and pennies be used to illustrate the basic principles of place value?

1-9 BASES OTHER THAN TEN

We have noted how the Babylonians used sixty as a base of their numeral system, and the Maya Indians used twenty. In modern literature on the subject there is to be found argument in support of the use of numbers other than ten as the base in the structure of our numeral system. Also, it is a well-known fact that electronic computation uses a binary number scale (base two) and that significant modifications of the straight binary code are used in many types of computer coding. Except for the use of two and certain powers of two as bases in electronic computation, sixty in measures of time, twenty in expressions such as "three score years and ten," and twelve in certain forms of measurement, our numeral system quite likely will retain ten as its base. Nevertheless, profit can be derived from a study of numeral systems constructed on bases other than ten. We shall pursue in some detail here an examination of systems using bases twelve and five, respectively. In this analysis we shall assume place value and the use of the multiplicative and additive principles, as in our decimal system. Fur-

† *Ibid.*, p. 23.

thermore, while we recognize the possibility of inventing new number symbols and names, we shall yield to the simple pattern of using, where possible, those already established, and shall not invent unless it becomes necessary to do so.

1-10 BASE TWELVE

10 (One-oh) represents the base

In the title to this section, why is the word "twelve" written instead of the symbol 12? The answer is simple. It is merely because the symbol **12†** (call it "one-two") in a duodecimal (base **twelve**) system of numeration is no longer the symbol for the concept of the number which can be used to tell how many strokes there are in Fig. 1-6. When we say we are going to use "twelve" as a base, we mean that we plan to group in "groups of twelve" and in such a grouping there are **10** (read this "one-oh"; it is no longer "ten") strokes in Fig. 1-6. There is one group of "twelve strokes" and zero strokes left over. Thus the symbol **12** (one-two) in base **twelve** means 1 group of twelve and 2 more (1 of the base and 2 more). If **10** (one-oh) no longer means ten, how can we write the number ten in this base? Since ten is smaller than twelve, it must be represented by a digital symbol, and so we are in need of a one-figure symbol which can occupy only one position in the formation of numerals to represent numbers larger than the base. We shall use **t** and read it "tē" just as we read the letter. Similarly, we shall use the letter **e** to represent eleven, since eleven is also less than twelve and we need a one-figure symbol and a name by which

Digital symbols in base twelve

to identify it. Thus, in a numeration system of base **twelve** our digit symbols will be **0, 1, 2, 3, 4, 5, 6, 7, 8, 9, t, e.** To be able to count and compute satisfactorily in this system, now that we have satisfactory digital symbols, we still have to invent number names. For example, what is the number **ee?** We shall avoid this difficulty simply by calling each digit separately in

Reading numerals in base twelve

order to read any numeral. For example, just as we read the symbol **12** as "one-two," we shall read the symbol **ee** as "e-e." For what does the symbol **3t4** stand? It means **300 + t0 + 4,** or **3 × 100 + t × 10 + 4 × 1.**

† In the remainder of this section and throughout the book boldface will be used to indicate numerals which are to be interpreted in bases other than ten. In all cases not printed in boldface the base is to be understood to be ten.

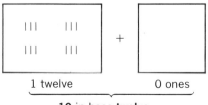

1 twelve   0 ones

10 in base twelve   FIGURE 1-6   **10** in base **twelve**

This means **3** × (twelve × twelve) + **t** × (twelve) + (**4** × one). We read it "three-te-four." In terms of our definitions "one dozen" = twelve units and "one gross" = twelve dozen, the numeral **3t4** might also be read **"three gross te** dozen **four."**

1-11 BASE FIVE

In order to operate efficiently in base **five** the only invention necessary, under our present agreement, will be that of number names. As digital symbols we can use **0, 1, 2, 3, 4.** In Fig. 1-7a the days cf the month of July are shown enumerated in the familiar way using the decimal system. In Fig. 1-7b the corresponding numerals are shown as they would be written in a system using **five** as the base. Note that the symbols 0, 1, 2, 3, and 4 are used as digital symbols here as well as in base ten and base **twelve.** While they have the same numerical significance in the languages of all three bases, they differ considerably in the meanings they carry, depending upon the base orientation of the numeral in which they occur. 23 as a

Digital symbols in base five

July

S	M	T	W	T	F	S
1	2	3	4	5	6	7
8	9	10	11	12	13	14
15	16	17	18	19	20	21
22	23	24	25	26	27	28
29	30	31				

Base ten
(a)

July

S	M	T	W	T	F	S
1	2	3	4	10	11	12
13	14	20	21	22	23	24
30	31	32	33	34	40	41
42	43	44	100	101	102	103
104	110	111				

Base **five**
(b)

FIGURE 1-7   The days of the month of July enumerated in base ten and in base **five**

a b c d e f g h i j	ten ⎫	
k l m n o p q r s t	ten ⎭ 2 × ten ⎫	
	+ ⎬ = 26	
u v w x y z	six } 6 ones ⎭	

FIGURE 1-8 The number of letters in the English alphabet as enumerated in base ten

symbol in base *ten* means 2 tens + 3 ones; as a symbol in base **twelve** it means **2** twelves + **3** ones; and as a symbol in base **five** it means **2** fives + **3** ones.

1-12 EQUIVALENT NUMERALS IN DIFFERENT BASES

How many letters in our alphabet?

How many letters are there in our alphabet? As an aid in answering this question, we might follow some pattern of arranging the letters in readily countable groups, as shown in the colored blocks of Figs. 1-8 to 1-10. If we arrange them in groups of ten, there are 2 groups of ten each and 6 more. We say there are 26 letters in the alphabet, because the symbol 26 in base ten means 2 tens and 6 ones (see Fig. 1-8). If the letters of the alphabet are grouped in groups of twelve, there are 2 groups of twelve and 2 more. We shall write this **22—twelve** and call it "two-two." There are 26—ten or **22—twelve** letters in our alphabet (see Fig. 1-9).

What symbol should be used if the letters were grouped by fives? There is one group of five × five, no groups of five, and one more (see Fig. 1-10). The symbol in base **five** for the number of letters in our alphabet is **101.** The symbol to be used to represent a quantitative concept depends on the base that is used for the purpose of grouping:

$$26\text{—ten} = \textbf{22—twelve} = \textbf{101—five}$$

What base is used if one says there are **35** (three-five) letters in our alphabet?

a b c d e f g h i j k l	twelve ⎫	
m n o p q r s t u v w x	twelve ⎭ 2 × twelve ⎫	
	+ ⎬ **22—twelve**	
y z	two} 2 ones ⎭	

FIGURE 1-9 The number of letters in the English alphabet as enumerated in base **twelve**

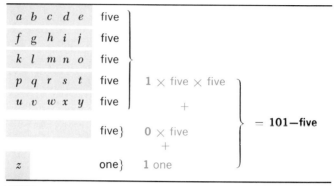

FIGURE 1-10   The number of letters in the English alphabet as enumerated in base **five**

In Fig. 1-11 the symbols **10, 100, 1000, 34,** and **4302** are interpreted for any base, for base ten, for base **twelve,** and for base **five.** It should be studied very carefully, as it carries significant implications for work not only in base ten but in whatever base one might choose to use.

EXERCISES

1   Use the indicated base to write the numeral which tells how many letters there are in the English alphabet: (a) **three;** (b) **six;** (c) **seven;** (d) **eight;** (e) **two.**

2   Why are these equivalent numerals? (a) **2314—five** and **334—ten;** (b) **302—five** and **65—twelve.**

3   First use base **five,** then base **twelve,** then base **eight,** and finally base **two** to count and record the number of words in each of the following quotations:

(a) . . . mathematics may be defined as the subject in which we never know what we are talking about, nor whether what we are saying is true. *Bertrand Russell*

(b) Pure mathematics is a collection of hypothetical, deductive theories, each consisting of a definite system of primitive, *undefined,* concepts or symbols and primitive, *unproved,* but self-consistent assumptions (commonly called axioms) together with their logically deducible consequences following by rigidly deductive processes without appeal to intuition . . . . *G. D. Fitch*

(c) Arithmetic has a very great and elevating effect, compelling the soul to reason about abstract number, and if visible or tangible objects are obtruding upon the argument, refusing to be satisfied *Plato*

(d) Leibniz saw in his binary arithmetic the image of creation. He imagined that Unity represented God, and zero the void; that the Supreme Being drew all beings from the void, just as unity and zero express all numbers in his system of numeration. *Laplace*

Symbol	Meaning			
	Any base	Base ten	Base twelve	Base five
10	1(base) + 0 ones	1(ten) + 0 ones	1(twelve) + 0 ones	1(five) + 0 ones
$100 = 10^2$	1(base × base) + 0(base) + 0 ones	1(ten × ten) + 0(ten) + 0 ones	1(twelve × twelve) + 0(twelve) + 0 ones	1(five × five) + 0(five) + 0 ones
$1,000 = 10^3$	1(base × base × base) + 0(base × base) + 0(base) + 0 ones	1(ten × ten × ten) + 0(ten × ten) + 0(ten) + 0 ones	1(twelve × twelve × twelve) + 0(twelve × twelve) + 0(twelve) + 0 ones	1(five × five × five) + 0(five × five) + 0(five) + 0 ones
34 = 3(10) + 4	3(base) + 4 ones	3(ten) + 4 ones	3(twelve) + 4 ones	3(five) + 4 ones
4302 = $4(10^3)$ + $3(10^2)$ + 0(10) + 2	4(base × base × base) + 3(base × base) + 0(base) + 2 ones	4(ten × ten × ten) + 3(ten × ten) + 0(ten) + 2 ones	4(twelve × twelve × twelve) + 3(twelve × twelve) + 0(twelve) + 2 ones	4(five × five × five) + 3(five × five) + 0(five) + 2 ones

FIGURE 1-11 Interpretation of certain symbols in different bases

4 Since twelve dozen = one gross and twelve gross = one great gross, what symbol in base **twelve** could be used to indicate each of the following?
(a) Eight dozen, four
(b) Ten gross
(c) Five great gross, three gross, two
(d) Two great gross, eleven gross, ten dozen, nine
(e) Ten great gross, eleven gross, ten dozen, eleven
(f) Eleven great gross
(g) Ten great gross, ten dozen

5 Use the duodecimal system of numeration to record each of these measures: (a) two feet, three inches; (b) eight feet, ten inches; (c) four feet; (d) eleven feet, eleven inches.

6 Explain why **23104** means

$$2(10^4) + 3(10^3) + 1(10^2) + 0(10) + 4(1)$$

in any system of numeration with a base larger than four.

7 Write the symbol in base **five** for: (a) the number of fingers you have on both hands; (b) the total number of fingers and toes you have on both hands and both feet.

8 Explain why 10—ten means the same as **20—five.**

9 (a) What symbol in a vigesimal scale might be used to express "three score years and ten"? (b) What is the symbol in the numeration system used by the Maya Indians?

10 In a sexagesimal scale what might be the symbol for: (a) five hours, six minutes, four seconds; (b) eight hours; (c) ten hours, eleven minutes, five seconds ($Hint:$ all digits must be one-figure symbols); (d) forty-five hours, thirty minutes, fifty seconds.

11 If **b** is the base of a given numeral system, how many digital symbols will be necessary if we assume that place value is to be used in writing all numerals?

12 What are some of the relative advantages and disadvantages in using ten, **twelve, twenty,** and **sixty** as bases in systems of numeration?

1-13 REVIEW EXERCISES

1–16 Write a carefully constructed answer to each of the guideline questions of this chapter.

In Exercises 17 to 22 convert each numeral to an equivalent decimal numeral.

17 (a) (b)

(c) (d)

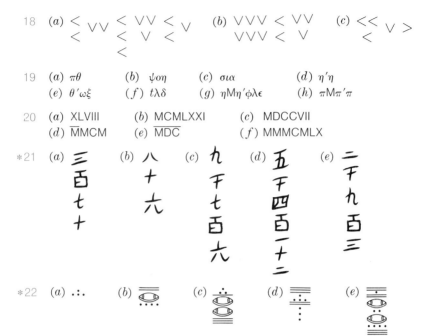

18 (a) < VV < VV < V   (b) VVV < VV   (c) << V >
    < VV < V < V   VVV < V   < V >
    <

19 (a) πθ   (b) ψοη   (c) σια   (d) η'η
    (e) θ'ωξ   (f) τλδ   (g) ηΜη'φλε   (h) πΜπ'π

20 (a) XLVIII   (b) MCMLXXI   (c) MDCCVII
    (d) $\overline{\text{MMCM}}$ (e) $\overline{\text{MDC}}$ (f) MMMCMLX

*21 (a)   (b)   (c)   (d)   (e)

*22 (a) .·.   (b)   (c)   (d)   (e)

*23 Convert each of these decimal numerals to Chinese-Japanese numerals:

     (a) 50   (b) 79   (c) 256   (d) 9,307
     (e) 8,008   (f) 5,020   (g) 4,671   (h) 8,888

*24 Convert each of these symbols to Egyptian numerals, to Greek numerals, to Roman numerals, and to Mayan numerals:

     (a) 2,025   (b) 8,063   (c) 15,300   (d) 25,000
     (e) 176,201   (f) 200,500   (g) 1,516,218   (h) 9,720,852

25   Simplify each expression:

     (a) $5^2 \times 5^3 \times 5$ (b) $10^7 \times 10^0 \times 10^2$ (c) $\dfrac{5^9}{5^3}$

     (d) $\dfrac{10^2}{10^3}$ (e) $\left(\dfrac{4}{5}\right)^3$ (f) $\left(\dfrac{5}{9}\right)^0$

     (g) $(10)^2$ (h) $\dfrac{6^0}{6^2}$ (i) $\dfrac{4^0}{5^2}$

26   Show first the two expanded forms and then the decimal numeral for the number six million, five hundred two thousand, eight hundred thirty-four.

27   Follow the example of Figs 1-8 to 1-10 to construct the numeral used to represent the number of letters in the English alphabet using each of these bases: (a) **eight**; (b) **two**; (c) **thirteen**; (d) **nine**; (e) **fifteen.**

28   Use each of the bases of Exercise 27 to make a calendar for the month

of (a) August; (b) February (not leap year). Let each month begin on a Sunday.

*INVITATIONS TO EXTENDED STUDY

1   History has revealed the use of different bases in the number languages of early civilization. What were some of the advantages and disadvantages of these early numeration systems?

2   Look up information on the numeration systems used by different tribes of early American Indians. What reasons would you give as to why some one of these systems did not prevail over the Hindu-Arabic system?

3   Look into the evolution of the shape of our numerals and our number names. Why do differences still exist?

4   Investigate the use of two as a base in electronic computers.

5   If any number **b** is used as the base for a system of numeration, distinct number names are needed for the digital symbols **0, 1, 2, 3, . . . , b − 1** and for the various powers of **b,** namely, **b, b^2, b^3, b^4, ...** How many different number names would be needed to name all the numerals from zero up to and including the numeral equivalent to 1,000—ten in the respective numeration systems for **b = two, five, eight, twelve, twenty,** or **sixty?** For example, if **b** = ten, we need names for 0, 1, 2, 3, 4, 5, 6, 7, 8, 9, 10, 100, and 1,000.

DEVELOPING A
BASIC LANGUAGE

GUIDELINES FOR CAREFUL STUDY

The study of numbers consists almost completely of an analysis of ideas expressed symbolically. Accordingly, a language which can be used efficiently in such a study must be characterized by clarity and effectiveness in the symbolism of its fundamental numeration system.

In this chapter certain concepts and symbols are introduced and discussed. They are of basic significance for intelligent comprehension not only of the content of this chapter but also of that of the entire book. For this reason, clear understanding, which results in accurate interpretation and ready use of each concept and symbol discussed, is very essential.

The following questions can serve you well as guidelines for careful study of the content of this chapter.

1 What is meant when one speaks of a set of elements?
2 What is the meaning of each of the symbols $\in$, $\subset$, and $\subseteq$?
3 What is the meaning of each of these symbols $\notin$, $\not\subset$, and $\not\subseteq$?
4 What are the accepted symbols for indicating a set?
5 What is meant by a well-defined set?
6 What is a set builder?
7 What is the roster symbol for a set?
8 What is a variable?
9 What is the domain, or replacement set, for a variable?
10 What is the inclusion relation?
11 What are the properties of the equivalence relation?
12 What is meant by and what is the symbol for each of these concepts: intersection and union of two sets, universal set, empty set, complement of a set, relative complement of one set with respect to another set, disjoint sets, equal sets?
13 What are Venn diagrams, and how may they be used in the study of set operations?
14 When applied to sets and operations on sets, what are the characteristics of each of these properties: closure, commutative, associative, and distributive?
15 What is the identity element with respect to union of sets? With respect to intersection of sets?
16 What is an ordered pair?
17 When are two ordered pairs said to be equal?
18 What is the cartesian product of two sets?
19 What is a relation?
20 What is meant by the relation of one-to-one correspondence between the elements of two sets?
21 What is meant by the cardinal number of a set?
22 What is meant by $n(S)$ where S represents a set?
23 When are two sets said to be equivalent?

24   What is the distinction between the cardinal and ordinal concepts of number?

25   What is meant by place value, or positional value?

26   What is meant by the base of a numeration system?

INTRODUCTION

What is number? It would be a rather difficult task to give an answer to this question that would be satisfactory from a logical point of view or even intelligible from a utilitarian point of view. In spite of these logical and utilitarian difficulties, it is possible to use our quantitative experiences to shape an intuitive concept of number which will serve as a satisfactory foundation upon which to build for an understanding of the number systems of elementary mathematics. It is the purpose of this chapter to lay the first stones in the structure of this foundation.

In order to pursue this development in an intelligent manner, it is first desirable to introduce certain terms and symbols which can be used advantageously to induce clarifications and avoid ambiguities of discussion.

2-1   THE CONCEPT OF SET

The symbolism and language of sets provide effective media for the discussion of the concepts and techniques important in the study of number and its uses.

Names for collections of objects

In the contemplation of his environment man observes and studies objects, either as individuals or as members of some *collection* or *family* of objects. The teacher has concern for his *class* as well as for each pupil. The grocer buys canned goods by the *case* and, in general, sells them by the single can. The hunter attempts to flush a *bevy* of quail in order to concentrate upon and bring down one for his game bag. The big-game hunter may seek a *herd* of elephants or a *pride* of lions so that he may be successful in bagging an individual animal as the trophy of his hunt. The fisherman seeks the habitat of a *school* of fish before dropping his line. A bridge *hand* of 13 cards is dealt from a *deck* of 52 cards. The tennis player concentrates on winning the individual game in order that he may win enough games to claim first the *set* and then the *match*.

The italicized words in the above paragraph are simply different names used to refer to an identifiable collection of objects. There are many other such words: assemblage, assembly, congress, congregation, flock, troop, posse, ensemble, club, to list a few. Whatever the terminology to specify a particular collection may be, all designated useful collections will have one characteristic in common, namely:

Common characteristic of all useful collections

It will, at least in theory, be possible to identify any given object as either a member of the specified collection or not a member.

Possibly the simplest term, and certainly the most generally used to designate clearly defined collections of objects, is that of *set*. The term "set" is mathematically undefined, but we do appeal to an intuitive understanding of its connotation and of such synonymous concepts as "collection" or "class" to provide the comprehension necessary for its intelligent use.

There are some sets for which the descriptions or definitions are so ambiguous and imprecise that it is difficult to select those elements which belong to them. For other sets this selection is rather simple and exact. In this book we shall be interested only in sets of this second type. Such a set is said to be *well defined* since its elements are so clearly and distinctly recognizable that there is no ambiguity or doubt in selecting them.

Well-defined set

Example For the set of "all interesting books," or the set of "all funny jokes," it would be difficult to identify the elements. Books which are interesting to one person or jokes funny to a certain individual might not be so to another. In contrast, there can be no ambiguity or doubt in selecting the elements of such well-defined sets as the set of "names of the five Great Lakes of the United States," or the set of "all whole numbers greater than 1 and less than 10." While the term "set" is undefined, the collection to which it refers is well defined.

2-2 SET NOTATION

The customary practice among mathematicians is to use capital letters to represent sets and small letters, among other symbols, to represent elements of a set. A convenient symbolic way to say that a is an element of set P is: $a \in P$. This symbol is read "a is an element of the set P," "a is a member of the set P," "a is contained in P," or "a is in P." If b is not in the set P, then we write $b \notin P$, and say "b is *not* an element of set P."

Methods for indicating sets

The three most generally accepted forms for indicating a set are:

1 $S = \{s_1, s_2, s_3, \ldots\}$. This symbol is read "$S$ is the set whose elements are s_1, s_2, s_3, and so on." The three dots (. . .) are used in any case to indicate an omission, whether the omission be of an unending number or of a fixed number of elements. For example, $N = \{1,2,3,4, \ldots\}$ is the symbol for the unending set of counting numbers, otherwise called natural numbers (see Chap. 3); the set $D = \{1,2,3, \ldots, 9\}$ could be used to indicate the set whose elements are the one-digit counting numbers 1, 2, 3, 4, 5, 6, 7, 8, 9. This is called the *roster* form for indicating a set.

Roster

2 $S = \{s \mid s$ has certain prescribed characteristics$\}$. This symbol is read "S is the set of all elements s such that s has certain prescribed characteristics." Of course different prescriptions will distinguish different sets. For example, the two sets N and D can be represented by $N = \{n \mid n$ is a counting number$\}$ and $D = \{d \mid d$ is a one-digit counting number$\}$. These two symbols are read: N is the set of *all* n such that n is a counting number,

and D is the set of *all* d such that d is a one-digit counting number. Note that the vertical bar in the symbol is read "such that." Such a symbol is *Set builder* called a *set selector*, or *set builder*. It indicates the pattern or formula for selecting the elements of the set.

3 $S =$ the set all of whose elements are s_i or even "S is the set all of whose elements are s_i." This is the least formal of the three and merely makes a statement which is sufficiently precise to identify the elements of the set. For example, $N =$ the set of all counting numbers and D is the *Description* set of all one-digit counting numbers. This is called the *description* form for indicating a set.

Attention is called to the fact that, in reading each of these forms, the symbol "$=$" is read simply as "is." Sometimes, in these situations, the symbol is read "is defined as."

In listing the elements of a set it is customary to list each distinct element only once. For example, although each of the letters a, i, and n occurs twice in the word "Indiana," the roster symbol for the set of all letters in the word "Indiana" is $I = \{a,d,i,n\}$. Note that the set symbol uses *only* curly brackets: { }.

Example In writing $I = \{a,d,i,n\}$ we have used the *roster* method for defining the set I. From this definition we can see that $a \in I$, $d \in I$, $i \in I$, and $n \in I$. This last statement may be abbreviated as a, d, i, $n \in I$, which is read "$a,d,i,$ and n are elements of I." It is also clear that $b \notin I$.

The set I may be *described* by the statement: The set I is the set of letters used in spelling the word "Indiana." The *set builder* form for the same set might be written as $I = \{* \mid * \text{ is } a,d,i, \text{ or } n\}$, or $I = \{* \mid * \text{ is one of the letters used in spelling the word "Indiana"}\}$. The first of these symbols is read: I is the set of *all* $*$ (star) such that $*$ is either a, d, i, or n; the second symbol is read: I is the set of *all* $*$ such that $*$ is one of the letters used in spelling the word "Indiana." Any symbol may be used in the set builder, so long as the same symbol is used both before and after the vertical bar.

Variable The symbol used on each side of the vertical bar in a set builder is called a *variable*. It is used to represent any element which can be selected from *Domain* a given set. This set is called the *domain*, or *replacement set*, for the variable.

Example In the set $N = \{n \mid n \text{ is a counting number}\}$, n is the variable and its domain is the set of all counting numbers. In the set builder for the set I of the previous example $*$ is the variable. Its domain is $\{a,d,i,n\}$ What is the variable and its domain in the set builder for the set D of page 33?

Given $E = \{i \mid i = 2n \text{ where } n \text{ is a counting number}\}$. In this set builder there are two variables i and n. The domain for n is the set of all counting numbers, and the domain for i is the set of even counting numbers, or $\{2,4,6,8, \ldots\}$.

Universe of discourse

Universal set

For any particular discussion there exists a "universe of discourse" which is the all-inclusive set of elements clearly identified by the specific orientation of the discussion. This set will be known as the *universal set* and, in general, will be designated by the symbol U. All other sets entering into the discussion will be *subsets* of U. In fact, in general, they will be *proper subsets* of U.

Subset and proper subset

Definition 2-1 The set P is a *subset* of the set Q if and only if every element of P is also an element of Q. The symbol $P \subseteq Q$ is to be read "P is a subset of Q." P is a *proper subset* of Q if and only if the elements of P are also elements of Q and there is at least one element of Q which is not an element of P. The symbol $P \subset Q$ is to be read "P is a proper subset of Q."

Example In any discussion concerning the correct spelling of words of the English language the universal set U is the English alphabet. The 26 letters of the alphabet are the elements from which selections are made to form combinations which we recognize as correctly spelled words. The letters of each word are then elements of a proper subset of U.

Relation

The symbols $\subseteq$ and $\subset$ indicate what we shall call a *relation* between two sets. At this point it will be sufficient to state that the term relation simply implies the consideration of two elements or objects which are linked to each other in some specified manner. Later, it will be desirable to give a more precise definition of the concept of relation. In the symbol $P \subseteq Q$ the set on the left (P) is linked to the set on the right (Q) by the phrase "is a subset of." From Definition 2-1 it should be evident that any set is a subset of itself ($P \subseteq P$). Since P is seen here both on the left and on the right of the symbol, the relation $\subseteq$ may be said to have the effect of reflecting the image of the object. Any relation which has this property is said to be

Reflexive

reflexive. Is the relation $\subset$ reflexive? Why?

Transitive

A property which is common to both relations $\subseteq$ and $\subset$ is that of being *transitive*. This means that each relation has the property of passing on successively from element to element. It should be clear from the definition that if $P \subseteq Q$ and $Q \subseteq R$, then $P \subseteq R$. Why? In other words, the relation "is a subset of" passes on successively from P to Q to R. Likewise, if $P \subset Q$ and $Q \subset R$, then $P \subset R$. Why?

Most frequently in discussions concerned with the relationships existing between two sets, we are not too greatly concerned with whether or not the proper-subset relation holds, but merely with which set, if either, is a subset of the other. In other words, are the elements of the set P also

Contains

elements of the set Q? If the answer to this question is yes, then we say "P is a subset of Q," "P is contained in Q," or "Q contains P," and we write $P \subseteq Q$. There are times when it is convenient to use $Q \supseteq P$ to symbolize "Q contains P."

If we can establish for two given sets P and Q that $P \subseteq Q$ and also $Q \subseteq P$, then we say that set $P =$ set Q. Since all elements of P are included in Q and all elements of Q are included in P, the two sets must have exactly the same elements.

*Equality of
two sets*

Definition 2-2 Set $P =$ set Q if and only if $P \subseteq Q$ and $Q \subseteq P$.

The full implication of this definition is, of course, that two sets are identical if every element of each is an element of the other. Note that, in this case, the symbol "$=$" is read "is the same as."

Example Although in the word "entertainment" each of the letters e, n, and t occurs three times, if $E = \{*\,|\,* \text{ is a letter in the word "entertainment"}\}$, then it may be written in roster form as $E = \{e,n,t,r,a,i,m\}$. Similarly the set $M = \{*\,|\,* \text{ is a letter in the word "maintainer"}\}$, in roster form, is $M = \{m,a,i,n,t,e,r\}$. Examination of these two sets reveals the fact that $E \subseteq M$ and $M \subseteq E$, whence $E = M$. Exactly the same letters are used in spelling the two words.

*Equality of
sets is
reflexive*

*Equality of
sets is
transitive*

*Equality of
sets is
symmetric*

From the definition of equality of two sets it follows directly that any set is equal to itself, $P = P$. This is simply saying that the elements of P *are the same* as the elements of P. Thus, the relation of "equality" between sets is *reflexive*. Furthermore, if the elements of P are the same as the elements of Q, and the elements of Q are the same as the elements of R, then it is true that the elements of P are the same as the elements of R. From these facts it follows that the relation "equality" between sets is also a *transitive* relation. Thus, for sets, the relations $=$ and $\subseteq$ have the reflexive and transitive properties in common. Similarly, it is intuitively acceptable to say that if the elements of P are the same as the elements of Q, then the elements of Q are the same as the elements of P. These facts may be stated symbolically in the form "if $P = Q$, then $Q = P$," and the relation "equality" between sets is said to be *symmetric*. Since this relation is reflexive, symmetric, and transitive it is said to be an *equivalence relation* in accordance with this definition:

*Equivalence
relation*

Definition 2-3 A relation between two sets, or any two elements, is said to be an *equivalence relation* if and only if it is reflexive, symmetric, and transitive.

Symbolically we may write that any relation R between two elements a and b is an equivalence relation if and only if it has these three properties:

Properties of the equivalence relation

Reflexive a R a. (a is R-related to a.)
Symmetric If a R b, then b R a.
Transitive If a R b and b R c, then a R c.

Example In this example the letters a, b, c represent people, S represents the relation "is shorter than," and T represents the relation "is as tall as." The relation T is an equivalence relation while S is not.

The following properties of each relation support this statement.

Since any person is as tall as himself, we have a T a and T is reflexive.

If a is as tall as b, then b is as tall as a. This may be written: if a T b, then b T a, and we see that T is symmetric.

If a is as tall as b and b is as tall as c, then a is as tall as c. Symbolically, if a T b and b T c, then a T c, and T is transitive.

Since T is reflexive, symmetric, and transitive, it is an equivalence relation.

On the other hand, no person is shorter than himself, so a $ a, or a is not shorter than a. For this reason S is *not* reflexive and, therefore, not an equivalence relation.

Furthermore, if a S b, then b $ a, and S is not symmetric.

While S is neither reflexive nor symmetric it is transitive, since if a S b and b S c, then a S c.

Order

A relation such as S, which is neither reflexive nor symmetric but is transitive, may be used to *order* a set of elements. If Carol S Sharon, Sharon S Diana, Diana S Rick, and Rick S David, then Carol, Sharon, Diana, Rick, and David can be arranged in that order from shortest to tallest.

Example There are both boys and girls enrolled in the sixth grade of Valley Elementary School. Let

V = the set of all pupils of Valley Elementary School
C = $\{c \mid c$ is a child enrolled in the sixth grade of Valley Elementary School$\}$ (this is read C is the set of all c such that c is a child enrolled in the sixth grade of Valley Elementary School)
B = $\{b \mid b$ is a boy in the sixth grade of Valley Elementary School$\}$
G = $\{g \mid g$ is a girl in the sixth grade of Valley Elementary School$\}$
P = the set of all boys and girls attending school at Valley Elementary School
T = $\{t_1, t_2\}$ (where t_1 and t_2 represent one each of a set of twins, a brother and sister, enrolled in the sixth grade of Valley Elementary School)

1 $V = P$. Since any pupil of Valley Elementary School is either a boy or a girl attending the school, $V \subseteq P$. Also, since any boy or girl attending the school is a pupil of the school, $P \subseteq V$.

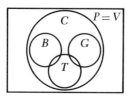

FIGURE 2-1 Venn diagram

2   $C \subset V$. Each child of the sixth grade is a pupil in the school, but there are pupils in the school who are not enrolled in the sixth grade. Why is it also true that $C \subset P$?

3   $B \subset C$ and also $G \subset C$. Why?

4   $B \subset C$ and $C \subset V$; therefore $B \subset V$. This is merely the symbolic way to say that any boy enrolled in the sixth grade of the school is a child enrolled in the sixth grade of the school and is therefore one of the pupils enrolled in the school.

5   T is not a subset of B ($T \not\subset B$) since one of the elements of T is not an element of B. Similarly, $T \not\subset G$. Why is $T \subset C$?

The entire context of this example is portrayed in a vivid manner in the diagram of Fig. 2-1.

The rectangle represents the set V which is the universal set of the example, and, since $P = V$, it also represents the set P. The set C, which is a proper subset of V, is represented by a circle contained entirely within the rectangle. Thus the interior of the rectangle is represented as containing all elements of the set V (all pupils of Valley Elementary School), and the interior of the circle as containing all elements of the set C (all children enrolled in the sixth grade of Valley Elementary School). The sets B, G, and T, which are proper subsets of C, are represented by circles, each of which is contained within the circle C. Since B and G have no elements in common, the circles representing these two sets do not overlap (do not have any portion in common). The set T has one element in common with set B and one element in common with set G. For this reason the circle for T does overlap the circle for B and also the circle for G.

Venn diagrams   Diagrams such as that of Fig. 2-1 are called *Venn diagrams*. They can be used very effectively to give vivid portrayal of indicated relationships between sets. Frequently the picture can be made even clearer and more expressive by the proper use of shading or color.

You should now be able to answer the first eleven of the guideline questions of this chapter. If you are not able to do so you should restudy very carefully pages 32–37 before attempting the exercises which follow.

EXERCISES

1   Which of these sets are well-defined sets?
   (*a*) The set of all letters of the English alphabet
   (*b*) The set of all letters in the word "Mississippi"

(c) The set of all handsome boys

(d) The set of all girls in the junior class of Blackford High School

(e) The set of all baseball teams in the United States

(f) The set of all baseball teams in the National League

(g) The set of all baseball teams in the Western Division of the American League

(h) The set of all days of the week

(i) The set of the names of all months in the year with exactly 31 days

(j) The set of all counting numbers

(k) The set of all good movies

2 Given the sets

$$S = \{2,4,6,8\}$$

and $\quad\quad D = \{d \mid d \text{ is a one-digit counting number}\}$

(a) For what values of the variable d is it true that $d \in S$?

(b) For what values of the variable d is it true that $d \notin S$?

Which of these statements are true and which are false?

(c) $6 \in S$	(d) $\{8\} \notin S$	(e) $\{2,3,4\} \subseteq S$
(f) $\{6\} \subseteq S$	(g) $6 \in \{6\}$	(h) $\{2,4,6,8\} \subset S$
(i) $\{6\} \subset S$	(j) $\{4\} \in S$	(k) $\{2,4,6\} \subset S$
(l) $\{8,2,6,4\} \subseteq S$	(m) $8 \subset S$	(n) $\{4,2,6\} \subseteq S$
(o) $\{2,3,4,5\} \nsubseteq S$	(p) $\{4,8\} \nsubseteq S$	

3 If $E = \{e \mid e \text{ is an even number between 1 and 9}\}$, and $S = \{2,4,6,8\}$, which of these statements are true and which are false?

(a) $E \subseteq S$ (b) $S \subset E$ (c) $E = S$ (d) $S \nsubseteq E$ (e) $E \not\subset S$

4 Which of the following words *cannot* be spelled if only the letters in the set $\{e,n,s,t\}$ can be used? Tens, send, Tennessee, tents, cent, tense, tension, set, nets, settee, setter, sentence, sennet, sent, senescent, sennit, sense

5 These sets are stated in the description form. Write the roster symbol and the set-builder symbol for each set.

(a) V is the set of vowels used to spell "equation."

(b) W is the set of whole numbers larger than 5 and less than 20.

(c) O is the set of odd numbers between and including 1 and 15.

(d) L is the set of letters used to spell the word "college."

(e) M is the set of names of the months in a year.

(f) P is the set of vowels used to spell "ambiguous."

(g) F is the set of names of all months of less than 30 days.

(h) E is the set of all numbers of the form $2k$ where the domain of k is $\{1,2,3,4,5\}$.

6 If S is the set of letters used in spelling "simp" and $M = \{l \mid l \text{ is a letter in the word "Mississippi"}\}$, what relation exists between sets S and M?

7 Write in roster form the symbol for the set C where C is the set of all letters common to the set B of letters used in spelling "bric-a-brac," and the set A of letters used in spelling "abracadabra."

8 Write in roster form the set K of all letters which are either in set A or in set B or in both sets A and B of Exercise 7.

9 Write in description form and in the set-builder form each of these sets, given in roster form.
   (a) $I = \{1,2,3,4,5\}$
   (b) $N = \{1,2,3,4,5, \ldots\}$
   (c) $R = \{2,4,6,8,10, \ldots, 18\}$
   (d) $S = \{$Atlantic, Arctic, Indian, Pacific$\}$
   (e) $U = \{$Utah$\}$
   (f) $T = \{$April, June, September, November$\}$

10 Write in description form and in roster form each of these sets, given in set-builder form.
   (a) $G = \{2k + 1 \mid k \in \{1,2,3,4, \ldots, 9\}\}$
   (b) $A = \{* \mid *$ is the name of a state in the United States which starts with the letter $A\}$
   (c) $B = \{n \mid n$ is an even number larger than 1 and smaller than 25$\}$
   (d) $D = \{d \mid d$ is the name of a day of the week$\}$
   (e) $C = \{\# \mid \#$ is the name of a state in the United States which starts with the letter $C\}$ ($\#$ may be read "sharp.")

11 Are the inclusion relations $\subseteq$ and $\subset$ between sets equivalence relations? Why?

12 What relation exists between sets P and C where $P = \{u \mid u$ is a letter used to spell "participator"$\}$ and C is the set of letters used to spell "capacitor"? Give reasons to support your answer.

13 In this exercise the symbols refer to sets defined in Exercises 5, 9, and 10. Which of these statements are true?

(a) $C = U$	(b) $U = F$	(c) $H = V$	(d) $P = V$
(e) $P \subset H$	(f) $O \subseteq G$	(g) $R \subseteq B$	(h) $H \subseteq V$
(i) $B \subseteq R$	(j) $R \subset B$	(k) $R \subset N$	(l) $F \subset M$
(m) $T \subset M$	(n) $W \subseteq N$	(o) $E \subset R$	(p) $O \subseteq W$
(q) $C \subseteq A$	(r) $F \subseteq T$	(s) $G \subset N$	(t) $V \subset H$

14 Use Definition 2-2 to construct the argument for proving that the relation $=$ between two sets is an equivalence relation.

*15 Which of these relations are equivalence relations? In each case give the argument to support your answer.
   (a) "Is not equal to," applied to numbers
   (b) "Is a sister of," applied to people
   (c) "Has the same length as," applied to skirts
   (d) "Lives within a mile of," applied to people
   (e) "Is equal to," applied to numbers
   (f) "Is east of," applied to cities in the United States
   (g) "Was born in the same town as," applied to people

(h) "Is taller than," applied to buildings
(i) "Has the same shape as," applied to geometric figures
(j) "Is a son of," applied to people
(k) "Is included in," applied to sets
(l) "Has the same area as," applied to farms
(m) "Is on the same side of Main Street," applied to buildings
(n) "Intersects," applied to straight lines
16 (a) There are five subsets of $L = \{a,d,h,i,n,o,w\}$, each of which contains only those letters needed to spell the name of a state of the United States. List in roster form each of these subsets and indicate the state to which it applies.
     (b) There is only one letter common to these five subsets. What is it?

2-3   OPERATIONS ON SETS

There are three basic operations used in working with sets. They are *complementation*, *union*, and *intersection*. Each of these operations will now be defined and also illustrated by an appropriate Venn diagram. In Figs. 2-2 to 2-6, circles and color are used to symbolize sets of elements selected from the universal set U, which, in turn, is represented by the rectangle.

Complement   **Definition 2-4**   When the universal set U is clearly defined, the complement of P is the set of all those elements of U which are not elements of P. The symbol $\sim P$ is read "the complement of P," sometimes abbreviated to "not P." (See Fig. 2-2.)

Example   If U is the set of all letters of the English alphabet and P is the set of all letters used to spell the word "mathematics," then $\sim P$ is the set of all letters of the English alphabet not used in spelling the word "mathematics."

$$U = \{a,b,c,d,e,f,g,h,i,j,k,l,m,n,o,p,q,r,s,t,u,v,w,x,y,z\}$$
$$P = \{a,c,e,h,i,m,s,t\}$$
$$\sim P = \{b,d,f,g,j,k,l,n,o,p,q,r,u,v,w,x,y,z\}$$

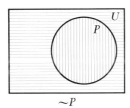

FIGURE 2-2   The rectangle represents U. The vertical lines represent P. The horizontal lines represent $\sim P$, that is, all those elements of U which are not elements of P.

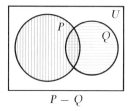

$P - Q$

FIGURE 2-3 The rectangle represents U. The vertical lines represent P and the horizontal lines represent Q. The portion of P containing vertical lines *only* represents $P - Q$, that is, those elements of P which are not also elements of Q. The portion of Q containing *only* horizontal lines represents $Q - P$, or those elements of Q which are not also elements of P.

At times one is concerned not so much with a clear-cut definition of the universal set as with those elements of one set P which are not elements of another set Q.

Relative complement **Definition 2-5** The *relative complement of set Q in set P* consists of all elements of set P which are not also elements of set Q. The symbol $P - Q$ is to be read "the relative complement of Q in P" or, in abbreviated form, "P minus Q." (See Fig. 2-3.)

Notice that the diagram emphasizes the fact that the relative complement of Q in P has meaning even when $Q \not\subset P$ (Q *is not a proper subset of P*). What is the diagram when $Q \subset P$?

Example Given $T = \{1,2,3,4,5,6,7,8,9\}$, $S = \{3,5,7,9,10,12,13\}$, and $R = \{2,4,6,8\}$.

$T - S = \{1,2,4,6,8\}$ $S - T = \{10,12,13\}$

$S - R = \{3,5,7,9,10,12,13\} = S$ $R - S = \{2,4,6,8\} = R$

$T - R = \{1,3,5,7,9\}$ $R - T$ contains no elements since $R \subset T$.

These same sets T, S, and R will be used to illustrate the concepts of union and intersection of two sets, which are now defined.

Union of two sets **Definition 2-6** The *union* of two sets P and Q consists of all those elements which are either in P or in Q or in both P and Q, with no duplication of elements. The symbol $P \cup Q$ is read "P union Q" or, more simply, "P cup Q." (See Fig. 2-4.)

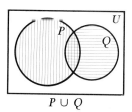

$P \cup Q$

FIGURE 2-4 The rectangle represents U. The vertical lines indicate elements of P. The horizontal lines indicate elements of Q. The set $P \cup Q$ is represented by lines in *either* direction.

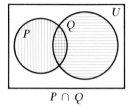

$$P \cap Q$$

FIGURE 2-5 The rectangle represents U. The vertical lines indicate elements of P. The horizontal lines indicate elements of Q. The set $P \cap Q$ is represented by crossing lines.

Intersection of two sets **Definition 2-7** The *intersection* of two sets P and Q consists of all those elements which are common to both P and Q. The symbol $P \cap Q$ is to be read "P intersection Q" or more simply, "P cap Q." (See Fig. 2-5.)

Example For the sets T, S, R, of the preceding example we have

$T \cup S = \{1,2,3,4,5,6,7,8,9,10,12,13\}$ $\qquad$ $T \cap S = \{3,5,7,9\}$

$T \cup R = \{1,2,3,4,5,6,7,8,9\} = T$ $\qquad$ $T \cap R = \{2,4,6,8\} = R$

$S \cup R = \{2,3,4,5,6,7,8,9,10,12,13\}$ $\qquad$ $S \cap R$ contains no elements since S and R have no elements in common.

In this example the intersection of sets T and S and the intersection of sets T and R are sets with easily identified elements. The intersection of sets S and R, however, presents a problem. The two sets have no elements in common, so there are no elements which we can identify as belonging to the set which is their intersection. In order that there will be no exception to the concept of the intersection of two sets it thus becomes necessary *Empty or null set* to define the *empty set, or null set, as that set which has no elements.* The symbol ∅ is read "the empty set" or "the null set"; the two expressions are used interchangeably. We may now write $S \cap R = \varnothing$, which is read "the intersection of sets S and R is the empty set." Similarly, in the example following Definition 2-5, we have $R - T = \varnothing$, or the relative complement of T in R is the empty set.

The sets S and R are disjoint sets in accordance with the following definition.

Disjoint sets **Definition 2-8** Two nonempty sets P and Q are said to be *disjoint sets* if their intersection is the empty set, that is, if they have no elements in common. (See Fig. 2-6.)

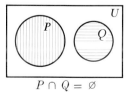

$$P \cap Q = \varnothing$$

FIGURE 2-6 The rectangle represents U. The vertical lines indicate elements of P. The horizontal lines indicate elements of Q. The set $P \cap Q$ is ∅ since there is no portion of U which has crossing lines. The sets P and Q have no elements in common. They are *disjoint sets*.

What sets of Fig. 2-1 are disjoint sets?

Attention is called to the fact that the set $\{0\}$ is not the null set. It is not empty since it contains one element, the element zero. Also the set $\{\varnothing\}$ is not the null set, since it contains the element $\varnothing$. This new set $\varnothing$ allows the use of the complement of a set with no exceptions. For example, $\sim U$ is $\varnothing$. Similarly, $\sim \varnothing$ is U. Although we shall make no effort to prove the statement, it is possible to prove that the empty set is a subset of every set. Symbolically, this statement may be written $\varnothing \subseteq P$ for every set P. Furthermore, $\varnothing$ introduces no exceptions when subjected to the set operations of union, intersection, and relative complementation.

∅ ⊆ P for all P

Now that the empty set has been introduced as a well-defined set for which the set operations also are well defined, it becomes desirable to specify that, *unless otherwise stated, all sets used in subsequent discussions, examples, and exercises will be assumed to be nonempty.*

If you have a clear comprehension of the content of Secs. 2-1 to 2-3, you should have no difficulty in answering the first thirteen guideline questions (page 31). You should review these questions and study this example before attempting the exercises which follow.

Example Consider the following sets:

U (the universal set) is the set of all letters of the English alphabet.

$$A = \{a,b,c,j,k,l,s,t,u\}$$
$$B = \{j,k,s,t\}$$
$$C = \{l,m,n,u,v,w\}$$
$$D = \{a,b,c,l,u\}$$
$$T = \{l \,|\, l \text{ is a letter in the word Tennessee}\}$$

S is the set whose elements are the letters e, n, s, t.

$\varnothing$

1 All sets, including U, are subsets of U, and $\varnothing$ is a subset of all sets, including $\varnothing$.

2 When any element from U is being considered, one can determine clearly and convincingly whether or not it is a member of any selected subset of U. For example, a is a member of A $(a \in A)$ but a is not a member of B $(a \notin B)$.

3 B is a proper subset of A $(B \subset A)$, since all elements of B are also elements of A and there is at least one element in A which is not in B.

4 C is not a subset of A $(C \not\subseteq A)$. Although the elements l and u which are members of C are also members of A, there are other elements of C which are not elements of A, namely, m, n, v, w.

5 $C \cap A = \{l,u\}$, since l and u are the only elements which the two sets have in common.

6 $A \cup B = A$, since $B \subset A$.

7 $A - B = D$, since the elements of D are left in A after the elements of B have been removed.

8 $B \cap C = \varnothing$, since B and C have no elements in common. They are disjoint sets.

9 $S = T$ since $T = \{t,e,n,s\}$, because the set selector selects *only* the letters used in the word Tennessee and says nothing about how many times any one letter is used nor anything about the arrangement of the letters. It follows then that $S \subseteq T$ and $T \subseteq S$.

10 $S - T = \varnothing$.

EXERCISES

1 Use the form $S = \{s_1,\ s_2,s_3,s_4,\ \ldots\}$ to describe each set.
  (*a*) $I = \{i \,|\, i$ is a counting number$\}$
  (*b*) $S = \{s \,|\, s$ is an integer greater than 5 but less than 20$\}$
  (*c*) $G = \{g \,|\, g$ is a state of the United States one of whose boundaries lies along the Gulf of Mexico$\}$
  (*d*) $L = \{l \,|\, l$ is the name of one of the Great Lakes of the United States$\}$
  (*e*) $E = \{e \,|\, e$ is a positive even integer$\}$
  (*f*) $M = \{m \,|\, m$ is a state of the United States bordering on Mexico$\}$

2 Write a symbol which will be the set selector for each of these sets:
  (*a*) $O = $ the set of all odd-counting numbers
  (*b*) $T = \{2,4,6,8,10,12\}$
  (*c*) $R = \{$Oregon, Hawaii, Alaska, California, Washington$\}$
  (*d*) $V = \{a,e,i,o,u\}$
  (*e*) $A = \{3, 1 + 2, 2 + 1, 1 + 1 + 1\}$
  (*f*) $B = \{5,7,9,11,13,15,17,19\}$
  (*g*) $P = \{1,3,5,7,9\}$
  (*h*) $Z = $ the set of all integers

3 Given the two sets

$$M = \{0,2,4,6,8\} \qquad \text{and} \qquad N = \{0,1,3,5,7,9\}$$

Is this a true statement: $\varnothing = M \cap N$? Give reasons to support your answer.

4 Identify all subsets of $R = \{0,1,2,3\}$.

5 Are these true or false statements about the sets of Exercises 1 and 2?

  (*a*) $I = Z$ (*b*) $O \subset I$ (*c*) $E \cap T = \varnothing$
  (*d*) $E \cup T = E$ (*e*) $M \cap R = \varnothing$ (*f*) $P \cup O = O$
  (*g*) $8 \notin E$ (*h*) $9 \in O$ (*i*) $M \not\subseteq G$
  (*j*) $B \not\subseteq S$ (*k*) $P - B = \{1,3\}$ (*l*) $P \not\subseteq O$
  (*m*) $G = R$ (*n*) $P \cap B = \{5,7,9\}$ (*o*) $P \cap T = \varnothing$

6 What is the roster symbol for each of these sets? The letters represent sets of Exercises 1 and 2.

  (*a*) $B \cap P$ (*b*) $O \cup E$ (*c*) $G \cap M$
  (*d*) $B - P$ (*e*) $B \cup S$ (*f*) $B \cap S$

7 Consider these sets:

$$F = \{\text{apple, peach, pear, orange}\}$$
$$V = \{\text{onion, carrot, potato, bean}\}$$
$$C = \{\text{apple, pear, carrot}\}$$
$$D = \{\text{apple, peach, orange}\}$$
$$E = \{\text{onion, carrot}\}$$

Which of the following statements are true and which are false?

(a) $F = V$ (b) $C \subseteq F$ (c) $E \subset V$
(d) $V \subset V$ (e) $D \cup C = F$ (f) $C \cap D = E$
(g) $F - D = \{\text{pear}\}$

8 If U is the universal set, what is the set $A \cup U$? What is the set $A \cap U$?
9 What is the set $U - A$? What is the set $A - U$?
10 What is the set $A \cup \sim A$? What is the set $A \cap \sim A$?
*11 Draw Venn diagrams to show each of the following sets:

(a) $(A \cup B) \cup C$ (b) $A \cap (B \cap C)$
(c) $A \cup (B \cap C)$ (d) $(A \cup B) \cap C$
(e) $(A \cup B) \cap (A \cup C)$ (f) $(A \cap B) \cup (A \cap C)$
(g) $A - B$ when $A \cap B = \varnothing$ (h) $\sim(A \cup B)$
(i) $\sim(A \cup B)$ (j) $\sim[(A \cap B) \cup C]$

*12 Use Venn diagrams to illustrate the truth of the following statements:

(a) $A \cup B = B \cup A$
(b) $A \cap B = B \cap A$
(c) $(A \cup B) \cup C = A \cup (B \cup C)$
(d) $(A \cap B) \cap C = A \cap (B \cap C)$
(e) $A \cup (B \cap C) = (A \cup B) \cap (A \cup C)$
(f) $A \cap (B \cup C) = (A \cap B) \cup (A \cap C)$

13 Let $C = $ the set of all the different letters in the word Cincinnati, and $M = $ the set of all the different letters in the word cinema. Are these sentences true or false? Give reasons for each answer.

(a) $C - M = M - C$ (b) $C = M$
(c) $C \cup (C - M) = C$ (d) $M \cap (C - M) = M$

14 Give six examples of the empty set.
15 Is the following statement true or false? The empty set has a subset but it has no proper subset. Give reasons to support your answer.
16 What is the resulting set in each of the following cases where A represents a nonempty set?

(a) $\varnothing \cup \varnothing$ (b) $\varnothing \cap \varnothing$ (c) $\varnothing - \varnothing$ (d) $A - \varnothing$
(e) $\varnothing - A$ (f) $A - A$ (g) $A \cap \varnothing$ (h) $A \cup \varnothing$

*17 Given the universal set $U = \{a,b,c,d,e,f\}$. In each case use the given information to identify the elements of the set P. Write P in roster form.

(a) $P \cup Q = U,$ $Q = \{a,b,c\},$ $P \cap Q = \varnothing$
(b) $P \cup Q = U,$ $Q = \{a,b,c\},$ $P \cap Q = \{a\}$
(c) $P = \sim Q,$ $Q = \{b,e,f\}$
(d) $P \cup Q = U,$ $P \cap Q = \{d,e\},$ $Q - P = \{a,b,c\}$
(e) $P \cap Q = \varnothing,$ $P \cup Q = \{a,b,c,d,f\},$ $P \cup (Q \cup R) = U$
       $Q \cap R = \varnothing,$ $P \cap R = \{a,c\},$ $Q \cup R = \{a,c,d,e,f\}$

For Exercises 18 to 29 there are corresponding Venn diagrams. Each diagram is shaded to represent the result of an operation or a combination of operations discussed in this section. The problem is to select from the given list of 16 indicated operations the proper one for each Venn diagram. In some cases one diagram will have more than one label. The given list of operations is:

(a) $Q \cap P$ (b) $P - Q$
(c) $\sim(\sim P)$ (d) $(P \cup Q) \cap (P \cup R)$
(e) $Q \cap \sim P$ (f) $\sim(P \cup Q)$
(g) $P \cup (Q \cap R)$ (h) $P \cap (Q - R)$
(i) $\sim \varnothing$ (j) $Q - P$
(k) $P \cup (Q \cup R)$ (l) $(P \cap Q) \cup (P \cap R)$
(m) $Q \cup P$ (n) $\sim(P \cap Q)$
(o) $P \cap (R - Q)$ (p) $P \cap (Q \cap R)$

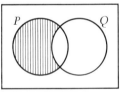

EXERCISE 18

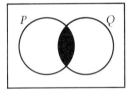

EXERCISE 19

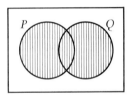

EXERCISE 20

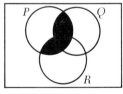

EXERCISE 21

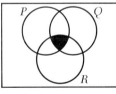

EXERCISE 22

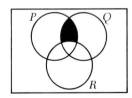

EXERCISE 23

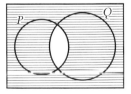

EXERCISE 24

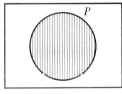

EXERCISE 25

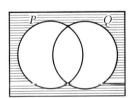

EXERCISE 26

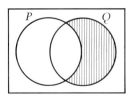

EXERCISE 27

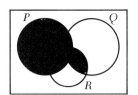

EXERCISE 28

EXERCISE 29

*30 Use the sets U, P, and Q as pictured in the figure and write in roster form the symbol for each of these sets.

(a) $P \cup Q$ (b) $P \cap Q$

(c) $\sim(P \cup Q)$ (d) $\sim(P \cap Q)$

(e) $P - Q$ (f) $\sim P$

(g) $Q - P$ (h) $(P - Q) \cup (Q - P)$

(i) $\sim Q$ (j) $\sim(P - Q)$

(k) $\sim(Q - P)$ (l) $\sim(P - Q) \cap \sim(Q - P)$

(m) $\sim P \cap Q$ (n) $\sim U$

(o) $\sim P \cap \sim Q$ (p) $(P - Q) \cap (Q - P)$

(q) $\sim \varnothing$ (r) $P \cap \sim Q$

(s) $\sim P \cup \sim Q$ (t) $\sim(P - Q) \cup \sim(Q - P)$

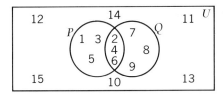

EXERCISE 30

*31 Given the sets P, Q, and R of the figure. Write the symbol for each of these sets in terms of operations on P and Q; P and R; Q and R; or P, Q, and R. Example:

$$Q - P = \varnothing \text{ (why?)} \quad P \cap Q = \{2,3\} \quad (P \cap Q) \cap R = \{3\}$$

(a) $\{1,2,3,4,5\}$ (b) $\{3,4\}$ (c) $\{1,4\}$ (d) $\{2,3,4\}$

(e) $\{1,2,3,4\}$ (f) $\{5\}$ (g) $\{4,5\}$ (h) $\{2,3,5\}$

(i) $\{1,2\}$ (j) $\{2\}$ (k) $\{1\}$ (l) $\{4\}$

EXERCISE 31

32 Given the universal set

$$U = \{a,b,c,d,e,f,g,h,i\}$$
$$A = \{a,e,i\}$$
$$B = \{e,f,g,h,i\}$$

Show that these are true equalities:

(a) $(A \cap B) \cup (A \cap \sim B) = A$
(b) $(A \cap B) \cap (A \cup \sim B) = A$

33 Given that the universal set U is the set of all letters of the English alphabet, P is the set of letters used to spell the word "elementary," and Q is the set of letters used to spell the word "education." Use the roster method to represent each of these sets:

(a) $P \cap Q$ (b) $P \cup Q$ (c) $\sim P$ (d) $\sim Q$
(e) $\sim P \cap \sim Q$ (f) $P - Q$ (g) $Q - P$ (h) $\sim P - \sim Q$
(i) $\sim Q - \sim P$ (j) $\sim P \cup \sim Q$

34 Use a set-builder symbol to represent each of these sets:

(a) $A \cup B$ (b) $A \cap B$ (c) $\sim A$ (d) $A - B$
(Hint: $A \cup B = \{e \mid e \in A \text{ or } e \in B\}$.)

35 (a) In each case draw a Venn diagram to illustrate the truth or falsity of the statement.

$$(P - Q) \cup (Q - P) = (P \cup Q) - (P \cap Q)$$
$$(P - Q) \cup (Q - P) = \sim(P \cap Q)$$

(b) Give an argument to support your answer.

*2-4 SETS WHOSE ELEMENTS ARE SETS

So far the discussion has been concerned only with sets whose elements have been primarily numerals representing counting numbers, letters of the alphabet, names of people or places. It is frequently the case that the elements of a given set are themselves sets. For example, in Exercise 4 on page 45, you were asked to identify all the subsets of $R = \{0,1,2,3\}$. The answer to this exercise is given here as the set S, where

$$S = \{\varnothing, \{0\}, \{1\}, \{2\}, \{3\}, \{0,1\}, \{0,2\}, \{0,3\}, \{1,2\}, \{1,3\},$$
$$\{2,3\}, \{0,1,2\}, \{0,1,3\}, \{0,2,3\}, \{1,2,3\}, R\}$$

The set operations also apply to sets of this type. The following example shows the operation tables for $\cup$ and $\cap$ defined on the set $T = \{\varnothing, A, \sim A, U\}$ where the elements are related subsets of the universal set U.

Example   The two tables of this example sum up the results of the operations ∪ and ∩ on the set T whose elements are ∅, a given set A and its complement, and the universal set U. $T = \{\emptyset, A, \sim A, U\}$.

∪	∅	A	$\sim A$	U
∅	∅	A	$\sim A$	U
A	A	A	U	U
$\sim A$	$\sim A$	U	$\sim A$	U
U	U	U	U	U

∩	∅	A	$\sim A$	U
∅	∅	∅	∅	∅
A	∅	A	∅	A
$\sim A$	∅	∅	$\sim A$	$\sim A$
U	∅	A	$\sim A$	U

These tables are to be read in the manner indicated here for the row in each table with A in the left-hand column.

$$A \cup \emptyset = A \qquad A \cup A = A \qquad A \cup \sim A = U \qquad A \cup U = U$$
$$A \cap \emptyset = \emptyset \qquad A \cap A = A \qquad A \cap \sim A = \emptyset \qquad A \cap U = A$$

EXERCISES

1   For the set $Q = \{1,2\}$ write the roster symbol for the set Y whose elements are all the subsets of Q. Construct the operation tables for ∪ and ∩ defined on Y.

2   If $P = \{1,2,3\}$, the set X of all subsets of P is $X = \{\emptyset, A, B, C, D, E, F, P\}$ where $A = \{1\}$, $B = \{2\}$, $C = \{3\}$, $D = \{1,2\}$, $E = \{1,3\}$, and $F = \{2,3\}$. Complete these operation tables for ∪ and ∩ defined on X.

∪	∅	A	B	C	D	E	F	P
∅								
A							P	
B								
C								
D								
E			E					
F								
P								

∩	Ø	A	B	C	D	E	F	P
Ø								
A							Ø	
B								
C								
D								
E					A			
F								
P								

EXERCISE 2

Four operations are completed for you: $A \cup F = P$, $E \cup C = E$, $A \cap F = \varnothing$, and $E \cap D = A$. What can you say immediately about $F \cup A$, $C \cup E$, $F \cap A$, and $D \cap E$?

*3 Construct operation tables for $\cup$ and $\cap$ defined on set S of the first paragraph of this section.

2-5 PROPERTIES OF THE SET OPERATIONS

Clear understanding of the basic properties which characterize the different set operations will provide a strong foundation for subsequent study of number and its use. This section, therefore, will be devoted to a closer analysis of the operations defined in the previous section in order to determine just what these properties are and how they govern the use of each operation. No attempt will be made to argue their validity, but appeal will be made to definitions, illustrative examples, and Venn diagrams to give support to their intuitive acceptance.

Unary operation   Complementation is a *unary operation* since, by definition, it involves one and only one set. (See Fig. 2-2.)

Example If $A = \{a,b,c,d\}$ and $U = \{a,b,c,d,e,f,g,h\}$ then $\sim A = \{e,f,g.h\}$.

Binary operation   In contrast union, intersection, and relative complementation are *binary operations* since, by definition, they each involve two sets. This fact is illustrated in Figs. 2-3 to 2-5.

Whether the operation is the unary operation of complementation or either of tho binary operations of union and intersection, the result is, in every case, a set of elements from the same universal set. For this reason, each *Closure* operation is said to have the property of *closure*.

An examination of Figs. 2-4 and 2-5 reveals that the same result is obtained whether we speak of $P \cup Q$ or $Q \cup P$, and also whether we speak of $P \cap Q$ or $Q \cap P$. Notice that in the two operations $P \cup Q$ or $Q \cup P$, the order of the sets has been changed without bringing about any change in the elements of the resultant set. The same thing is true for $P \cap Q$ and $Q \cap P$. For this reason the binary operations of union and intersection of *Commutative* sets are said to be *commutative*. The binary operation of relative comple-

property mentation is *not* commutative. Figure 2-3 illustrates very vividly that the set $P - Q$, in general, is not the same as the set $Q - P$.

Since $\cap$ and $\cup$ are binary operations, a problem is created when there are three or more sets to be combined under either operation. For example, what does $A \cup B \cup C$ or $A \cap B \cap C$ mean? It is not too difficult to argue that it is immaterial whether one first finds $A \cup B$ and then takes the union of this set with C, or if one takes the union of A with the set $B \cup C$. In other words $(A \cup B) \cup C$ is the same set as $A \cup (B \cup C)$. For this reason

Associative the binary operation $\cup$ is said to be *associative* and either form may be

property used to find the union of three or more sets. Similarly, the operation $\cap$ is associative. (See Exercises 12*c* and *d*, page 46.)

Example   Consider the sets

$$A = \{1,2,3,4,5,6\}$$
$$B = \{3,5,6,7,8\}$$
$$C = \{1,3,6,8,9,10\}$$

$A \cup B = \{1,2,3,4,5,6,7,8\}$	$A \cap B = \{3,5,6\}$
$(A \cup B) \cup C = \{1,2,3,4,5,6,7,8,9,10\}$	$(A \cap B) \cap C = \{3,6\}$
$B \cup C = \{1,3,5,6,7,8,9,10\}$	$B \cap C = \{3,6,8\}$
$A \cup (B \cup C) = \{1,2,3,4,5,6,7,8,9,10\}$	$A \cap (B \cap C) = \{3,6\}$
$(A \cup B) \cup C = A \cup (B \cup C)$	$(A \cap B) \cap C = A \cap (B \cap C)$

The binary operation of relative complementation does not have the property of being associative.

Example   Consider the same sets of the previous example.

$A - B = \{1,2,4\}$	$(A - B) - C = \{2,4\}$
$B - C = \{5,7\}$	$A - (B - C) = \{1,2,3,4,6\}$
$(A - B) - C \neq A - (B - C)$	

The two binary operations of $\cup$ and $\cap$ can be combined in a very

Distributive important way by what is known as the *distributive property*. (See Exer-

property cises 12*e* and *f*, page 46.)

$$A \cap (B \cup C) = (A \cap B) \cup (A \cap C)$$
or
$$A \cup (B \cap C) = (A \cup B) \cap (A \cup C)$$

In the first case we say that "intersection is distributive over union," and in the second case, that "union is distributive over intersection." It is also true that

$$(B \cup C) \cap A = (B \cap A) \cup (C \cap A)$$

and
$$(B \cap C) \cup A = (B \cup A) \cap (C \cup A)$$

Example Again using the same sets A, B, and C of the two previous examples, we have

$$(B \cup C) = \{1,3,5,6,7,8,9,10\}$$
$$A \cap (B \cup C) = \{1,3,5,6\}$$
$$A \cap B = \{3,5,6\} \qquad A \cap C = \{1,3,6\}$$
$$(A \cap B) \cup (A \cap C) = \{1,3,5,6\}$$
$$A \cap (B \cup C) = (A \cap B) \cup (A \cap C)$$

Also
$$B \cap C = \{3,6,8\}$$
$$A \cup (B \cap C) = \{1,2,3,4,5,6,8\}$$
$$A \cup B = \{1,2,3,4,5,6,7,8\}$$
$$A \cup C = \{1,2,3,4,5,6,8,9,10\}$$
$$(A \cup B) \cap (A \cup C) = \{1,2,3,4,5,6,8\}$$
$$A \cup (B \cap C) = (A \cup B) \cap (A \cup C)$$

Use these same sets to illustrate that

$$(B \cup C) \cap A = (B \cap A) \cup (C \cap A)$$
$$(B \cap C) \cup A = (B \cup A) \cap (C \cup A)$$

are true statements.

By definition $A \cup \varnothing = A$ and $A \cap U = A$. Thus, the union of any set *Identity element* A with the empty set is always A. For this reason $\varnothing$ is called the *identity element* for the operation of *union*. Similarly, the intersection of any set A with the universal set is always A, and U is the *identity element* for *intersection*.

EXERCISES

Given the sets $P = \{a,b,c,d,e,f\}$
$$V = \{a,e,i,o,u\}$$
$$Q = \{b,d,o,u\}$$
$$U = \text{the set of letters of the English alphabet}$$
$$\varnothing = \text{the empty set}$$

Use these sets to illustrate the properties of closure, associativity, commutativity, and identity as they apply to the binary operations of $\cup$, $\cap$, and relative complementation.

2-6 THE CARTESIAN PRODUCT

Ordered pair

Another method for combining the elements of two sets is that of *cartesian product*. When two sets are combined by union or intersection the elements of the resulting set are selected elements of the given sets. This is not the case when the cartesian product is formed. Before defining this new concept it is desirable to state what is meant by an ordered pair of elements. Such pairs of numbers, or numbers and letters, are used in forming grids for graphs and maps. They simplify the task of locating points on a graph and cities or other locations on a map. A symbol of the form (x, y) is called an *ordered pair*† simply because a pair of symbols, x and y, is used, and one of them, x, is written first and called the *first element*, while the other, y, is written second and called the *second element*. It is important to note that the ordered pair (y,x) has a different ordering of its components x and y from that of the ordered pair (x,y). Two ordered pairs are *equal* if and only if they have the same first element and the same second element:

Equality of ordered pairs

$$(x, y) = (a,b) \quad \text{if and only if } x = a \text{ and } y = b$$

Example $(1,2) \neq (2,1)$ and $(a,b) = (1,2)$ if and only if $a = 1$ and $b = 2$.

$\{1,2\}$ is the same as $\{2,1\}$, whose two elements are 1 and 2.
$\{1,2\}$ does not mean the same as $(1,2)$.
$\{(1,2)\}$ is a set whose one element is the ordered pair $(1,2)$.
$\{(1,2), (2,1)\}$ is a set of two distinct elements $(1,2)$ and $(2,1)$.

The parentheses are an essential part of the symbol for an ordered pair. Just listing two symbols such as 1,2 or a,b does *not* indicate an ordered pair.

Cartesian product

Definition 2-9 The *cartesian product*, $P \times Q$, of the two sets P and Q is the set of all possible ordered pairs (p,q) such that $p \in P$ and $q \in Q$. The symbol $P \times Q$ is read the cartesian product of sets P and Q, the cross product of P and Q, or, still more briefly, P cross Q.

$$P \times Q = \{(p,q) \mid p \in P, q \in Q\}$$

From the nature of the definition it should be clear that it is possible to have the cross product of a set with itself. Thus $P \times P = \{(p,q) \mid p, q \in P\}$.

Example Consider the sets $P = \{a,b,c\}$ and $Q = \{1,2\}$.

$$P \times Q = \{(a,1), (a,2), (b,1), (b,2), (c,1), (c,2)\}$$

---

† A more formal and rigorous definition may be found in Norman T. Hamilton and Joseph Landin, "Set Theory: The Structure of Arithmetic," p. 44, Allyn and Bacon, Inc., Boston, 1961.

$Q \times P = \{(1,a), (2,a), (1,b), (2,b), (1,c), (2,c)\}$
$P \times Q \neq Q \times P$ but they have the same number of elements.
$P \times P = \{(a,a), (a,b), (a,c), (b,a), (b,b), (b,c), (c,a), (c,b), (c,c)\}$
$Q \times Q = \{(1,1), (1,2), (2,1), (2,2)\}$

From the definition and the example it can be observed that the cartesian product of two sets results in a set of ordered pairs, the elements of which bear some relation to each other. In the set $P \times Q$ of the example, the only observable relation is that each element of Q is associated as a second element to an element of P as the first element. This is quite a different relation from that expressed by the set $Q \times P$. In either case the relation is said to be defined *on the sets P and Q*. The same type of relation is pictured both in $P \times P$ and in $Q \times Q$ between the respective elements of each set. There are, however, other relations expressed by certain subsets of $P \times P$ and $Q \times Q$. For example, the set $\{(a,a), (b,b), (c,c)\}$, which is a subset of $P \times P$, defines the relation "is the same as" on the set P. a is the same as a, b is the same as b, and c is the same as c. Similarly, the set $\{(1,1), (2,2)\}$, a subset of $Q \times Q$, defines the same relation on the set Q. Two other subsets from $Q \times Q$ are $\{(2,1)\}$ and $\{(1,2)\}$. The relation "is greater than" on the set Q is expressed by $\{(2,1)\}$, since 2 is greater than 1. This same set can also be interpreted as expressing the relation "is an integral multiple of" since 2 is an integral multiple of 1. The set $\{(1,2)\}$ defines the relation "is less than" *on the set Q*.

The previous paragraph has provided an illustration for the more precise definition of the concept of *relation*.

Relation **Definition 2-10** A *relation* is a set of ordered pairs. More precisely, a relation on sets P and Q is any subset of their cartesian product $P \times Q$.

Domain and range **Definition 2-11** In a relation the replacement set for the first element is called the *domain* of the relation and the replacement set for the second element is called the *range* of the relation.

As has been mentioned previously, cartesian products can be very helpful in maps and various types of direction charts. In turn, grid diagrams can contribute a great deal to clearer comprehension of the relations associating elements of chosen subsets of cartesian products. The elements of set $P = \{1,2,3,4\}$ may be placed at equal intervals along a horizontal line and those of $Q = \{a,b,c\}$ at equal intervals on a vertical line, as in Fig. 2-7. If vertical and horizontal lines are then drawn as indicated in the figure, each point of intersection will represent an element of $P \times Q$. The elements of this set are all labeled in the figure. By placing the elements of Q along the horizontal line and those of P along the vertical, a diagram can be

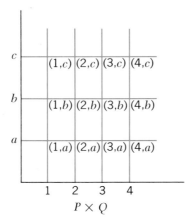

$P \times Q$

FIGURE 2-7

obtained in which each point of intersection represents an element of $Q \times P$. Draw this diagram. The diagrams for $P \times Q$ and $Q \times P$ emphasize even more vividly that $P \times Q \neq Q \times P$, but that they do have the same number of elements.

It is customary practice in drawing such diagrams of cartesian products to place the elements of the domain along the horizontal line and of the range along the vertical line.

Example In the diagram of Fig. 2-8 the open circles represent the relation "is a factor of" defined on $P \times Q$ where $P = \{1,2,3,4,5\}$ and $Q = \{2,4,6\}$. This subset is $\{(1,2), (1,4), (1,6), (2,2), (2,4), (2,6), (3,6), (4,4)\}$. While the domain of the relation $P \times Q$ is $\{1,2,3,4,5\}$ and the range is $\{2,4,6\}$, that is not the case for the relation "is a factor of." Its range is $\{2,4,6\}$ but its domain is $\{1,2,3,4\}$.

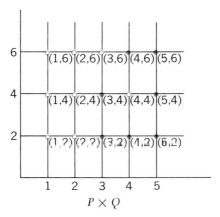

$P \times Q$

FIGURE 2-8

EXERCISES

For Exercises 1 to 9, $A = \{1,2,3\}$ and $B = \{a,b\}$.

1 Write the roster symbols for $A \times B$ and $B \times A$.
2 Does $A \times B = B \times A$? Why?
3 How many elements are there in $A \times B$? In $B \times A$?
4 Construct the diagram for $A \times B$. For $B \times A$.
5 Write the roster symbols for $A \times A$ and $B \times B$.
6 How many elements are there in $A \times A$? In $B \times B$?
7 Construct diagrams for $A \times A$ and $B \times B$.
8 Write the roster symbol for the subset of $A \times A$ which defines the relation "is equal to."
9 Write the roster symbol for the subset of $A \times A$ which defines the relation "is greater than."
10 If set M contains m elements and set N contains n elements, how many elements are there in $M \times N$?
11 Explain the difference in meaning of (a,b), $\{a,b\}$, and $\{(a,b)\}$.
12 What is $\{a,b\} \cap \{(a,b)\}$? $\{a,b\} \cup \{(a,b)\}$?
13 Given the sets $C = \{$Nashville, Atlanta, Sacramento, Austin, Columbia, Lincoln$\}$ and $S = \{$California, Missouri, Tennessee, Nebraska, Georgia, Texas$\}$. Write the roster symbol of the subset of $C \times S$ which defines the relation "is the capital of."
14 Given sets $T = \{$Dodgers, Braves, Tigers, Astros, Red Sox, Indians, Cardinals, Orioles$\}$ and $L = \{$American League, National League$\}$. Write the roster symbol of $T \times L$ which defines the relation "is the name of a team in the."
15 Write the roster symbol for the domain and range of each relation in Exercises 8, 9, 13, and 14.
16 Draw the diagram for $P \times P$ where $P = \{1,2,3,4\}$. Use this diagram as an aid in selecting the subsets of $P \times P$ which express these relations.

(a) "is equal to" $\quad(b)$ "is less than" $\quad(c)$ "is greater than"
(d) "is a multiple of" $\quad(e)$ "divides" $\quad(f)$ "is a factor of"

17 Write the roster symbol for the domain and range of each relation identified in Exercise 16.

2-7 THE CARDINAL NUMBER OF A SET

From history we learn that the shepherds of ancient times used an interesting technique for keeping records of their flocks of sheep. As the sheep filed into the pasture the shepherd would set aside one pebble and only one pebble for each sheep that passed. Furthermore, he was careful that no

Object-to-object correspondence

Tallying

pebble was placed in the pile unless it was placed there to represent a sheep. By using this technique he was assured of the fact that there was an object-to-object correspondence between the pebbles in the pile and the sheep in the flock. There were just as many sheep as there were pebbles, and as many pebbles as sheep.

One modern version of the shepherd's technique is the process of *tallying*, which is frequently used in counting votes in a class or school election. One and only one tally mark is made for each vote cast, and no mark is made which does not represent a vote. In this way there is an object-to-object correspondence between the register of tally marks and the total number of votes cast. When such a procedure is followed, there will be just as many tally marks as there are votes cast, and, conversely, there will be just as many votes cast as there are tally marks made.

If one enters a classroom and sees that each chair in the room is occupied by one and only one student and that there are no students for whom there are no chairs, one knows immediately that there exists an object-to-object correspondence between the chairs in the room and the students in the class. There will be just as many chairs as there are students, and, conversely, there will be just as many students as there are chairs.

The object-to-object correspondence of each of the above three illustrations is an example of what is known as a *one-to-one correspondence* between the elements of one set and the elements of a second set.

One-to-one correspondence

Definition 2-12 Two sets A and B are said to be in *one-to-one correspondence* with each other when the elements of the two sets are so paired that to each element of set A there corresponds one and only one element of set B, and to each element of set B there corresponds one and only one element of set A. (Such a correspondence may be indicated also by "1-1 correspondence.")

The correspondence of the first illustration is a one-to-one correspondence between the set (pile) of pebbles and the set (flock) of sheep; in the second illustration, the one-to-one correspondence is that between the set (register) of tally marks and the set (total number) of votes cast; and in the third illustration the one-to-one correspondence is that between the set (room) of chairs and the set (class) of students. In each case the two sets which are in one-to-one correspondence would be said to have the same *cardinal number*, or the same *cardinality*.

Cardinal number

Counting is a familiar process in which the set of cardinal numbers 1, 2, 3, 4, 5, 6, . . . is used. Just what is the cardinal number 1; what is the cardinal number 2; what is the cardinal number 3; . . . ? For the purpose of this discussion it is not essential that we get too involved in the philosophical subtleties involved in these questions.† Suffice it to say here that all sets which can be placed into one-to-one correspondence with the set $\{a\}$ have the cardinal number *one;* that is, they each contain one and only one element. Similarly, all sets which can be placed into one-to-one correspondence with the set $\{a,b\}$ have the cardinal number *two;* that is, they each contain two and only two elements, and so on. Thus, through the simple technique of establishing one-to-one correspondence between sets, the numbers 1, 2, 3, 4, 5, . . . , familiar to us as counting numbers, are seen to be cardinal numbers of sets.

The cardinal number of a set P is sometimes called the *count of the set* $P.$ It tells how many elements are contained in the set. This is the context in which 0 is identified as the cardinal number of the empty set. There are 0 elements in the empty set. If any given set is such that its count exists and is any one of the cardinal numbers 0, 1, 2, 3, . . . , n, the set is said

Finite or infinite

to be *finite;* otherwise it is said to be *infinite.* It will be convenient to use the symbol $n(P)$ to indicate the cardinal number of $P.$ For example, $n(\varnothing) = 0$ and $n(\{1,2,3,4\})$ is 4. Thus, we may refer to the set $C = \{0,1,2,3,4, . . .\}$ as the set of all cardinal numbers. Such numbers may be used to answer the question: How many elements in a given set? There should be no confusion in the fact that, although C is an infinite set, the use of any element of C, no matter how large, to designate the number of elements in any given set indicates that specific set is finite.

Example One pattern of one-to-one correspondence between the elements of $V = \{a,e,i,o,u\}$ and those of the set $N = \{1,2,3,4,5\}$ is $a \leftrightarrow 1$ (Read: "a corresponds to 1"), $e \leftrightarrow 2$, $i \leftrightarrow 3$, $o \leftrightarrow 4$, and $u \leftrightarrow 5$. Another pattern of one-to-one correspondence might be $a \leftrightarrow 5$, $e \leftrightarrow 2$, $i \leftrightarrow 1$, $o \leftrightarrow 4$, and $u \leftrightarrow 3$. In either case the correspondence reveals that $n(V) = 5$. There are 5 elements in the set V. Can you write other patterns showing a 1-1 correspondence between sets V and N?

Suppose three different people were given the assignment to count the chairs in some given classroom. While they might use three distinct techniques in making their individual counts, the number reported should be the same from all three. This illustrates a very important property of sets, namely:

The cardinal number of a set is a property of the set only, and is independent of the count used to obtain it.

This is intuitively reasonable and we shall accept it without proof, although

† A more thorough treatment may be found in Norman T. Hamilton and Joseph Landin, *op. cit*, pp. 74–106.

it is possible to derive it by rigorous argument. This property of a set, which is as real and significant as such properties as color, size, shape, sourness, sweetness, length, and width, might be called the *number property of the set.* We can thus say that number is that property of a set of objects which the set has in common with every other set with which it can be placed in a one-to-one correspondence. It should be noted that to say $n(A) = n(B)$ is not to imply that set $A =$ set B. For example, if $A = \{1,2,3,4,5\}$ and $B = \{a,b,c,d,e\}$, then $n(A) = n(B)$, but the elements of A and B are not identical, and $A \neq B$.

Number property of a set

Equivalent sets

Definition 2-13 Two sets which have the same cardinal number are *equivalent* sets.

2-8 THE CONCEPT OF ORDINAL NUMBER

We have seen that one important concept of number is that of the cardinal number of a set, which answers the question: How many? Another important notion is that of an *ordinal number,* which answers the question: What is the order of the elements in a set? When one states that there are seven contestants entered in a race, there is no particular concern over the order of their arrangement; only the cardinality of the set is of consequence. At the end of the race, however, there is no particular concern about "how many" entrants there were; the major concern is the order in which they finished. Who was first, who was second, and even, at times, who was last?

Ordinal number

What basic principles control the determination of order? To answer this question let us consider, for example, the sets A, B, C, and D which are all subsets of the same universal set. Furthermore, let the sets be such that $A \subset B$, $B \subset C$, and $C \subset D$. By the transitive property of the relation "is a proper subset of," it follows that $A \subset B \subset C \subset D$. Thus a specified ordered arrangement is determined for the four sets. When such an ordering has numerical significance the names "first," "second," "third," and "fourth" are associated with sets A, B, C, and D, respectively. Other titles such as "earliest," "tallest," or "shortest" may be given to a member of set A, depending on the criteria used in ordering the sets.

Example Suppose the seven contestants in a footrace are Bob, Tom, Joe, Jim, David, John, and Bill. The universal set for the race = {Bob, Tom, Joe, Jim, David, John, Bill}. If the sets of finishers, in their respective order of occurrence, are {David}, {David, Joe}, {David, Joe, Jim}, {David, Joe, Jim, Bob}, {David, Joe, Jim, Bob, John}, {David, Joe, Jim, Bob, John, Tom}, and {David, Joe, Jim, Bob, John, Tom, Bill}, then we say that David was *first;* Joe was *second;* Jim, *third;* Bob, *fourth;* John, *fifth;* Tom, *sixth;* and Bill, *seventh.* From another point of view we might say that David was the *fastest* runner, and Bill the *slowest* runner, in the race.

	1	2	3	4	5	6	7	8	9	R
Cincinnati Reds	0	1	0	0	0	0	2	0	0	3
Baltimore Orioles	2	0	1	0	1	4	1	0	✕	9

FIGURE 2-9   A baseball scoreboard

Example   A good illustration of the use of cardinal number, ordinal number, and placeholder is the box score of a recent World Series baseball game (Fig. 2-9). The numbers across the top line indicate the innings of play. They serve either as cardinal numbers or as ordinal numbers depending on the way in which they are used. One example of a cardinal use is the following: The box score shows that the two teams played a *nine*-inning game and not one of extra innings. An example of ordinal use is as follows: The Reds scored in the *second* and *seventh* innings, while the Orioles scored in the *first, third, fifth, sixth,* and *seventh* innings.

During their turn at bat in the first inning the Reds scored 0 runs and the Orioles scored 2 runs during their turn. The numerals 0 and 2 which occur in the column for the record of the first inning answer the question: How many runs did each team score in the first inning? They represent cardinal numbers. Similarly, each numeral appearing in color throughout the box score represents a cardinal number which answers the question: How many runs were scored? What does the symbol ✕ in the lower part of the ninth inning mean? The Orioles had already won the game, so there was no reason for them to make any additional effort to score. The ✕ is *Placeholder*   simply a *placeholder:* the frame might as well have been left empty except that it is the custom to write in a symbol. It would be entirely incorrect to replace any of the 0s by the symbol ✕. In the frame labeled *R*, the 3 and 9 indicate the total number of runs scored by each team.

*2-9   THE NEED FOR A SYMBOLIC LANGUAGE

Suppose that, for the moment, you remove from your thinking all number names and number symbols with which you are familiar. What are some of the major problems with which you are confronted if you have a need to count and record the results of your counting? For example, how would you answer the question: How many letters are there in the word "number"? You start with the letter *n*, but you have neither a name nor a symbol to record the result of your count. These must be invented or determined. If the set of elements to be counted is fairly large in content, it becomes desirable to use some pattern of grouping in order to simplify the system *Place value*   of symbols and names. The principle of *place value*, or *positional value*, can be used as a most effective aid in building symbols for large numbers. By place value we mean the value which a digit symbol acquires by virtue of its position in any given numeral. Place value has no significance at all, however, until some agreement has been reached as to the cardinal number

Base of a numeral system

of the set to be used in grouping. This number is called the *base* of the system. In the number system that we use the base is ten because in forming the symbols for larger numbers we group in terms of the set whose

Decimal numeral system

cardinal number is ten. Hence this system is called a *decimal* system, from the Latin word *decem*, meaning ten. Thus, if the symbol ☐☐☐☐ represents a four-digit number, the square on the extreme right is the ones place, the next to the left is the tens place, the next is the place for ten × ten, and the next is for ten × ten × ten. In other words, in the number 2,345 the 2 is in the thousands place, the 3 is in the hundreds place, the 4 is in the tens place, and the 5 is in the ones place.

Building a system of numeration

It is the purpose of this discussion to outline the salient problems in building a system of numeration. We shall use place value and the principles of multiplication and addition to obtain the symbol for any expressed quantitative concept such as how many letters there are in the word "number." To illustrate: In our system 2,345 means 2 thousands + 3 hundreds + 4 tens + 5 ones, or (2 × 1,000) + (3 × 100) + (4 × 10) + 5. We read the numeral by saying two thousand three hundred forty-five.

The only thing left for us to determine before we can begin our experiment of building a system of number symbols is the cardinal number of the set we shall use as the base. Let us choose as the base of our system the number which is the cardinal number of all sets which can be placed in one-to-one correspondence with the set $\{a,b,c,d,e\}$. If we were counting in the familiar decimal number system this number would be five, represented by the symbol 5. In the number system we are constructing we have neither the number name "five" nor the number symbol 5. We shall use the number

Base verto

name *verto* and the number symbol **l0** to indicate the base of our new number system. In Fig. 2-10 there are suggested number names and symbols to assist you in understanding just what takes place in the building of a number symbolism. Note that **l0** is the symbol for **verto,** that is, it is

l vert	l0 verto	l00 vertoo	l000 vertong
⊥ perp	⊥0 perpo	⊥00 perpoo	⊥000 perpong
△ trang	△0 trango	△00 trangoo	△000 trangong
☐ sar	☐0 saro	☐00 saroo	☐000 sarong

This system uses 0(zero) as it is used in the decimal system.
l0 **(verto)** represents the base.
l00 **(vertoo)** represents the base × the base.
l000 **(vertong)** represents the base × the base × the base.

FIGURE 2-10 Number names and symbols in base **verto**

the symbol for the base of the system. Similarly, **l00** is the notation for base $\times$ base, or **verto** $\times$ **verto,** and **l000** represents base $\times$ base $\times$ base, or **verto** $\times$ **verto** $\times$ **verto.**

Now to count the letters in the word "number," starting with n, we have **vert;** u is **perp;** m, **trang;** b, **sar;** e, **verto;** and r, **verto-vert (ll).** In the word "arithmetic" there are **perpo (⊥0)** letters, and in the words "elementary mathematics" there are **saro-vert (□l)** letters. Count the letters to check these statements. As a final check on your ability to count in this new system, count the words and then the letters in this entire sentence.†
To interpret symbols such as □□, △□△, and □⊥△l we use place value and the multiplicative and additive principles: □□ = □0 + □; △□△ = △00 + □0 + △; □⊥△l = □000 + ⊥00 + △0 + l. This last symbol is read **sarong perpoo trango-vert.**

This experiment may be pursued as far as one likes. It has been designed merely for the purpose of giving the reader some opportunity to understand some of the basic principles which have contributed to the evolution of our system of enumeration out of the confusion and awkwardness of more primitive systems.

EXERCISES

1 When are two sets of objects said to be in one-to-one correspondence with each other?

2 Without counting, use one-to-one correspondence to show that you have just as many fingers on your left hand as on your right hand.

3 Under what conditions can the empty chairs of a classroom be counted to determine the number of pupils absent from class?

4 Give four illustrations, other than those of the text, of one-to-one correspondence between sets.

5 What is meant by the "number property" of a set?

6 Can you cite evidence from the literature or from your experience of the existence of a number sense among animals, birds, or insects?

7 Give two illustrations, other than those in the text, of situations in which people might exhibit number sense without recourse to counting.

8 Team up with someone to try this experiment:
(a) Each present the other with groups of objects in disarray, and in varying numbers, for example,

```
XX X   X
XXX X  X
 X X X  X
  X X   X
```

the object being to determine the upper limit of immediate recognition.

---

† There are **saro-sar (□ □)** words and **trangoo-saro-trang (△ □ △)** letters in the entire sentence.

(b) Each present the other with groups of objects in various forms of orderly array, for example,

X X	X X	X X
X	X	X	X
X X	X X	X X

also to determine the upper limit of immediate recognition.

9 What are some illustrations of the use of tallying as an aid to counting?

10 Distinguish between the concepts of cardinal and ordinal numbers. Give illustrations.

11 What is the cardinal number of each of the following sets?
(a) All circles which are square
(b) All letters common to the words "bar," "call," and "saw"
(c) All occurrences of the letter "e" in the word "cough"
(d) The days in a week
(e) The minutes in an hour

12 Classify each of the following numbers as a cardinal number or an ordinal number.

First day of the week	Baseball score 3–0
32 children	6 months
24 hours in a day	7-day week
Page 75	December 25
Henry VIII	75 pages

13 List the sets which would show the set of racers = {Linda, Carol, Cathy, Sharon, Marly, Diana} finishing in this order: Linda, first; Sharon, second; Diana, third; Marly, fourth; Cathy, fifth; and Carol, sixth.

14 List the sets which would show the girls of Exercise 13 finishing in this order: Diana and Marly tied for first place; Cathy, third; Carol and Sharon tied for fourth place; and Linda, sixth.

*15 Write the numeral in the **verto** system for each of these numbers:
(a) **Perpoo, verto-trang**
(b) **Sarong, trangoo, vert**
(c) **Perpong, perpo-perp**

*16 Write the number name for each of these numerals:

(a) I⊥△   (b) △0□   (c) □△0   (d) □I△□

17 Prove this statement: 1-1 correspondence between sets is an equivalence relation.

18 Given sets $P = \{1,2,3,4,5,6\}$, $Q = \{7,8,9\}$, and $R = \{5,6,7,8,9\}$. Show that:
(a) $n(P \cup Q) = n(P) + n(Q) - n(P \cap Q)$
(b) $n(P \cup R) = n(P) + n(R) - n(P \cap R)$

19 Argue that for any two sets X and Y, it is always true that $n(X \cup Y) = n(X) + (Y) - n(X \cap Y)$.

20   Given sets $A = \{a,b,c,d,e\}$, $B = \{f,g,h,i\}$ and $C = \{c,d,e,f,g\}$. Show that:

(a) $n[A \cup (B - A)] = n(A \cup B)$

(b) $n[A \cup (C - A)] = n(A \cup C)$

*21   Argue that for any two sets M and N, it is always true that $n(M \cup N) = n[M \cup (N - M)]$.

2-10   REVIEW EXERCISES

1–26   Write a carefully constructed answer to each of the guideline questions of this chapter.

27   Given the sets

$$U = \{1,2,3,4, \ldots , 15\}$$
$$E = \{e \mid e \text{ is an even number between 1 and 15}\}$$
$$R = \{7,9,11,13\}$$
$$S = \{1,2,3,4,5,6\}$$

Write the roster symbol for each of these sets:

(a) $\sim E$ (b) $S \cap \sim E$ (c) $S \cup \sim E$

(d) $R \cap E$ (e) $R \cup E$ (f) $R \cap U$

(g) $S \cap \sim S$ (h) $S \cup \sim S$ (i) $E - S$

(j) $E - R$ (k) $S - E$ (l) $R - E$

(m) $\sim U$ (n) $\sim R - \sim S$ (o) $\sim S - \sim R$

*28   For the sets of Exercise 27 which of these statements are true, and which are false? In each case support your answer.

(a) $S \cap E = S - E$ (b) $n(R \cap S) = n(S \cap E)$

(c) $N[\sim(S \cup R)] = n(E - S)$ (d) $n(S) = n(E) - n(R)$

(e) $\sim(S \cup R) = E - S$ (f) $n(S - \{5,6\}) = n(R)$

(g) $\sim R - S = \sim(S \cup R)$ (h) $n(S - E) - n(S \cap E) = n(R) - n(E - S)$

(i) $n(R \cap E) = n(S \cap R)$

(j) $S - \{5,6\} = R$

29   Given the sets

$$P = \{a,b,e,i,m,n,o,r,u\}$$
$$Q = \{b,r,m,n\}$$
$$R = \{* \mid * \text{ is a letter in the word ``number''}\}$$
$$S = \{a,e,i,o,u\}$$

Which of these statements are true and which are false? In each case support your answer.

(a) $Q \subseteq P$ (b) $a \in Q$

(c) $b \in R$ (d) $Q \subseteq R$

(e) $n(Q) = n(S)$

(f) $R \cap P = R$

(g) $u \in R$

(h) $R \subseteq S$

(i) $R \subset P$

(j) $S \cup R = P$

(k) $o \notin R$

(l) $Q \cup R = P$

(m) $n(Q) + n(R) = n(P)$

(n) $Q \cap R = \varnothing$

(o) $P \cap S = S$

(p) $m \in P \cap Q$

(q) $R \not\subset Q$

(r) $n(Q) + n(S) = n(P)$

(s) $Q \not\subseteq R$

(t) $P \cap Q = P$

(u) Q and R are disjoint sets

(v) Q and S are disjoint sets

(w) P and S are disjoint sets

(x) $S \cup (Q \cap R) = (S \cup Q) \cap (S \cup R)$

(y) $S \cap (Q \cup R) = (S \cap Q) \cup (S \cap R)$

30 Complete each of these set-builder definitions of the given set.

(a) $\sim A = \{a \,|\, ?\}$

(b) $P \subseteq Q = \{* \,|\, ?\}$

(c) $\sim P \cap \sim Q = \{\# \,|\, ?\}$

31 Is either of these statements a true statement about sets A and B? Support your answer.

(a) If $A = B$, then A is equivalent to B.

(b) If A is equivalent to B, then $A = B$.

32 Why is it true that, if a one-to-one correspondence exists between the elements of two sets, the sets are equivalent?

33 Use at least two patterns for setting up a one-to-one correspondence between pairs of equivalent sets.

(a) $\{a,e,i,o,u\}$ (b) $\{2,4,6,8,10\}$ (c) $\{1,2,3,4\}$

(d) $(a,b,c,d,e,f\}$ (e) $\{g,o,l,d\}$

34 Use at least three patterns for setting up a one-to-one correspondence between pairs of *equal* sets.

(a) $\{1,3,5,7,9,11\}$

(b) $\{* \,|\, *$ is a letter in the word "marble"$\}$

(c) $\{9,7,11,1,5,3\}$

(d) $\{\# \,|\, \#$ is a letter in the word "rambler"$\}$

(e) $\{2,4,6,8,10,12\}$

35 Which sets of Exercise 34 are equivalent?

36 Given the sets $U = \{1,2,3,4,5,6,7,8,9,10\}$, $A = \{2,4,6,8,10\}$, $B = \{1,2,3,4,5,6\}$, and $C = \{1,3,5,7\}$. Use these sets to illustrate the properties of closure, associativity, commutativity, distributivity, and identity as they relate to union and intersection. For distributivity consider these cases: $A \cup (B \cap C), (B \cap C) \cup A, A \cap (B \cup C)$, and $(B \cup C) \cap A$.

37 Use the sets A, B, and C of Exercise 36 to examine the same properties of closure, associativity, and commutativity as they relate to relative complementation.

38   If $P = \{2,4,6,8\}$ and $Q = \{1,3,5,7,9\}$, draw the diagrams for $P \times Q$ and $Q \times P$.

39   Use the diagram for $P \times Q$ to select the subset which defines the relation "is less than."

40   What are the domain and range of the relation of Exercise 39?

41   Construct diagrams for $P \times P$ and $Q \times Q$, where P and Q are defined as in Exercise 38.

42   What is the subset of $P \times P$ which defines the relation "is equal to"? What are the domain and range of this relation?

*43   Use the sets $A = \{a,b,c\}$, $B = \{k,l,m\}$, and $C = \{l,m,n\}$ to illustrate the truth of each of these statements.

(a) $A \times (B \cap C) = (A \times B) \cap (A \times C)$
(b) $(B \cap C) \times A = (B \times A) \cap (C \times A)$
(c) $A \times (B \cup C) = (A \times B) \cup (A \times C)$
(d) $(B \cup C) \times A = (B \times A) \cup (C \times A)$

*44   Use Venn diagrams to illustrate the truth or falsity of each of these statements about sets P, Q, and R.

(a) $(P \cup Q) \cap (P \cup R) = P$ (b) $P \cup (Q - P) = P \cup Q$
(c) $P \cap (\sim P \cup Q) = Q$ (d) $\sim(Q - P) = P \cup (\sim P \cap \sim Q)$
(e) $P \cap (Q - R) = P \cap (R - Q)$
(f) $(P \cap Q) \cap (P \cap R) = P \cap (Q \cap R)$

*45   Write the numeral in the **verto** system for each of these numbers.

(a) **Vertong**   (b) **Perpo sar**
(c) **Perpong trang**   (d) **Saroo trango-perp**
(e) **Sarong vertoo saro-perp**

*46   Write the number names in the **verto** system which correspond to each of these numerals.

(a) $\perp$**O**   (b) $\triangle \perp$ (c) $\square\triangle$**O**
(d) $\perp\triangle$**I**   (e) **IO**$\triangle\perp$ (f) $\triangle\perp$**O**$\triangle$
(g) $\perp$**OO**$\perp$ (h) $\triangle$**O**$\triangle$**O**   (i) $\square\square\square\square$

*47   Use the **verto** system to count and record the number of words in these statements made about number by mathematicians interested in the history of number.

(a) God created the integers, the rest is the work of man. *Leopold Kronecker*

(b) In the history of culture the discovery of zero will always stand out as one of the greatest single achievements of the human race. *Tobias Dantzig*

(c) The grandest achievement of the Hindoos and the one which, of all mathematical investigations, has contributed most to the general progress of intelligence, is the invention of the principle of position in writing numbers. *Florian Cajori*

∗INVITATIONS TO EXTENDED STUDY

1 Investigate the significance of symbolism in such areas as highway travel, home life, business, industry, and religious and organization ceremony.

2 Use Venn diagrams to illustrate these relations between sets. They are known as De Morgan's laws.

$$\sim(A \cup B) = \sim A \cap \sim B \quad \text{and} \quad \sim(A \cap B) = \sim A \cup \sim B$$

3 Develop the argument needed to establish De Morgan's laws.

4 Develop the argument needed to establish the statements of Exercise 12 on page 46.

5 Use De Morgan's laws and the properties of Exercise 12 (page 46) to establish the truth of each of these equations for nonempty sets P, Q, and R.
(a) $P \cup (\sim P \cap Q) = P \cup Q$
(b) $[(P \cap Q) \cup (P \cap \sim Q)] \cup [\sim(\sim P \cup Q)] = P \cup \sim Q$
(c) $[P \cap (\sim P \cup Q)] \cup [Q \cap (Q \cup R)] \cup Q = Q$

6 Use Venn diagrams to illustrate the truth of each of these statements. For all sets P, Q, and R:
(a) $R - (P \cup Q) = (R - P) \cap (R - Q)$
(b) $R - (P \cap Q) = (R - P) \cup (R - Q)$
(c) $(P - Q) \cup (Q - P) = (P \cup Q) - (P \cap Q)$
(d) $(P - Q) \cup (Q - R) = (P \cup Q) - (R \cap Q)$

7 Invent number names and symbols for a number system using ℳ II as a base. Assume place value and the additive principle in constructing symbols for numbers larger than base.

THE NATURAL NUMBER SYSTEM

We now have a context within which to undertake a careful study of the number systems that serve as the foundation upon which the structure of elementary mathematics is built. The natural number system is the simplest and most fundamental of these number systems. Its basic concepts and principles are the concern of Chap. 3.

The first thirteen of the guideline questions constitute a review of that content of the two previous chapters which is pertinent to this chapter. It is important that the reader have ready answers to these questions. The remaining questions may serve as an outline for preview and later review of the present chapter. The guidelines for the study of this chapter are the following:

1 What is meant by the concepts of *place*, or *positional*, value and *digit* value in the structure of numerals?

2 What is a cardinal number?

3 When are the elements of two sets said to be in one-to-one correspondence with each other?

4 How is the relation of one-to-one correspondence between two sets used to determine the cardinality of a set?

5 Why does the cardinal number of the union of two disjoint sets indicate the total number of elements in the two sets?

6 What is an ordered pair of elements?

7 When are two ordered pairs equal?

8 What is the cartesian product of two sets?

9 What is a relation?

10 What are the domain and range of a relation?

11 What are the properties of an equivalence relation?

12 What are the characteristics of each of the following properties when applied to the operations of union and intersection of sets: (*a*) closure, (*b*) associative, (*c*) commutative, (*d*) distributive?

13 What is meant by the statement that U and $\emptyset$ are the identity elements for the respective set operations of intersection and union?

14 What is a number system?

15 What are the definitions of addition and multiplication as defined on the set of natural numbers?

16 What are the basic functions of the two operations?

17 When is an operation upon the elements of a set said to be well defined?

18 What do the words "closure," "associative," "commutative," "distributive," and "identity" mean when applied to addition and multiplication of natural numbers?

19 How does place value simplify the algorithms used for addition and multiplication?

20 What are some of the effective checks of addition and multiplication? Why do they work?

21 What are the properties which characterize the relation of equality between two natural numbers?

22 Is the relation of equality of natural numbers an equivalence relation? Why?

23 What is the principle of *trichotomy?*

24 Which of the properties of equality do not hold for the relations "is greater than" and "is less than" between two natural numbers?

25 What are the basic properties of the natural number system?

26 Why is it that a numeral such as 2,034 has no meaning until it is given orientation in the context of a pattern of grouping to be used as the base for the system of numerals?

INTRODUCTION

The concept of the cardinality of a set is an important capstone in the structure of modern civilization. Answers to such questions as How many?, How much?, How far?, and How long? are found by means of counting, the technique of determining the cardinal numbers of identifiable sets of objects. Thus the counting numbers have evolved as a natural means to aid civilized human beings in comprehending and communicating quantitative ideas. For the more efficient use of these numbers certain combinatorial operations have been defined. Intuition and experience have combined to formulate certain operational characteristics into fundamental properties or laws. Using these fundamental properties as a foundation, one may employ the techniques of logic to derive other properties, relations, and principles of operation which may be combined with them to shape the basic structure of the natural number system.

3-1 THE NATURAL NUMBERS

Natural numbers

Legend has it that Leopold Kronecker (1823–1891) made the statement, "Die ganzen Zahlen hat Gott gemacht, alles andere ist Menschenwerk."† Herein lies the foundation of the philosophy that associates the concept of natural numbers with the numbers 1, 2, 3, 4, 5, 6, 7, 8, 9, 10, 11, . . . , which are felt to have an existence independent of man. As was noted in Chap. 2, each such natural number n is the cardinal number of some clearly identifiable set of n elements, as well as of all other sets whose elements can be put in one-to-one correspondence with these n elements.

---

† "God created the integers, the rest is the work of man."

Characteristics of the set of natural numbers

In the use of these numbers man has observed certain fundamental characteristics which he has accepted on a purely intuitive basis. Since 1 is the cardinal number of the set $\{1\}$ and

$$\{1\} \subset \{1,2\} \subset \{1,2,3\} \subset \{1,2,3,4\} \subset \cdots \subset \{1,2,3,4, \ldots, n\}$$

it seems intuitively acceptable to say that 1 is a natural number and that each natural number (except 1) follows a natural number. Let us say this another way by stating that each natural number has a successor (a number which follows it) and that each natural number (except 1) is the successor of another natural number. Two other statements about natural numbers which are as readily acceptable as those already mentioned are (1) if two natural numbers a and b have successors which are equal, then $a = b$; (2) if a set S contains the natural number 1 and also contains the successor of any one of its members, which may be selected arbitrarily, then it contains all the natural numbers.

Principle of finite induction

This last statement is known as the *principle of finite induction*. The above statements are adapted from a list of five statements about natural numbers known as Peano's postulates. They were originally stated by G. Peano (1858–1932) and used as a basis for a critical study of the natural numbers and their properties.

3-2 THE NATURE OF A NUMBER SYSTEM

Structure of a number system

A number system consists of (1) a nonempty set of elements (defined or undefined); (2) at least one well-defined operation† upon these elements; (3) a set of postulates (acceptable assumptions) which govern the use of the operation, or operations; (4) an equivalence, or equality, relation; and (5) all the valid statements, or properties, which can be derived as logical consequences of the postulates and any previously derived statements.

The title of this chapter, The Natural Number System, indicates our intention to examine now the set of natural numbers $N = \{1,2,3,4, \ldots\}$, along with certain operations to be defined and postulates to be stated, in terms of its fundamental structure as a mathematical system. In the preceding section we have at least hinted at a basis for setting up a list of postulates. There still remains the necessity for specifying "at least one well-defined operation" and for providing a pattern for deriving all the consequential valid statements, or properties. For the natural number system there are two operations: *addition* and *multiplication*. We now need to define clearly what is meant by each of these two operations.

---

† An operation is said to be *well defined* on a set of elements if, no matter to what elements of the set the operation is applied, the result obtained has the characteristic that it is clearly and uniquely identifiable.

3-3 WHAT IS ADDITION?

The two sets $A = \{\triangle, I, \perp\}$ and $B = \{*, \#, @, \&, ?\}$ are disjoint sets. To find the total number of elements in the two sets, one can follow either of two plans:

1 Count the elements in one set, say A. When this count is completed continue in the natural sequence using the elements of the other set, B.

$$\{\triangle, I, \perp\} \quad \{*, \#, @, \&, ?\}$$
$$1 \ 2 \ 3 \quad\quad 4 \ 5 \ 6 \ 7 \ 8$$

2 Form $A \cup B$ and count the elements of this set.

$$A \cup B = \{\triangle, I, \perp, *, \#, @, \&, ?\}$$
$$1 \ 2 \ 3 \ 4 \ 5 \ 6 \ 7 \ 8$$

The results of these two patterns of counting will always be the same for any two disjoint sets. (See Exercises 18 and 19 of Sec. 2-9.) This fact provides the basis for defining the process of *addition* on the set of natural numbers. It is the basic process for combining the cardinal numbers of two disjoint sets for the purpose of determining the total number of elements in the two sets.

Definition of addition

Definition 3-1 If A and B are two disjoint sets, then the *sum* of the cardinal number of A and the cardinal number of B is the same as the cardinal number of the set $A \cup B$. The operation used to find this sum is indicated by the symbol $+$, and is called *addition*. If $A \cap B = \varnothing$, then $n(A) + n(B) = n(A \cup B)$.

Addition is a binary operation

The table of basic addition facts, which may be determined by counting, completes the definition. It is for this reason that we may think of addition as a shortened form of counting. For natural numbers addition is, thus, a *binary operation* which combines two numbers to give a *sum*, or *total*.

3-4 NAMES AND SYMBOLS USED IN ADDITION

In the annals of history there are to be found many different names for the process which we call addition. Some of the more frequently used names are *aggregation, summation, composition,* and *collection.* In fact, nowadays in some phases of more advanced mathematics the term *summation process* is used to indicate that addition is taking place.

Addend

The terms to be added are called *addends.* This name is derived from the Latin *numeri addendi,* meaning "numbers to be added." Interestingly enough, this name has been used frequently in the past to apply to all the numbers in a column to be added except one, which was called the *augend,* the number to be augmented. The word "augend" dropped from popular usage for a long period of time, but is appearing again in current literature relative to electronic computation.†

Sum

Two words were used rather generally by early writers to indicate the result of adding a group of numbers. These words were "sum," which applied only to addition, and "product," which apparently was used to represent the result of any form of computation.‡ Although other words and phrases were used, these two seemed to be the most popular. Usage has established "sum" as the name for the result obtained by the process of addition. The word "total" is also used fairly generally today.

Total

Symbols used to indicate addition

In the oldest known mathematical writing, the Ahmes papyrus (ca. 1650 B.C.), the symbol Λ (representing a pair of legs walking forward) was used to indicate the process of addition. In the writing of unit fractions (fractions with one as the numerator) the Egyptians used juxtaposition to indicate addition. For example, $\frac{1}{2} \frac{1}{18}$ represented $\frac{1}{2} + \frac{1}{18}$. This practice of using juxtaposition to indicate addition was also used in certain early Greek and Arab manuscripts. It is one which we still follow today in writing mixed numbers; for example, $4\frac{1}{3}$ is used to represent $4 + \frac{1}{3}$.

The symbol +

During the latter part of the fifteenth century the sign $\tilde{p}$ was introduced to indicate addition; it remained as an important symbol among Italian writers during the sixteenth century. A cross, much like our + sign, appeared in the Bakhshālī manuscript as a symbol to indicate subtraction. In 1429, Widman first used the + sign in print, and it was used to indicate addition. By the early part of the seventeenth century the use of the symbol "+," in the sense in which we use it today, had become rather well established. Some of the problems given by Widman, however, seem to indicate that the symbol might have been used in warehouses to denote excess weights. It apparently was used also as an abbreviation for "and," or to indicate error in computation involving the *rule of false position*. There is also evidence that the symbol might have found its origin in variations of the word *et* in Latin manuscripts.§ Attention should be called to the fact that, in the early use of the symbol, it seemed to be employed quite frequently as a descriptive label (as we use it today in directed numbers) rather than as a symbol of operation.

*3-5 THE ADDITION ALGORITHM

Just as our number system evolved out of a struggle for simplicity, compactness, and effectiveness in the recording of quantitative experience, so

---

† See, for example, C. B. Tompkins, Computing Machines and Automatic Decisions, *Twenty-third Yearbook of the National Council of Teachers of Mathematics*, p. 381, Washington, 1957.
‡ David Eugene Smith, "History of Mathematics," vol. II, pp. 89–90, Ginn and Company, Boston, 1925.
§ F. Cajori, "A History of Mathematical Notations," vol. I, pp. 229–236, The Open Court Publishing Company, La Salle, Ill., 1928.

did our computational algorithms emerge from a desire for speed, accuracy, and efficiency in carrying out numerical calculations. The concept of place value has made fully as significant a contribution to the refinement of computation as it has to the simplification of number symbolism.

Place value as an aid in computation

The numerals of the ancients did not adapt themselves very readily to calculation. For this reason different races, of necessity, had to devise computational aids to make up for their notational deficiencies. Three principal mechanical aids, all variants of the abacus, were used. They were the dust board, the table with loose counters, and the table with counters fastened to lines. It is of interest to note that these primitive "computing machines," which were based on the concept of place value, were used by races of people whose number notation had no semblance of the concept. The abacus is still an important aid to computation in some countries, and a high degree of proficiency in its use is developed.

Early devices used in addition

The process of adding two or more numbers is so elementary that one would not expect in the course of time to find much change in its techniques. The principal variations, particularly since the general acceptance of the Hindu-Arabic numerals, have been in the forms of recording the results. The Hindus seem to have used two methods: one was the method we now use, while the other, which they called the *inverse* or *retrograde*, consisted in adding from the left and recording the result for each column, blotting out (canceling in the illustration) as it was necessary to "carry" from one column to the next. Figure 3-1 shows the step-by-step analysis of this form of addition in finding the sum 2,642 + 1,756 + 833. The colored numerals show the complete analysis of each step, with each result being recorded in the colored block. On the right the algorithm is shown as actually used, except that the auxiliary digits 3, 1, and 2 would not have shown. Only the sum 5,231 would have shown in the end, as the canceled digits 3, 1, and 2 would have been blotted out as they were replaced.

Retrograde method of addition

In the sixteenth century Gemma Frisius introduced a method of addition

	2,642		2,642
	1,756		1,756
	833		833
Step 1	3̶	2 + 1 = 3	3̶,1̶2̶1
Step 2	2̶ 1̶	6 + 7 + 8 = 21; record 1	5 23
Step 3	5	Replace 3 by 3 + 2 = 5	
Step 4	1̶2̶	4 + 5 + 3 = 12; record 2	
Step 5	2	Replace 1 by 1 + 1 = 2	
Step 6	1̶1	2 + 6 + 3 = 11; record 1	
Step 7	3	Replace 2 by 2 + 1 = 3	

The sum is 5,231

FIGURE 3-1 Retrograde addition

	8,762
	3,256
	783
	652
Step 1	13
Step 2	24
Step 3	2 2
Step 4	11
Step 5	13,453

FIGURE 3-2 Algorithm of Gemma Frisius. Sums are recorded by columns from right to left.

which is still used by some today, particularly in the addition of long columns of numbers. This method is illustrated in Fig. 3-2. Note that he wrote the largest addend first, recorded the sum of each column in order from right to left, as shown by the colored numerals, and then added the partial sums.

A review of the variations in addition forms used by different races and individuals impresses one with the fact that the major source for differences and difficulties was an inadequate comprehension of the full significance of place value in a system of numeration. Our system is a decimal system; consequently, any natural number symbol is to be interpreted in terms of ones, tens, hundreds, thousands, and higher powers of 10. Hence, in any addition problem, place value becomes a very effective guide in arranging the addends so that it becomes rather simple to make sure that we are *combining* disjoint sets of like things, ones with ones, tens with tens, hundreds with hundreds, etc. It also serves very effectively as a guide in recording the sum. The sum of the ones column in Fig. 3-3 is 13, which by the form in which it is written means $1(10) + 3$. Therefore, we record the 3, since that is the only group of ones we have, and then we do the natural thing of combining the 1 ten with the other tens we have. This may be done mentally or, as is sometimes done, the 1 may be written at the top of the tens column. In either case, the number of tens is $1 + 6 + 5 + 8 + 5 = 25$. This we recognize as 2 hundreds and 5 tens. The 5 tens are recorded in the tens column and the 2 hundreds are com-

Significance of place value in addition

	Step 4	Step 3	Step 2	Step 1
8,762	8 (1,000)	+7 (100)	+6 (10)	+2 (1)
3,256	3 (1,000)	+2 (100)	+5 (10)	+6 (1)
783		7 (100)	+8 (10)	+3 (1)
652		6 (100)	+5 (10)	+2 (1)
13,453	1 (10,000) +3 (1,000)	+4 (100)	+5 (10)	+3 (1)

FIGURE 3-3 Use of place value in addition

bined with those represented by the given numerals to obtain 24 hundreds, which we write as 2 thousands and 4 hundreds. When the 2 thousands are totaled with those given we obtain 13 thousands and think of it as 1 ten thousand and 3 thousands. Our sum is thus 13,453. If one thinks in terms of the full significance of place value in number symbolism, the process of addition loses virtually all its major aspects of difficulty.

The properties of natural numbers under addition, which provide the authority for the procedures used in this section, will be discussed in the next section.

3-6 PROPERTIES OF THE NATURAL NUMBERS UNDER ADDITION

By Definition 3-1, addition, the process of finding the sum of two natural numbers, produces a unique, clearly identifiable natural number. It is a *Well-defined* *well-defined* operation, the result of which can be verified by counting the *operation* elements in the union of two disjoint sets. This fact leads us to examine the properties of sets under the operation of union for suggestions as to basic properties of numbers under the operation of addition.

A review of Sec. 2-5 will reveal that under the operation of union, sets have the properties of closure, commutativity, and associativity. There exist analogous properties of natural numbers under the operation of addition. Also, the empty set $\varnothing$ was shown to be an identity element for the union of sets since $A \cup \varnothing = A$ for all sets A. Since $n(\varnothing) = 0$, and 0 is not a natural number, there does not exist an analogous identity element for natural numbers under addition. (We shall return to this consideration in Chap. 4.) Furthermore, the distributive property mentioned in Sec. 2-5 has no application at this time since it relates the two set operations of union and intersection. We are concerned here *only* with union. In a later section of this chapter we shall examine some of the implications of distributivity as it relates to natural numbers.

Each natural number is the cardinal (counting) number of some identifiable nonempty set. Definition 3-1 specifies that $n(A) + n(B) = n(A \cup B)$ if *Closure under* $A \cap B = \varnothing$. This is simply saying that the sum of two natural numbers *addition* is a natural number. In other words, the natural numbers are *closed* under addition. (For an example of a set of numbers not closed under addition see Exercise 1, Sec. 3-17.) Also, we have seen that $A \cup B = B \cup A$ for any two sets A and B. This fact gives strong intuitive support for assuming that the order in which two natural numbers are added has no effect on the sum obtained. The sum of 3 and 5 is 8 regardless of whether we think *The* of adding 5 to 3 or 3 to 5; $3 + 5 = 5 + 3$. Since this is true for any two *commutative* natural numbers, we say that natural numbers are *commutative* under *property of* *addition* addition.

From its definition, addition is a binary operation; that is, it is a process that operates only on two numbers. Because of this fact there is a question as to how one might proceed to find the cardinal number $n(A \cup B \cup C)$, call it $a + b + c$, where A, B, and C are three *disjoint* sets. From the definition $n(A \cup B)$ is $n(A) + n(B)$, which is a natural number by the closure property. If $a = n(A)$ and $b = n(B)$, then $n(A \cup B)$ is the natural number $(a + b)$. It thus follows that, if $c = n(C)$, $n(A \cup B) + n(C)$ is the natural number $(a + b) + c$. By a similar argument it follows that $n(A) + n(B \cup C)$ is the natural number $a + (b + c)$. Furthermore, in Sec. 2-5 attention was called to the fact that $(A \cup B) \cup C = A \cup (B \cup C)$, which supports the assumption that in the addition of any three natural

The associative property of addition

numbers a, b, and c, the same sum $a + b + c$ is always obtained whether the grouping is $a + (b + c)$ or $(a + b) + c$. Hence natural numbers are said to be *associative* under addition.

These three properties which hold for the set N under the operation of addition have such far-reaching significance that it is important that they be listed for emphasis.

Under the operation of addition the set $N = \{1,2,3,4, \ldots\}$ satisfies these properties:

Properties of the natural numbers under addition

A-1 Closure For $a,b \in N$, it is true that $a + b$ is a unique element of N.

A-2 Commutative For $a,b \in N$, it is true that $a + b = b + a$.

A-3 Associative For $a,b,c \in N$, it is true that $(a + b) + c = a + (b + c)$.

Analysis of the addition process

Example Figures 3-4 to 3-6 complete the definition of addition for natural number systems whose numerals are written in the respective bases of ten, **three,** and **six.** In Fig. 3-4 the basic addition facts are recorded for a numeral system for natural numbers with ten as a base. As an example of how the table is to be used we find, in the row with 4 at the extreme left, the numeral 12 in the column headed by 8. This tells us that $4 + 8 = 12$. Similarly, 12 is found opposite 8 and under 4. Thus $8 + 4 = 4 + 8 = 12$. In the same manner the sum of any two one-digit numbers, with numerals written in base ten, can be read from the table.

Figure 3-5 may be used in a corresponding manner for numerals written in base **three.** The basic addition facts are few in this system.

$$1 + 1 = 2 \qquad 1 + 2 = 2 + 1 = 10 \qquad 2 + 2 = 11$$

In Fig. 3-6 we have the addition facts for a numeral system using **six** as a base. Opposite **4** and under **2** we find **10,** which tells us that $4 + 2 = 10$. Opposite **5** and under **4** we find **13,** so that $5 + 4 = 13$.

Each table contains only numerals representing numbers from the set of

+	1	2	3	4	5	6	7	8	9
1	2	3	4	5	6	7	8	9	10
2	3	4	5	6	7	8	9	10	11
3	4	5	6	7	8	9	10	11	12
4	5	6	7	8	9	10	11	12	13
5	6	7	8	9	10	11	12	13	14
6	7	8	9	10	11	12	13	14	15
7	8	9	10	11	12	13	14	15	16
8	9	10	11	12	13	14	15	16	17
9	10	11	12	13	14	15	16	17	18

FIGURE 3-4 Basic addition facts for natural numbers (base ten). In the row headed by 4 on the left we find 12 in the column headed by 8 above; thus $4 + 8 = 12$. Similarly $8 + 4 = 12$.

natural numbers as written in the particular system specified, and serves as an illustration of the closure property: The sum of two natural numbers is a natural number. We assume that this would continue to be true if each table could be extended to contain the sum of any two natural numbers of that particular system.

Examination of each table reveals the fact that the table has symmetry with respect to the dotted line which is a diagonal of the particular diagram (the same numerals are seen in corresponding positions on each side of the diagonal). This fact illustrates that addition in each system is commutative. Why? As in the case of closure, we assume that this characteristic would continue to be true if each table could be extended to contain the sum of any two natural numbers.

As an illustration of the associative property we raise the question of what is the sum of the four numbers **3 + 4 + 1 + 2** if the numerals represent numbers written in base **six.** Since addition is a binary operation we must group these numbers in such a way that we can combine them two at a time. We might have any one of four groupings without changing the relative order of the four numbers. Note that in each case the order is **3, 4, 1, 2.**

1 **(3 + 4) + (1 + 2)**, base **six**
From Fig. 3-6, first we have **3 + 4 = 11** and **1 + 2 = 3;** then **11 + 3 = 14 — six.**

+	1	2
1	2	10
2	10	11

FIGURE 3-5 Basic addition facts for natural numbers (base **three**). In the row headed by **1** on the left we find **10** in the column headed by **2** above; thus **1 + 2 = 10 — three.** Similarly, **2 + 1 = 10.**

+	1	2	3	4	5
1	2	3	4	5	10
2	3	4	5	10	11
3	4	5	10	11	12
4	5	10	11	12	13
5	10	11	12	13	14

FIGURE 3-6 Basic addition facts for natural numbers (base **six**). In the row headed by **3** on the left we find **11** in the column headed by **4** above; thus **3 + 4 = 11 − six.** Similarly **4 + 3 = 11.**

2   **3** +(**4 + 1**) + **2,** base **six**
First we have **4 + 1 = 5,** and the desired sum becomes **3 + 5 + 2.** This sum may be found by the grouping (**3 + 5**) + **2 = 12 + 2 = 14 − six,** or by **3** + (**5 + 2**) = **3 + 11 = 14 − six.**

3   **3 + (4 + 1 + 2),** base **six**
To find this sum we first must determine a sum for **4 + 1 + 2** which may be grouped either as (**4 + 1**) + **2,** giving the sum **5 + 2 = 11,** or as **4** + (**1 + 2**) = **4 + 3 = 11.**   Then   **3 + (4 + 1 + 2) = 3 + 11 = 14 − six.**

4   (**3 + 4 + 1**) + **2,** base **six**
First   we   have   **3 + 4 + 1 = (3 + 4) + 1 = 11 + 1 = 12,**   or   **3** + (**4 + 1**) = **3 + 5 = 12.** Whence it follows that (**3 + 4 + 1**) + **2 = 12 + 2 = 14 − six.**

It will be informative at this point to observe to just what extent the associative and commutative properties enter into finding the sum of any set of addends, for example those in the illustrations of Figs. 3-1 to 3-3. Rather than use the large addends of those illustrations it will suffice to analyze the process of finding the sum 57 + 83. This analysis, which is given in the following example, will be simpler to follow if the addition is accomplished with the numbers in horizontal rather than vertical array, although the details of the operation are the same. The steps are given in the left column, with their respective reasons given directly opposite in the right column.

Example   Find the sum of 57 and 83, base ten.

1	$57 + 83 = (50 + 7) + (80 + 3)$	Place value
2	$(50 + 7) + (80 + 3) = 50 + (7 + 80) + 3$	Associative property
3	$50 + (7 + 80) + 3 = 50 + (80 + 7) + 3$	Commutative property
4	$50 + (80 + 7) + 3 = (50 + 80) + (7 + 3)$	Associative property
5	$(50 + 80) + (7 + 3) = 130 + 10$	Addition
6	$130 + 10 = (100 + 30) + 10$	Place value

7	$(100 + 30) + 10 = 100 + (30 + 10)$	Associative property
8	$100 + (30 + 10) = 100 + 40$	Addition
9	$100 + 40 = 140$	Place value
10	$57 + 83 = 140$	The justification for this last step is found in a property (E-3) of equality, listed later

EXERCISES

For the set of natural numbers $N = \{1,2,3,4,5, \ldots\}$ cite the property, or properties, which support the truth of the statement in each exercise.

1 $3 + 5 \in N$ 2 $2 + 4 + 7 \in N$
3 $42 + 35 \in N$ 4 $4 + 3 + 2 + 6 + 1 \in N$
5 $2 + 8 = 8 + 2$ 6 $(5 + 4) + 9 = 5 + (4 + 9)$
7 $(3 + 8) + (7 + 4) = 3 + (8 + 7) + 4$
8 $(4 + 5) + 9 = 4 + (9 + 5)$
9 $56 + 73 = (50 + 70) + (6 + 3)$
10 $124 + 317 + 873 = (100 + 300 + 800) + (20 + 10 + 70) + (4 + 7 + 3)$
11 $(32 + 41) + 56 = (56 + 41) + 32$
12 For $a,b,c,d,e,f, \in N$
    $(a + b) + (c + d) + (e + f) = a + (b + c) + d + (e + f)$
13 $(8 + 7) + (3 + 4) + (2 + 6) = (8 + 2) + (7 + 3) + (4 + 6)$
14 If one has found the sum of a column of numbers by adding down the column, a check can be obtained by adding up the column.
15 It does not affect the sum of a set of one-digit addends, such as the given set, to follow this procedure:
    (a) Combine each set of addends whose sum is 10.

$$9 + 1 \qquad 4 + 6 \qquad 1 + 7 + 2$$

    (b) Find this sum.

$$10 + 10 + 10 = 30$$

    (c) To this sum add the sum of the remaining addends.

$$5 + 7 = 12 \qquad 30 + 12 = 42$$

$$
\begin{array}{r}
1 \\
5 \\
7 \\
4 \\
2 \\
9 \\
1 \\
6 \\
7 \\
\hline
42
\end{array}
$$

3-7 PROPERTIES OF EQUALITY

Up to this point we have, on occasion, used the symbol $=$ to express the relation of equality between two sets or two numbers. It is used to mean "is the same as." For example, in the discussion of Fig. 3-4 (page 80), there occurs the statement $4 + 8 = 12$, or $4 + 8$ is the same as 12. When we are dealing with numbers, this relation has five properties which distinguish it from other relations which may exist between numbers. In this context the numbers with which we are concerned are natural numbers. However, these properties of equality hold for the numbers of any of the number systems discussed in this text. They, will, therefore, be stated in that context.

The relation of equality between any two elements of the set $S = \{a, b, c, \ldots\}$ has these properties:

Properties of the relation of equality

E-1 Reflexive $a = a$. Any number is equal to itself.

E-2 Symmetric If $a = b$, then $b = a$. It is immaterial whether we read an equality from left to right or from right to left.

This is not always true for a relation between two numbers. For example, it is a true statement to say that 2 is less than 4, but it is not a true statement to say that 4 is less than 2.

E-3 Transitive If $a = b$ and $b = c$, then $a = c$.

E-4 Additive If $a = b$, then $a + c = b + c$, or $c + a = c + b$.

E-5 Multiplicative If $a = b$, then $ac = bc$, or $ca = cb$.

This last property, of course, has significance only in the context of the definition of multiplication, as stated later. (See pages 88 to 94.)

The transitive property and the fact that the result from any well-defined operation is a unique result together provide the authority for the *rule of substitution*.

Substitution rule for equals

Substitution rule A number may be substituted for its equal in any statement or operational process.

Example Since $3 + 4 = 7$ it follows that, in any situation where desirable, $3 + 4$ may be replaced by 7, or 7 may be replaced by $3 + 4$. As a consequence, we have, for example, $(3 + 4) + 5 = 7 + 5$.

In each of the four illustrations of grouping related to Fig. 3-6 the properties of substitution and transitivity provide the authority for the continued equalities. The reader should verify this statement.

EXERCISES

In these exercises different symbols for equality will be used: (1) = to mean "the same as" as defined for the relation between sets and between natural numbers; and (2) other symbols as defined in each particular case. In each exercise, present the argument sufficient to establish that the particular relation of equality is an equivalence relation.

1   $P = Q$ for any two sets P and Q.
2   $n(P) = n(Q)$ for any two sets P and Q.
3   If (a,b) and (c,d) are ordered pairs of natural numbers, then $(a,b) \doteq (c,d)$ if and only if $a = c$ and $b = d$. (Read $\doteq$ as "dot equal.")
4   If (a,b) and (c,d) are ordered pairs of natural numbers, then $(a,b) \ominus (c,d)$ if and only if $ad = bc$. (Read $\ominus$ as "circle equal.")
*5   If (a,b) and (c,d) are ordered pairs of natural numbers, then $(a,b)//(c,d)$ if and only if $a + d = b + c$. (Read $//$ as "slash equal.")
*6   If a and b are natural numbers, then $a \boxminus b$ if and only if $a = 2k + c$ and $b = 2n + c$ where c, k, and n are natural numbers. (Read $\boxminus$ as "square equal.")

3-8 CHECKING ADDITION

Commutative property provides a check

The surest check for the accuracy of an addition is to combine the addends in the reverse order to that used in finding the sum. If the sum is found by "adding down" the columns, then the check is established by "adding up," and vice versa. The commutative and associative properties of addition assure us that the two sums thus obtained must be the same.

Casting out nines

Probably the most easily used of the simple checks is that known as *casting out nines*. If the digits of a number are added and, if this sum is more than a one-digit number, the process is repeated until a one-digit number is obtained then the one-digit number so obtained is called the *excess of nines* in the original number. For example, in Fig. 3-7 the sum of the digits in 4,823 is $4 + 8 + 2 + 3 = 17$, and the sum of the digits in 17 is $1 + 7 = 8$. From this it follows that the excess of nines in 4,823 is 8. This simply means that if 4,823 is divided by 9 there will be a remainder of 8.

4,823	$4 + 8 + 2 + 3 = 17$	$1 + 7 = 8$	
5,736	$5 + 7 + 3 + 6 = 21$	$2 + 1 = 3$	
2,591	$2 + 5 + 9 + 1 = 17$	$1 + 7 = 8$	
3,425	$3 + 4 + 2 + 5 = 14$	$1 + 4 = 5$	
		24	$2 + 4 = 6$
16,575	$1 + 6 + 5 + 7 + 5 =$	24	$2 + 4 = 6$

FIGURE 3-7 Casting out nines is a quick check. It is not perfect, however, since transposition of digits will not change their sum.

$$
\begin{array}{r}
5\,000 \\
6\,000 \\
3\,000 \\
3\,000 \\
\hline
17\,000
\end{array}
$$

FIGURE 3-8 Approximation can help determine whether an obtained sum is in the right order of magnitude.

Why this works we shall leave as an open question to return to at a later time (Sec. 5-4). The process of casting out nines can be made still simpler. Look at the digits in a number and discard any group whose sum is 9. Then add the remaining digits. For example, in 4,823, $4 + 2 + 3 = 9$. When these digits are discarded, the digit 8 is left. It is less than 9 and is the excess of nines in the number. In 5,736 the sum of the digits 3 and 6 is 9, and so they are discarded. Then $5 + 7 = 12$ and $1 + 2 = 3$, the excess of nines in 5,736.

The check of casting out nines in addition then is in accordance with the following statement, which will be proved later (Sec. 5-4):

The excess of nines in the sum of two or more numbers is the same as the excess in the sum of the excesses of each of the numbers.

In the example of Fig. 3-7 the excess of nines in the sum 16,575 is 6. The sum of the excesses in each of the numbers is 24 in which the excess is 6.

The check of addition by casting out nines is by no means an absolute check. If the excesses in the two sums indicated are not the same, then one knows there is an error somewhere. However, errors can be made and not be detected by this process. For example, suppose that either of the incorrect sums 16,665 or 16,755 had been obtained in the example. In either case the excess of nines is 6, so that the check would fail. In spite of this weakness, this method of checking is recognized as worthy of use because of its simplicity.

Another quick and effective check, especially against the making of absurd errors, is the method of estimating the answer by approximation. For example, in Fig. 3-8 the first addend of Fig. 3-7 is close to 5,000, the second to 6,000, and the other two to 3,000. Thus a quick approximation to the sum is 17,000.

EXERCISES

1 Give a brief description of the use of the abacus in addition.
2 What is meant by the statement that addition of natural numbers is a well-defined process?

3　What is the basis of the procedure called "carrying" in addition?

4　Use the following sums to illustrate the importance of place value in addition:

(a)	75	(b) 2,519	(c) 295	(d) 4,123
	213	3,428	76	125
	42	4,632	81	73
	56	5,678	324	3,982

(e) 2,526,114	(f) 721,836	(g) 4,271,856
3,142,878	1,856,417	232,431
4,243,851	9,682,563	2,682

5　Use the algorithm of Gemma Frisius (Fig. 3-2) and that for retrograde addition (Fig. 3-1) to find these sums:

(a) 649	(b) 167	(c) 843	(d) 967	(e) 34
278	423	217	868	833
976	275	586	492	25
685	274	965	438	766

(f) 452	(g) 777	(h) 764	(i) 427	(j) 1,623
689	333	493	276	2,346
928	499	876	634	763
54	563	798	895	489
76	256	365	467	68

(k) 9,412	(l) 6,878	(m) 21,594	(n) 24,355
6,754	272	96,633	67,932
3,854	565	97,851	43,661
9,499	39	88,563	42,876
6,483	236		

(o) 2,871,663	(p) 8,762,534	(q) 32,756,823
8,684,236	61,892	15,862,154
3,667,126	14,512,476	4,627,693
4,685,263	3,898,663	83,162,435

6　Use Exercises 4a and 5a to illustrate how the associative and commutative properties of addition support each algorithm used.

7　Use the three addition checks suggested in the text for checking the sums of Exercises 4 and 5.

∗ 8　The addition facts for the **verto** system are given in Fig. 3-9. How does an examination of the table illustrate the fact that the **verto** system is closed under addition?

∗ 9　Does an examination of Fig. 3-9 show that addition in the **verto** system is commutative? Why?

+	I	⊥	△	□
I	⊥	△	□	10
⊥	△	□	10	11
△	□	10	11	1⊥
□	10	11	1⊥	1△

FIGURE 3-9 Addition facts for **verto** system. In the row headed by △ on the left we find 1⊥ in the column headed by □ above. △ + □ = 1⊥. Similarly, □ + △ = 1⊥.

∗10 Find the sums:

    (a) (⊥ + □) + I; ⊥ (□ + I)

    (b) (△ + △) + □; △ + (△ + □)

    (c) (⊥I + △□) + △⊥; ⊥I + (△□ + △⊥)

∗11 What property of addition does each pair of sums illustrate for the **verto** system?

∗12 Find these sums:

    (a) I⊥△ + □⊥⊥ + △⊥□

    (b) △△△ + ⊥I□ + ⊥I⊥

    (c) I△△□ + ⊥△I□ + △□I□

13 Construct addition tables for base **five** and base **eight.**

14 How can the tables of Exercise 13 be used to illustrate the fact that addition in each of the respective number systems has the properties of closure and commutativity?

15 Construct examples based on each table of Exercise 13 to illustrate the associative property for addition with each of the respective bases.

16 Find these sums in base **five:**

    (a) 14 + 23 + 31 + 42

    (b) 2133 + 4143 + 4424

    (c) 142 + 321 + 413 + 224 + 34

    (d) 2341 + 4121 + 1424 + 3124

17 Find these sums in base **eight:**

    (a) 14 + 23 + 31 + 47

    (b) 321 + 475 + 432 + 135

    (c) 7643 + 1525 + 347 + 6445

    (d) 1634 + 2746 + 3124 + 5447

    (e) 5224 + 7162 + 1335 + 2172

    (f) 45 + 3423 + 712 + 1377 + 2564

18 Use the table for base **three** (Fig. 3-5) to find these sums:

    (a) 111 + 112 + 11 + 111 + 111

    (b) 11111 + 11211 + 1221 + 11221 + 111

    (c) 121112 + 111 + 111111 + 1211 + 11211

    (d) 111 + 112111 + 12 + 1111 + 121121 + 11211

19 Use the table for base **six** (Fig. 3-6) to find these sums:

    (a) 123 + 234 + 512 + 343

    (b) 4412 + 3145 + 2325 + 1245

(*c*) 3421 + 54243 + 121 + 3542 + 45

(*d*) 5435 + 12335 + 341 + 2435 + 52434

20 Construct examples based on Exercises 18 and 19 to show how the commutative property provides a check for addition.

21 Construct addition tables for bases **seven, eleven,** and **twelve.**

22 Find these sums in base **seven.** Use the commutative property to provide a check for each sum.

(*a*) 16 + 52 + 34 + 44 + 6 + 12

(*b*) 264 + 312 + 451 + 325

(*c*) 2136 + 5432 + 616 + 4542

(*d*) 41343 + 31264 + 52361 + 41421

(*e*) 123 + 4662 + 36 + 55231 + 6126

23 Find these sums in base **eleven.** Use the commutative property to provide a check for each sum.

(*a*) 189 + 34 + 798 + 259

(*b*) 6889 + 7182 + 3672 + 9419

(*c*) 561 + 8197 + 99 + 7652

(*d*) 4842 + 8467 + 7165 + 329

24 Why is casting out **fives** in base **six** similar to casting out nines in base ten?

25 Check each sum of Exercise 19 by casting out **fives.**

26 Casting out what numbers in bases **seven** and **eleven** is similar to casting out nines in base ten?

27 Use the techniques of Exercise 26 to check the sums in Exercises 22 and 23.

28 Find these sums in base **twelve.** Use the commutative property to provide a check for each sum.

(*a*) 14 + 23 + 31 + 4t

(*b*) 52 + 78 + 67 + 83

(*c*) 769 + 234 + 189 + 487

(*d*) 472 + 569 + 3te + e4e + t5e

(*e*) 1778 + te82 + ette + 36te

(*f*) e56e + 8182 + ttte + tete

*29 Use the sum 87 + 72 + 58 in base ten to illustrate how the associative and commutative laws justify the procedure followed in finding such sums.

3-9 WHAT IS MULTIPLICATION?

Multiplication is a short form of addition

From the most elementary point of view, multiplication of natural numbers is a short form of addition. It is the process which is used when we desire to determine how many objects there are in several disjoint sets when the sets all have the same cardinal number. For example, suppose we desire to know how many letters of the alphabet there are in the three sets

$A = \{a,b,c,d\}$, $B = \{e,f,g,h\}$, and $C = \{i,j,k,l\}$. These are disjoint sets and, by the definition of addition, $n(A \cup B \cup C) = n(A) + n(B) + n(C) = 4 + 4 + 4 = 12$. Since $n(A) = n(B) = n(C) = 4$, the answer to this problem can be found also by the process of multiplication. Thus $n(A \cup B \cup C)$ is the product 3×4. The table of basic multiplication facts, which complete the definition of the operation, makes it possible for us to know that the result of the indicated multiplication 3×4 is the *product* 12. As in the case of addition, each fact of the table may be established by

Multiplication is unique

counting or, even better for multiplication, by addition. Just as the sum in addition is unique, so is the product in multiplication unique.

A less restricted definition of the process of multiplication may be obtained from the following consideration. The elements of each of the sets A, B, and C may be placed in one-to-one correspondence with each other and with the elements of $Q = \{1,2,3,4\}$. Since this is true we may use the diagram of Fig. 3-10 to illustrate the operation discussed in the previous paragraph. This diagram is seen to be that of the cartesian product $P \times Q$ where $P = \{A,B,C\}$ and $Q = \{1,2,3,4\}$. In the diagram the ordered pairs $(A,1)$, $(A,2)$, $(A,3)$, and $(A,4)$ are elements of $P \times Q$ with first element A; similarly for the ordered pairs whose first elements are B and C.

In Sec. 2-6 attention was called to the fact that, while $A \times B \neq B \times A$, it is true that $n(A \times B) = n(B \times A)$ for any two sets A and B. Also, in the exercises of that section it was found by actual count in several cases that $n(A \times B)$ was the same as $n(A)$ times $n(B)$. It is to be observed that in the example being discussed here, $n(P \times Q) = 12$ and $n(P)$ times $n(Q) = 12$. This discussion has presented a specific illustration of the manner in which the less restricted definition of multiplication, which follows, incorporates the definition previously discussed.

General definition of multiplication

Definition 3-2   For any two sets A and B the *product* of the cardinal number of A and the cardinal number of B is the same as the cardinal

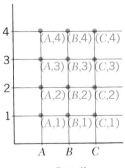

4   $(A,4)$ $(B,4)$ $(C,4)$

3   $(A,3)$ $(B,3)$ $(C,3)$

2   $(A,2)$ $(B,2)$ $(C,2)$

1   $(A,1)$ $(B,1)$ $(C,1)$

    A B C

$P \times Q$

$P = \{A,B,C\}$; $Q = \{1,2,3,4\}$ **FIGURE 3-10**

number of the cartesian product of the two sets. The operation used to find this product is indicated by the symbol $\times$ and is called *multiplication*.
$$n(A) \times n(B) = n(A \times B) = n(B \times A) = n(B) \times n(A).\dagger$$

One of the immediate advantages of this definition of multiplication lies in the fact that, while it is equivalent to the definition given for the case in which P and Q are disjoint sets, it no longer requires that the sets be disjoint. In fact, it allows for the product in the extreme case in which $P = Q$. For example, consider $A = \{1,2\}$ and $B = \{1,2\}$. The set of ordered pairs $A \times B$ is the set $\{(1,1),(1,2),(2,1),(2,2)\}$, and it is evident that $n(A \times B) = n(A) \times n(B)$. Incidentally, an important point implied by the above discussion is the fact that when the product of two natural numbers is obtained, the order in which they are multiplied is immaterial. In other words, it seems intuitively acceptable to assume that *multiplication of natural numbers is commutative*. Attention is also called to the fact that centuries of experience in the use of the multiplication process have given mathematicians strong intuitive background for asserting that *multiplication of natural numbers is associative as well as commutative*. In symbols we say that for all natural numbers a, b, and c it is true that $(a \times b) \times c = a \times (b \times c)$. For example, from the definition $(2 \times 3) \times 5 = 6 \times 5 = 30$, and $2 \times (3 \times 5) = 2 \times 15 = 30$. Thus $(2 \times 3) \times 5 = 2 \times (3 \times 5)$.

Multiplication is commutative and associative

Multiplication of natural numbers is thus a well-defined binary operation which gives a *unique* result known as a *product*.

Product

3-10   NAMES AND SYMBOLS USED IN MULTIPLICATION

Names used in multiplication

The terms *multiplicand, multiplier,* and *product* come from the Latin expressions *numerus multiplicandus* (number to be multiplied), *numerus multiplicans* (multiplying number), and *numerus productus* (number produced). Since the word *numerus* was dropped occasionally by Latin writers, the terms finally became the single words in technical translations. As was pointed out in an earlier section, the word "product" was rather generally used to indicate the result obtained from any form of computation, although in recent years it has been used almost exclusively to indicate the result obtained from the process of multiplication. In more modern usage the two words "multiplier" and "multiplicand" are being replaced by the one word "factor." This is due to the fact that multiplication is commutative and any two factors combine to give the same product, independent of which is considered as the multiplier and which as the multiplicand.

†On occasion $a \cdot b$ and ab are used to indicate the same product as $a \times b$. Juxtaposition as in ab cannot be used with numerals to indicate a product. For example, 34 means $(3 \times 10) + 4$ and not 3×4.

Factor   A *factor* of a natural number n is any one of two or more natural numbers whose product is n.

The symbol ×   The symbol ×, most commonly used in arithmetic to indicate multiplication, was developed in England in the early seventeenth century. Cajori was of the opinion that the first use of the "St. Andrew's cross" as a symbol for multiplication is to be attributed to W. Oughtred.† Other methods of indicating multiplication include the practice of Diophantus of using no symbol at all, juxtaposition in the Bakhshālī manuscripts, the dot of Bhāskara, ideograms by the Babylonians and Egyptians, and the use of M by such mathematicians as Stifel, S. Stevin, and Descartes.‡ Although some of the Bhāskara manuscripts contained an unexplained use of the dot as a symbol for multiplication and similar unexplained uses occurred in manuscripts of Thomas Harriot (1631) and Thomas Gibson (1655), the actual introduction of the use of this symbol is attributed to G. W. Leibniz (1646–1716), who, in writing to John Bernoulli, stated: "I do not like × as a symbol in multiplication, as it is easily confounded with x, . . . often I simply relate two quantities by an interposed dot and indicate multiplication by ZC · LM."§ Modern notation uses both symbols to indicate multiplication.

*3-11 THE MULTIPLICATION ALGORITHM

Little is known about the multiplication algorithm used in ancient times. There is rather strong evidence, however, that the Egyptians and many generations of their successors used the method of duplation (doubling). This method is used in Fig. 3-11 to find the product 43×23. The multiplicand 23 is written opposite the number 1. As 1 is doubled, so is 23.

Method of doubling

†F. Cajori, *op. cit.*, p. 251.
‡*Ibid.*, p. 250.
§*Ibid.*, p. 267.

$$
\begin{array}{lll}
1\checkmark & \checkmark\,23 & \\
2 \times 1\ \ = 2\checkmark & \checkmark\,46 = 2 \times 23 & \text{Step 1} \\
2 \times 2\ \ = 4 & 92 = 2 \times 46 & \text{Step 2} \\
2 \times 4\ \ = 8\checkmark & \checkmark\,184 = 2 \times 92 & \text{Step 3} \\
2 \times 8\ \ = 16 & 368 = 2 \times 184 & \text{Step 4} \\
2 \times 16 = 32\checkmark & \checkmark\,736 = 2 \times 368 & \text{Step 5} \\
\end{array}
$$

$1 + 2 + 8 + 32 = 43$

$23 + 46 + 184 + 736 = 989 = 43 \times 23$

FIGURE 3-11 Multiplication by doubling. In each step both multiplier and multiplicand are doubled.

√ 43	23 √	
√ 21	46 √	Step 1
10	92	Step 2
√ 5	184 √	Step 3
2	368	Step 4
√ 1	736 √	Step 5
	989	

FIGURE 3-12 Multiplication by doubling and halving. In each step the multiplicand is doubled but the multiplier is halved; remainders are discarded.

This process is repeated, as shown by the colored numerals, until the next double in the column headed by 1 would be greater than 43. Then the numbers in this column whose sum is 43 are found and checked (✓) along with the numbers opposite them. The sum of these corresponding numbers (those checked) in the column headed by 23 gives the desired product. It is to be noted that this process reduces the necessary multiplication facts down to only those in which 2 is the multiplier. It results from the following generalized definition of multiplication which is of historical significance:

Multiplication is the process of finding a third number related to one of two given numbers in the same ratio as the second number is related to 1.

Basically similar to the duplation method is the technique of finding a product sometimes referred to as the "Russian peasant method." In this method duplication is accompanied by mediation (halving), as illustrated in Fig. 3-12. Both the multiplier and multiplicand are written down. As the multiplicand is doubled, the multiplier is halved. In the halving process any remainder is discarded. Thus, the process will terminate ultimately with 1 in the multiplier column. To get the product, add those numbers in the multiplicand column which correspond to the odd numbers in the multiplier column. Note that the same numbers are checked in the multiplicand column in Figs. 3-11 and 3-12.

The forms used through the ages have varied so greatly that it would be tedious to present all of them. The number receiving significant attention had increased to eight by the time Pacioli published his "Suma" in 1494.†

†David Eugene Smith, *op. cit.*, p. 107.

7 5 9			
3 6			
4 5 5 4	= 6 × 759		Step 1
2 2 7 7	= 30 × 759		Step 2
2 7 3 2 4	= 4554 + 22770		Step 3

FIGURE 3-13 Early form of multiplication

					3		
					6		
			7	5	9		
Step 1				5	4		$= 6 \times 9$
Step 2			2	7			$= 30 \times 9$
Step 3			3	0			$= 6 \times 50$
Step 4		1	5				$= 30 \times 50$
Step 5		4	2				$= 6 \times 700$
Step 6	2	1					$= 30 \times 700$
Step 7	2	7	3	2	4		$=$ Total

$$
\begin{array}{rll}
 & 7\ 5\ 9 & \\
\hline
6 \times 759 = & 4\ 5\ 5\ 4/6 & \text{Step 1} \\
30 \times 759 = & 2\ 2\ 7\ 7/3 & \text{Step 2} \\
\hline
\text{Total} = & 2\ 7\ 3\ 2\ 4 & \text{Step 3}
\end{array}
$$

FIGURE 3-14 Scacchera method of multiplication

Of these eight, only four are of interest to us in this discussion. They picture something of the evolution of form through which the multiplicative process passed in the struggle for the most effective use of place value to produce efficiency of operation. Interestingly enough, one of the forms is the one which we use today, with the slight variations shown in the example (Fig. 3-13). It is rather evident that the extra lines were drawn as an aid in keeping the digits aligned.

The *scacchera* (chessboard) method was quite popular for many years. There are variations in the relative positions of the two factors and the partial products. Two distinct forms of this particular method are illustrated in Fig. 3-14, in which we find the same product, 36 × 759.

The *gelosia* method, so called because of the resemblance of its framework to latticework, simplifies multiplication in that it records the result of each multiplication and thus does away with adding unseen numbers in obtaining the partial products. One of its several different forms is illustrated in Fig. 3-15 in finding the product 269 × 8,678. After each product is recorded in its proper block, the final product is then found by summing along each diagonal, starting at the extreme right. This product is 2,334,382.

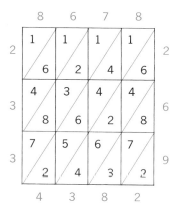

FIGURE 3-15 Gelosia (lattice) multiplication. In each block the colored line separates the tens digit from the ones. For example, in the lower right corner 72 = 9 × 8. The adding is along the diagonals. 8 + 7 + 3 = 18, and the 1 ten of this sum is added with the digits in the next diagonal array: 1 + 6 + 4 + 2 + 6 + 4 = 23.

234	
482	
468	= 2 × 234 Step 1
1872	= 80 × 234 Step 2
936	= 400 × 234 Step 3
112788	= 93,600 + 18,720 + 468

FIGURE 3-16 Algorithm used today

The *repiego* method broke the multiplier down into simple factors and then multiplied the multiplicand by these factors in order. For example, to find the product 36 × 759, first multiply 759 by 9 and then multiply this result by 4. Multiplication is *associative:* (4 × 9) × 759 = 4 × (9 × 759). Of course, other factors could be used. This is an effective method for oral multiplication.†

Each of these algorithms depends heavily on a property which multiplication has when used in conjunction with addition. This new property, called the *distributive* property, states that for any three natural numbers (a,b,c) it is true that $a \times (b + c) = (a \times b) + (a \times c)$. For example, 5 × (8 + 2) = (5 × 8) + (5 × 2). This property, which is easily verified in this case since 5 × 10 = 50 and 40 + 10 = 50, is displayed as a property relating multiplication and addition of natural numbers which is intuitively acceptable. This is used in the method of multiplication by duplation, as shown in Fig. 3-11. Since 43 = 1 + 2 + 8 + 32, the product 43 × 23 is written as (1 + 2 + 8 + 32) × 23, which is the same as 23 × (1 + 2 + 8 + 32) = (23 × 1) + (23 × 2) + (23 × 8) + (23 × 32). Each of these products was found by doubling so that 23 × 43 = 23 + 46 + 184 + 736 = 989.

Multiplication is distributive over addition

The algorithm which we use today also depends on the distributive property. In the product shown in Fig. 3-16, the partial products are placed in their respective positions due to place value and the distributive property of multiplication. The colored numerals illustrate the fact that the product is found by taking the product 234 × (400 + 80 + 2). Thus the 1,872 is in reality 18,720, since it is the product 234 × 80. Likewise, 936 actually represents 234 × 400 = 93,600. Place value makes it unnecessary to write the zeros. What other properties of multiplication are used in finding this product?

3-12 PROPERTIES OF NATURAL NUMBERS UNDER MULTIPLICATION

Attention has been called already to the fact that multiplication of natural numbers is associative, is commutative, and also is distributive over addition. It also has the property of closure and one additional property, the existence of an identity, not yet mentioned.

† *Ibid.*, pp. 106–128.

The set $N = \{1,2,3,4, \ldots\}$ of natural numbers satisfies these properties under the operation of multiplication:

Properties of the natural numbers under multiplication

M-1 **Closure** For $a,b \in N$ it is true that $a \times b$ is a unique element of N.

M-2 **Commutative** For $a,b \in N$ it is true that $a \times b = b \times a$.

M-3 **Associative** For $a,b,c \in N$ it is true that $(a \times b) \times c = a \times (b \times c)$.

M-4 **Distributive** For $a,b,c \in N$ it is true that $a \times (b + c) = (a \times b) + (a \times c)$ and $(b + c) \times a = (b \times a) + (c \times a)$.

M-5 **Identity** There exists a unique natural number 1 such that $1 \times a = a \times 1 = a$ for any natural number a.

Property M-4 states that multiplication distributes over addition; that is, in any case of a product involving the multiplication of a sum of two natural numbers by a third natural number, one may either add first and then multiply or multiply first and then add. The identification of 1 as the identity element for multiplication will increase in significance as we work further with natural numbers and later as we extend our considerations to other number systems.

EXERCISES

For $N = \{1,2,3,4,5, \ldots\}$ cite the property, or properties, which support the truth of the statement in each exercise.

1 $4 \times 7 \in N$
2 $5 \times 6 \times 8 \in N$
3 $21 \times 132 \in N$
4 $8 \cdot 17 \cdot 16 \cdot 52 \in N$
5 $3 \times 8 = 8 \times 3$
6 $(6 \times 9) \times 7 = 6 \times (9 \times 7)$
7 $5 \times (3 + 4) = (5 \times 3) + (5 \times 4)$
8 $(4 \cdot 8) \cdot (7 \cdot 3) = 4 \cdot (8 \cdot 7) \cdot 3$

In Exercises 9 to 12 how is the substitution rule used? What other properties are used?

9 $26 \times 4 = (20 + 6) \times 4 = (20 \times 4) + (6 \times 4)$
10 $23 \times 30 = (23 \times 3) \times 10$
11 $6 \times 42 = 3 \times (2 \times 42) = 2 \times (3 \times 42)$
12 $25 \times 37 = (20 + 5) \times (30 + 7)$
$= [(20 + 5) \times 30] + [(20 + 5) \times 7]$
$= [(20 \times 30) + (5 \times 30)] + [(20 \times 7) + (5 \times 7)]$
$= (600 + 150) + (140 + 35)$
$= 750 + 175$

3-13 CHECKING MULTIPLICATION

Commutative property provides a check

One of the surest checks in multiplication is to interchange multiplicand and multiplier and redetermine the product. The rule for casting out nines is

Casting out nines

The excess of nines in the product is equal to that of the product of the excesses in the multiplicand and multiplier.

For example, in Fig. 3-17 the excess of nines in 481 is 4 since $8 + 1 = 9$. The excess in 719 is 8 since 9 may be discarded and $7 + 1 = 8$. The excess in 345839 is 5 since 9 may be discarded, $4 + 5 = 9$, $3 + 8 + 3 = 14$, and $1 + 4 = 5$. The product of the excesses is $4 \times 8 = 32$, and $3 + 2 = 5$. As in addition, the check by casting out nines is in no sense an absolute check; its principal value lies in the ease and simplicity of its use. Another check, valuable because of its simplicity, is that of estimating the answer. In the example of Fig. 3-17, 481 is near 500 and 719 is near 700, so that a very satisfactory check on the reasonableness of the answer is $500 \times 700 = 350,000$.

EXERCISES

Follow these instructions for Exercises 1 to 24: (*a*) Estimate each product. (*b*) Compute each product. (*c*) Check each product by casting out nines. (*d*) Check each product by interchanging factors and computing the new product.

1	28	2	86	3	723	4	236
	37		71		3		7

5	345	6	487	7	721	8	642
	73		29		89		76

9	623	10	856	11	926	12	734
	327		133		472		678

13	8,264	14	7,534	15	6,418	16	9,273
	6		7		23		18

```
481      4
719      8
4329     8 × 4 = 32, 3 + 2 = 5
481
3367
345839   3 + 8 + 3 = 14, 1 + 4 = 5
```

FIGURE 3-17 Checking multiplication by casting out nines

| 17 | 4,273 | 18 | 8,129 | 19 | 7,246 | 20 | 6,749 |
| | 321 | | 762 | | 832 | | 914 |

| 21 | 1,234 | 22 | 3,286 | 23 | 6,731 | 24 | 8,162 |
| | 2,143 | | 1,784 | | 4,215 | | 7,256 |

*25 Use the product 74 × 63 to indicate how the associative, commutative, and distributive properties justify the procedure followed.

3-14 THE POSTULATES OF THE NATURAL NUMBER SYSTEM

The set of natural numbers $N = \{1,2,3,4, \ldots\}$, for which we now have two well-defined operations, addition ($+$) and multiplication ($\times$), satisfies the following postulates.

For any $a,b,c \in N$:

Postulates of the natural number system

N-1 Closure If $a + b = x$ and $a \times b = y$, then x and y are unique elements of N.

N-2 Associative $(a + b) + c = a + (b + c)$; $(a \times b) \times c = a \times (b \times c)$.

N-3 Commutative $a + b = b + a$; $a \times b = b \times a$.

N-4 Distributive $a \times (b + c) = (a \times b) + (a \times c)$, and $(b + c) \times a = (b \times a) + (c \times a)$.

N-5 Multiplicative identity There exists a unique natural number 1 such that $1 \times a = a \times 1 = a$ for any $a \in N$.

N-6 Trichotomy For any two elements a and b of N exactly one of these relations holds: $a = b$, $a + c = b$, or $a = b + c$ for some $c \in N$.

N-7 Cancellation If $a + c = b + c$, then $a = b$; if $a \times c = b \times c$, then $a = b$.

N-8 Finite induction If $P \subseteq N$ such that
  1   $1 \in P$, and
  2   $(k + 1) \in P$ whenever $k \in P$
then $P = N$.

Law of trichotomy

Postulates N-1 to N-5 will be recognized as those which have been listed as satisfied by the set N under addition (A-1 to A-3) and multiplication (M-1 to M-5). Postulate N-6 is frequently referred to as the *law of trichotomy*. Another useful form for stating this same law is provided by Definition 3-3.

Definition 3-3 If, for $a,b \in N$, it is true that $a + c = b$ for some $c \in N$, then a is said to be *less than b*$(a < b)$, or b is *greater than a*$(b > a)$.

Natural numbers are ordered

Inequality

In accord with this definition we may say that if any two natural numbers a and b are selected at random, we know that one and only one of three relations exists between them: $a = b$, $a > b$, or $a < b$. These last two relations are called *inequalities*. They do not satisfy all the properties of equality. For example, no natural number is greater than itself ($a \not> a$ for $a \in N$). We may say this another way: there exists no natural number a to which a natural number can be added so that the sum is a. This means that the relation "is greater than," applied to natural numbers, does not have the reflexive property.

Since the principle of finite induction (N-8) is of great significance in the study of the natural number system, it is listed here for completeness. In the development of this text we shall have only limited occasion to refer to it.

These eight postulates can be shown to be equivalent to the Peano postulates mentioned earlier. From them and the properties of equality (page 83), all the remaining properties of natural numbers can be derived.†

EXERCISES

1 What is meant by the statement that multiplication of natural numbers is a well-defined process?
2 How are addition and multiplication related to each other?
3 What is the basis of the procedure called "carrying" in multiplication?
4 First estimate, and then compute each product:

 (a) 48   (b) 437   (c) 539   (d) 8,124
    32   546   753   723

 (e) 719   (f) 9,317   (g) 5,214   (h) 6,639
    624   421   89   546

†Howard Eves and Carroll V. Newsom, "An Introduction to the Foundations and Fundamental Concepts of Mathematics," pp. 207–216, rev. ed., Holt, Rinehart and Winston, Inc., New York, 1965.

(*i*) 8,764	(*j*) 5,623	(*k*) 16,856	(*l*) 25,124
3,216	2,184	4,282	3,324

5   Check each product in Exercise 4 first by interchanging multiplier and multiplicand and then by casting out nines.

6   Use Exercises 4*b* and 4*h* as a means for illustrating the importance of place value in multiplication.

* 7   Follow these steps in considering the sum of the two columns of figures of this exercise.

(*a*) Make a guess as to which sum you think is the larger sum.

(*b*) Find the sum of each column.

(*c*) Use the *properties* of addition and multiplication to justify the result you obtain.

578	467
578	467
578	467
578	467
78	467
78	67
8	67
	7

8   Supply the reasons for each step in the conversion of this exercise in addition to one in multiplication.

$32 + 28 + 20 + 16$

$\qquad = (4 \times 8) + (4 \times 7) + (4 \times 5) + (4 \times 4)$ Why?

$(4 \times 8) + (4 \times 7) + (4 \times 5) + (4 \times 4)$

$\qquad = 4 \times (8 + 7 + 5 + 4)$ Why?

$4 \times (8 + 7 + 5 + 4) = 4 \times 24$ Why?

$4 \times 24 = 96$ Why?

$32 + 28 + 20 + 16 = 96$ Why?

9   In each exercise first rewrite each addend as a product with the encircled number as a factor; then convert the addition exercise into one in multiplication for finding the total for the numbers given.

(*a*) ⑤	(*b*) ⑩	(*c*) ②	(*d*) ⑪
15	10	8	22
35	70	6	44
25	90	4	11
55	20	16	55
45	80	14	88
5	30	12	66

(e) ⑧	(f) ⑨	(g) ⑦	(h) ③
72	45	49	15
48	81	63	21
32	18	35	42
56	54	14	27
24	63	7	36
16	99	21	24
64	36	42	39

10   Construct the table of basic multiplication facts for base ten, base **verto,** base **eight,** and base **twelve.**

11   What do the tables of Exercise 10 illustrate about the properties of closure and commutativity in each system?

12   Construct examples based on the tables of Exercise 10 to illustrate the fact that multiplication in each system is associative.

*13   Find these products and check:

(a) IΔ
　　 Δ⊥

(b) □I⊥
　　 ΔI□

(c) ΔI□⊥
　　 ⊥I□

14   Use the duplation method to find the product 77 × 29. Explain how this illustrates the distributive law of multiplication over addition.

15   Use the duplation and mediation method to find the product 45 × 95.

16   Use the repiego method to find each of these products:

(a) 63 × 36   (b) 85 × 28   (c) 272 × 54

17   How are the products of Exercise 16 illustrations of the associative property of multiplication?

18   Use the gelosia method to find each of these products:

(a) 839   (b) 3,737   (c) 7,986
    76   865   873

19   Find each of these products in base **eight:**

(a) **24 × 72**   (b) **576 × 327**   (c) **746 × 2756**

20   Check the products of Exercise 19 by casting out **sevens.**

21   Find each of these products in base **twelve.**

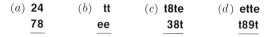

(a) **24**   (b) **tt**   (c) **t8te**   (d) **ette**
    78   **ee**   **38t**   **t89t**

22   Check the products of Exercise 21 by casting out **e's.**

23   Use these products in base ten to illustrate how the distributive property is used in finding such products.

(a) 247   (b) 3,145   (c) 24,562
      8   6   9

(d) 236 (e) 3,624 (f) 2,526
    74   527   374

∗24 Use each of these products to show in detail how the associative, commutative, and distributive properties of addition and multiplication are used in finding such products.

(a) 24 × 63 (b) 3 × 5,438

∗**3-15 THE NATURE OF PROOF**

Deduction and the nature of implication

The process used in deriving the properties of natural numbers from the basic postulates is that of *deduction*, which is the process of reasoning by *implication*. If we have two propositions, represented by the symbols p and q, which are so related to each other that it is absolutely impossible for q to be false if p is true, then we say that p implies q and write it symbolically as $p \rightarrow q$. By a "proposition" we shall mean a sentence which is stated so precisely that it has the property of being either true or false. No effort will be made here to go beyond the intuitively acceptable connotations of the terms "true" and "false." Furthermore, there is no requirement that we must know which of the labels applies to the sentence in order to be able to classify it as a proposition.

Example Consider these sentences:

1 Christmas Day fell on Tuesday in 1970.
2 Christmas Day will fall on Tuesday in 1979.
3 Christmas Day will fall on Tuesday in 2963.
4 Christmas Day will fall on Tuesday.

The first two sentences make statements which can be checked readily by reference to the appropriate calendars. Statement 1 is false and statement 2 is true. While the truth or falsity of statement 3 cannot be so readily checked, it does have the property of being either true or false. These three sentences are, therefore, bona fide propositions. Statement 4, however, does not meet the conditions necessary to qualify as a proposition. Although the sentence makes a complete statement, there is no way in which its truth or falsity may be judged.

Nature of a deductive system

In any deductive system there are (1) undefined terms, (2) defined terms, (3) assumed propositions (*postulates* or *axioms*), and (4) derived propositions (*theorems*). In the natural number system we have undefined terms such as *set* and *number* which we use to build a vocabulary of such terms as *addition, multiplication, closure, associative, commutative, distributive, identity, equal, is less than,* and *is greater than*. The postulates of the system are the propositions listed as N-1 to N-8 and E-1 to E-5. We are to use all this information as the context within which all the remaining properties of natural numbers can be derived as provable propositions (theorems). Within the limitations of this book we shall not be able to exhaust the possibilities of all such theorems. We shall prove a few and state as exercises a few others for the reader to prove.

All theorems follow a pattern of statement which is of the form "if p then q," where p and q are propositions. This form of statement is synonymous *p implies q* with "p implies q" ($p \rightarrow q$), since the statement "if p then q" is simply saying that if proposition p is assumed or known to be true, then it must follow that proposition q is true. Thus every theorem has two distinct parts: (1) *Hypothesis and conclusion* the *hypothesis*, or that which is given as true, and (2) the *conclusion*, or the proposition which follows as a logical consequence of the hypothesis.

Actually, in a deductive pattern of reasoning we are not so much concerned with whether the propositions involved are true or false; rather, we are concerned with whether or not the argument is a valid one. In other words, we are concerned not so much with whether the conclusion we can *Valid conclusion* draw is true as with the fact that it is valid. A *valid conclusion* is one which follows through a chain of implications from the hypothesis. Whether a *True conclusion* conclusion is true or false is most frequently determined by a check against experience or previously established knowledge.

*3-16 SOME DERIVED PROPERTIES OF THE NATURAL NUMBER SYSTEM

Two familiar properties, used frequently with numbers, are informally stated as follows: (1) *If equals are added to equals, the sums are equal;* and (2) *If equals are multiplied by equals, the products are equal.* These are theorems which will now be stated formally as Theorems 3-1 and 3-2 and proved for natural numbers. In each case the arguments will be presented, in parallel columns, for a special case and for the general case, with the supporting reason for each step appearing in the middle column.

Theorem 3-1 If $a,b,c,d \in N$ such that $a = b$ and $c = d$, then $a + c = b + d$.

As a special case we shall use:

Theorem If $1,2,3,4,7 \in N$ such that $1 + 2 = 3$ and $3 + 4 = 7$, then $(1 + 2) + (3 + 4) = 3 + 7$.

Hypothesis	$1,2,3,4,7 \in N$	*Hypothesis*	$a,b,c,d \in N$
	$1 + 2 = 3, \; 3 + 4 = 7$		$a = b, \; c = d$
Conclusion	$(1 + 2) + (3 + 4) = 3 + 7$	*Conclusion*	$a + c = b + d$

Proof

Statements	*Reasons*	*Statements*
1 $1 + 2 = 3$; $3 + 4 = 7$	Hypothesis	$a = b$, $c = d$
2 $(1 + 2) + (3 + 4) = 3 + 7$	Substitution	$a + c = b + d$

In step 2 the substitution rule has been used to replace the natural number $(1 + 2)$ by 3 and the natural number $(3 + 4)$ by 7 in the sum $(1 + 2) + (3 + 4)$.

In step 2 the substitution rule has been used to replace the natural number a by b and the natural number c by d in the sum $a + c$.

Theorem 3-2 If $a,b,c,d \in N$ such that $a = b$ and $c = d$, then $ac = bd$.

As a special case we shall use:

Theorem If $1,2,3,4,7 \in N$ such that $1 + 2 = 3$ and $3 + 4 = 7$, then $(1 + 2) \cdot (3 + 4) = 3 \cdot 7$.

Hypothesis $1,2,3,4,7 \in N$		*Hypothesis* $a,b,c,d \in N$
$1 + 2 = 3$; $3 + 4 = 7$		$a = b$, $c = d$
Conclusion $(1 + 2) \cdot (3 + 4) = 3 \cdot 7$		*Conclusion* $ac = bd$

Proof

Statements	*Reasons*	*Statements*
1 $1,2,3,4,7 \in N$	Hypothesis	$a,b,c,d \in N$
2 $1 + 2 = 3$	Hypothesis	$a = b$
3 $(1 + 2) \cdot (3 + 4) = 3 \cdot (3 + 4)$	E-5	$ac = bc$
4 $3 + 4 = 7$	Hypothesis	$c = d$
5 $(1 + 2) \cdot (3 + 4) = 3 \cdot 7$	Substitution rule	$ac = bd$

In step 5 the substitution rule is used to justify replacing $(3 + 4)$ by its equal 7 and c by its equal d, each in step 3. The suggestion for the multiplication of both sides of the equality stated in step 2 came from the fact that the conclusion called for the product $(1 + 2) \cdot (3 + 4)$ in the specific case and ac in the general case. E-5 furnished the authority for the multiplication of one side of an equality, provided the other side is multiplied by the same number. Inspection of the result suggested the use of the substitution rule.

One more illustration of the use of the postulates and definitions of the natural number system will be given. This theorem will extend the additive property to the relation of inequality. You should realize that any proposition established for the relation "is less than" is, by the same argument, true for "is greater than." This is true since $a < b$ states both propositions: a is less than b and b is greater than a.

Theorem 3-3 If a, b, c are natural numbers and $a < b$, then $a + c < b + c$.

For comparison purposes we shall follow the practice of stating the same theorem for a particular case: If 3, 8, 4 are natural numbers such that $3 < 8$, then $3 + 4 < 8 + 4$.

| *Hypothesis* | 3,8,4 $\in$ N | $a,b,c \in N$ |
| | $3 < 8$ | $a < b$ |

| *Conclusion* | $3 + 4 < 8 + 4$ | $a + c < b + c$ |

Proof

Statements	*Reasons*	*Statements*
1 $3 < 8$	Hypothesis	$a < b$
2 There exists a natural number k (5 in this case) such that $3 + 5 = 8$.	Definition 3-3	There exists a natural number k such that $a + k = b$.
3 $(3 + 5) + 4 = 8 + 4$	E-4	$(a + k) + c = b + c$
4 $(3 + 5) + 4 = (3 + 4) + 5$	N-2; N-3	$(a + k) + c = (a + c) + k$
5 $(3 + 4) + 5 = 8 + 4$	Substitution rule; steps 3,4	$(a + c) + k = b + c$
6 $3 + 4 < 8 + 4$	Definition 3-3	$a + c < b + c$

These few theorems have been presented merely to illustrate the fact that the properties of natural numbers can be developed as logical consequences of the basic postulates stated. The set N of natural numbers, the well defined operations of addition and multiplication, the set of postulates N-1 to N-8, and all the properties of natural numbers which follow as logical consequences from the definitions and these eight postulates constitute the number system we call the *natural number system*.

EXERCISES

1   The relations "less than" ($<$) and "greater than" ($>$) are called *inequalities* because they do not have all the properties E-1 to E-5 of equality. Which of these properties do these relations have?

2   Use the substitution rule to prove each form of the multiplicative property of equality. Theorem 3-4: If $a,b,c \in N$ such that $a = b$, then $ac = bc$ and $ca = cb$.

* 3   Prove Theorem 3-5: If $a,b,c \in N$ such that $a < b$, and $b < c$, then $a < c$.

* 4   Prove Theorem 3-6: If $a,b,c,d,e,f \in N$ such that $a = b, c = d$, and $e = f$, then $a + c + e = b + d + f$.

* 5   Prove Theorem 3-7: If $a,b,c,d,e,f \in N$ such that $a = b, c = d$, and $e = f$, then $ac + e = bd + f$.

* 6   Prove Theorem 3-8: If $a,b,c,d \in N$ such that $a < b$ and $c < d$, then $a + c < b + d$.

* 7   Prove Theorem 3-9: If $a,b,c \in N$ such that $a < b$, then $ac < bc$.

For the natural numbers a, b, c, d, and e, state the four theorems each of which has one of the following propositions as its conclusion. Prove each theorem.

* 8 $(a + b) + (c + d) + e = (a + c) + b + (e + d)$.

* 9 $(a + b)(c + d) = (ac + ad) + (bc + bd)$.

*10 $(a + b)(c + d + e) = (ae + be) + (c + d)(a + b)$.

*11   When $a > b$, then $ac + ad > bc + bd$.

3-17   TWO PARTITIONS OF THE SET OF NATURAL NUMBERS

A set is partitioned if its elements have been separated into two or more clearly defined disjoint subsets in accordance with some characterizing principle, definition, or law. For example, the students in a given school may be partitioned into the two disjoint subsets designated as male and female. Another partitioning of students which exists in most high schools and colleges is accomplished in accordance with the accumulation of specified amounts of academic credit. The disjoint subsets of this partition usually are designated as freshmen, sophomores, juniors, and seniors. In an elementary school such a partitioning usually is by grades: kindergarten, first grade, second grade, and so on.

Two partitions of N   There are two important partitions of the set N of natural numbers which are of interest at this point: (1) the prime numbers and the composite numbers, and (2) the even natural numbers and the odd natural numbers.

The prime numbers and the composite numbers   If a natural number a can be written in the form bc where b and c are both natural numbers

Factor and multiple

other than 1, it is called a *composite number*. The numbers b and c are said to be *factors* of a. Furthermore, a is a *multiple* of either factor.

Prime number

Definition 3-4 A *prime number* is any natural number greater than 1 which has no factors other than itself and 1.

This definition partitions the set N of natural numbers into the three disjoint sets:

1 The set whose only element is 1, $\{1\}$
2 The set of prime numbers, $\{p \mid p \text{ is a prime number}\}$
3 The set of composite numbers, $\{c \mid c \text{ is a composite number}\}$

Relatively prime natural numbers

Two natural numbers which have no common factor other than 1 are said to be *relatively prime*. Thus 33 and 70 are relatively prime, since $33 = 3 \times 11$ and $70 = 2 \times 5 \times 7$, and it is evident that they have no common factor other than 1.

From the theory which has developed as a result of the study of primes and their use in factorization problems there are three theorems of significance to the development here. They will be stated as postulates since their proofs are of a rather abstract nature.

Postulate 3-1 There are infinitely many primes.

Fundamental theorem of arithmetic

Postulate 3-2 The fundamental theorem of arithmetic Except for the ordering of the factors, any natural number greater than 1 is either a prime number or may be written as the product of primes in one and only one way.

Example The composite number 105 may be factored into primes as $3 \times 5 \times 7, 3 \times 7 \times 5, 5 \times 3 \times 7, 5 \times 7 \times 3, 7 \times 3 \times 5,$ or $7 \times 5 \times 3$. While these are six different arrangements, they are of the same three primes 3, 5, and 7.

The prime number 17 has no factors other than itself and 1, so that it has only one prime factor, namely 17.

Postulate 3-3 If the prime p is a factor of the composite number ab, then it must be a factor of either a or b and, if a and b are relatively prime, p is a factor of exactly one of them.

Example If by any means one knows that 17 is a factor of 1,615 and also that $1,615 = 5 \times 323$, then by Postulate 3-3 it follows that 17 is a factor of 323 since it is not a factor of 5. On the other hand, if one knows that 12 is a factor of 1,020 and also that $1,020 = 6 \times 170$, it does not follow that 12 must divide 170 since it does not divide 6. The difference lies in the fact that 17 is a prime number while 12 is not.

When two or more natural numbers are used in a computation it frequently becomes desirable to find either the *greatest common factor* (*g.c.f.*) of the numbers or their *least common multiple* (*l.c.m.*).

Greatest common factor (g.c.f.)   **Definition 3-5**   The *greatest common factor* (*g.c.f.*) of two or more natural numbers is the largest natural number which is a factor of each number.

Least common multiple (l.c.m.)   **Definition 3-6**   The *least common multiple* (*l.c.m.*) of two or more natural numbers is the smallest natural number of which each of the given numbers is a factor.

Example   Find the g.c.f. and l.c.m. of 12, 30, and 210.

Both of the numbers (g.c.f. and l.c.m.) to be found are composed of factors of the three given numbers. This makes it necessary that each of the given numbers be expressed as a product of its prime factors.

$$12 = 2 \times 2 \times 3$$
$$30 = 2 \times 3 \times 5$$
$$210 = 2 \times 3 \times 5 \times 7$$

Since the greatest common factor of the given numbers must be the largest number which is a factor of all three, it must contain any prime which occurs as a common factor. An inspection of the factored form of the numbers reveals that 2 and 3 are the only such primes. For this reason the g.c.f. of 12, 30, and 210 is 2×3 or 6.

Since the least common multiple of the three numbers must be the smallest number of which each of the given numbers is a factor, it must contain all the prime factors of each number. If 12 is to be a factor, the unknown number must contain the factors $2 \times 2 \times 3$. For the number to contain both 12 and 30 as factors its prime factorization must contain $2 \times 2 \times 3 \times 5$. Finally, to contain 210 also as a factor the prime factorization must be $2 \times 2 \times 3 \times 5 \times 7$. The l.c.m. of 12, 30, and 210 is, therefore, $2 \times 2 \times 3 \times 5 \times 7 = 420$.

The theory of prime numbers and its applications have occupied an important place in the world of mathematics and still do. Mathematicians have engaged in a great deal of research and expended a vast amount of *Sieve of Eratosthenes* effort in search of techniques which will discover primes. One of the first and simplest of all such techniques is the sieve of Eratosthenes (ca. 230

	2	3	~~4~~	5	~~6~~	7	~~8~~	~~9~~	~~10~~
11	~~12~~	13	~~14~~	~~15~~	~~16~~	17	~~18~~	19	~~20~~
~~21~~	~~22~~	23	~~24~~	25	~~26~~	27	~~28~~	29	~~30~~
31	~~32~~	~~33~~	~~34~~	35	~~36~~	37	~~38~~	39	~~40~~

/ indicates a number with 2 as a factor
— indicates a number with 3 as a factor
\ indicates a number with 5 as a factor

FIGURE 3-18 Eratosthenes sieve.

B.C.). This is a scheme to select all primes less than a selected natural number n. In Fig. 3-18 it is used to select all primes less than 40. All integers beginning with 2 and up to the selected number (40) are written in an array such as the rectangular array of Fig. 3-18. Then starting from 2, strike out every second number thereafter; next start with 3 and strike out every third number thereafter; then with 5, the next number not already stricken, and strike every fifth number thereafter; then repeat the process with 7, and so on. Continue until there are no further numbers less than n to be stricken. The numbers which remain will be all the primes less than n. From Fig. 3-18 it is seen that the primes less than 40 are 2, 3, 5, 7, 11, 13, 17, 19, 23, 29, 31, 37.

This process soon becomes very cumbersome and unwieldy. In the present day of the rapid computing machine, remarkable success is being achieved in determining prime numbers and in the factorization of composite numbers of tremendous size.

The even naturaı numbers and the odd natural numbers If a number has 2 as one of its factors, then it is said to be an *even number;* otherwise it is an *odd number.* Thus the set N may be separated into the two following disjoint subsets:

1 The set O of odd natural numbers

$$O = \{1,3,5,7,9,11,13, \ldots\}$$

2 The set E of even natural numbers

$$E = \{2,4,6,8,10,12,14,16, \ldots\}$$

There are many occasions in which these two subsets of N have siɑnifıcaııce. ɪney are cited here to illustrate an interesting and important property of infinite sets. We may describe the two subsets O and E of N by the set selectors

$$O = \{2n - 1 \mid n \text{ is a natural number}\}$$
$$E = \{2n \mid n \text{ is a natural number}\}$$

$2n$	n	$2n - 1$
2	1	1
4	2	3
6	3	5
8	4	7
10	5	9
12	6	11
14	7	13
16	8	15
18	9	17

FIGURE 3-19 Correspondence between sets of natural numbers

The table of Fig. 3-19 shows how to each natural number n there corresponds one and only one element in each of the sets O and E as determined by their respective set selectors, and to each element in either O or E there corresponds one and only one element of N. Thus we have not only a one-to-one correspondence between the elements of O and the elements of E, but also a one-to-one correspondence between the elements of either of these sets and the elements of N. In other words, the elements of N can be put into one-to-one correspondence with the elements of a subset

Characteristic of an infinite set of elements

of N other than N itself. This is a property of infinite sets which does not hold for finite sets. For example, suppose we consider the set A = the set of all letters of the English alphabet. There are 26 elements in this set; it is a finite set. No subset of A, except A itself, can be selected such that its elements can be placed in one-to-one correspondence with the elements of A. This property of infinite sets was first observed by Galileo and pointed out in a work published in 1638. It is used today as a characteristic property which distinguishes infinite sets from finite sets.

EXERCISES

1 Which of these sets are closed under addition and multiplication?
  (a) The set of all prime numbers
  (b) The set of all composite natural numbers
  (c) The set of all odd natural numbers
  (d) The set of all even natural numbers

2 Complete these operation tables for the set $\{E,O\}$ where O = the set of all odd natural numbers and E = the set of all even natural numbers and, for example, $E + E$ means the sum of two even natural numbers.

+	E	O
E		
O		

×	E	O
E		
O		

EXERCISE 2

3 Find the prime factors of each of these numbers: (*a*) 140; (*b*) 184; (*c*) 1,980; (*d*) 7,350.

4 If two primes differ by 2, they are called *twin primes*. What are the pairs of twin primes less than 100?

5 Use Eratosthenes's sieve to find all prime numbers less than 200.

6 Why is 2 the only even prime number?

7 What is the greatest common factor of 15 and 77? What are these two numbers called?

8 What is the greatest common factor of any two prime numbers? Why?

* 9 A convenient scheme for finding the l.c.m. of two or more numbers is illustrated here with the numbers 18, 30, 45, and 126. Why does it work?

(*a*) Arrange the given numbers in a row with a place provided for recording selected prime factors.

(*b*) Select a prime factor of *at least one* of the given numbers and list it in the place provided. (In the given problem 2 is selected.)

(*c*) When the selected prime *is* a factor of a given number write the other factor immediately below the number. When the selected prime *is not* a factor of a given number, repeat the number in the line immediately below it. (2 is a factor of 18, 30, and 126; the other factors are 9, 15, and 63, respectively. 2 is not a factor of 45.)

(*d*) Repeat this process until the row of numbers contain only 1s. It is important that only a prime factor be used in each step.

(*e*) The lowest common multiple of the given numbers is the product of the listed prime factors. (In the illustration the l.c.m. is $2 \times 5 \times 3 \times 3 \times 7 = 630$.)

Prime factors		*Numbers*		
2	18	30	45	126
5	9	15	45	63
3	9	3	9	63
3	3	1	3	21
7	1	1	1	7
	1	1	1	1

The l.c.m. = $2 \times 5 \times 3 \times 3 \times 7 = 630$.

10 On a sheet of paper construct a table similar to that shown on page 111. Use the letters *n*, *o*, and *e* to indicate whether the given numerals represent *no* number (*n*), an *odd* number (*o*), or an *even* number (*e*). As an illustration, the table is marked for the numeral 31.

11 On a sheet of paper construct another table for the same set of bases used in Exercise 10. Use the numerals 2, 10, 11, 23, 32, 47, 54 as the given numerals, and indicate which of them represent *no* number (*n*), which represent a *prime* number (*p*), and which represent a *composite* number (*c*).

	Base					
Numeral	**two**	**three**	**five**	**seven**	**ten**	**twelve**
1						
2						
10						
11						
31	n	n	e	e	o	o
43						
44						

EXERCISE 10

In Exercises 12 to 16 first find the greatest common factor of the given numbers and then the least common multiple.

12 24, 28, 100
13 192, 108, 234
14 13, 17, 23
15 78, 105, 110
16 315, 462, 693, 1155
∗17 Let the symbol $F(a,b)$ mean "find the greatest common factor of a and b" where $a,b \in N$. This means that the symbol for the operation of finding the g.c.f. is $F(\ ,\ )$.
  (a) Is N closed under the operation $F(\ ,\ )$?
  (b) Is N commutative under the operation $F(\ ,\ )$?
  (c) Is N associative under the operation $F(\ ,\ )$?
∗18 Let the symbol $M(a,b)$ mean "find the least common multiple of a and b" when $a,b \in N$.
  (a) Is N closed under the operation $M(\ ,\ )$?
  (b) Is N commutative under the operation $M(\ ,\ )$?
  (c) Is N associative under the operation $M(\ ,\ )$?

3-18 REVIEW EXERCISES

1–26 Write a carefully constructed answer to each of the guideline questions of this chapter.

In Exercises 27 and 28 show how the associative, commutative, and distributive properties are used in carrying out each operation.

27 Find the sum and check by reversing the order.
  (a) $2 + 3 + 4 + 5 + 6 + 7 + 9 + 1$
  (b) $25 + 36$
  (c) $213 + 456$

28 Find the product and check by interchanging factors: (*a*) 6 × 32; (*b*) 48 × 76.

29 First make an estimate of each sum and then find the sum.

(*a*) 82	(*b*) 462	(*c*) 2,136
79	756	8,756
48	613	5,968
35	824	8,956
91	598	1,339

30 Find the sum and check by casting out nines.

(*a*) 4,235	(*b*) 51,623	(*c*) 486,725
8,714	42,876	623,524
3,253	78,962	213,672
9,127	87,643	965,342

31 Find the sum for each base indicated.

(*a*) **Eight**	(*b*) **Five**	(*c*) **Twelve**
231	**2,143**	**2,856**
452	**4,211**	**e,t12**
673	**3,422**	**t,tet**
114	**1,234**	**e,et9**

(*d*) **Six**	(*e*) **Seven**	(*f*) **Nine**
1,542	**2,365**	**5,478**
2,132	**456**	**6,851**
4,123	**213**	**782**
542	**4,651**	**643**

32 First estimate the product and then compute the product.

(*a*) 28	(*b*) 234	(*c*) 7,213
46	672	426

33 Find the product and check by casting out nines.

(*a*) 273	(*b*) 567	(*c*) 428
28	321	379

34 Find the product for each base indicated.

(*a*) **Eight**	(*b*) **Five**	(*c*) **Twelve**
427	**234**	**tet**
35	**123**	**28t**

(d) **Six**	(e) **Seven**	(f) **Nine**
352	643	872
413	135	534

*35 None of these relations is an equivalence relation. In each case give an example of each property which is not satisfied. The first four relations apply to natural numbers and the last two to sets.

(a) "Is relatively prime to"

(b) "Is a factor of"

(c) "Is a multiple of"

(d) "Is not equal to"

(e) "Is contained in the same set as"

(f) "Contains elements in common with"

36 Can a one-to-one correspondence be set up between the elements of the set of all natural numbers less than 20 and the elements of the set of all odd natural numbers less than 20?

*37 Construct an Eratosthenes sieve to find all prime numbers less than 300.

*38 What are the pairs of twin primes less than 300?

39 Find the greatest common factor and the least common multiple of these sets of numbers.

(a) 195, 210, 450   (b) 464, 546, 630

(c) 17, 23, 37   (d) 21, 26, 55

(e) 34, 39, 51, 65   (f) 114, 190, 266, 418

*40 On a sheet of paper construct a table like that in the figure. Use the letters n, p, c, o, and e to indicate whether the given numerals represent *no* number (n), a *prime* number (p), a *composite* number (c), an *odd* number (o), or an *even* number in the respective bases indicated at the top of each column. The entries for the numeral 35 serve as an illustration.

	Base					
Numeral	**Two**	**Four**	**Seven**	**Eight**	**Ten**	**Twelve**
3						
11						
35	n	n	c,e	p,o	c,o	p,o
36						
57						
111						
121						
214						

*INVITATIONS TO EXTENDED STUDY

1   What is meant by the contrapositive or the inverse of a given theorem?

2   Why is proving the contrapositive of a theorem equivalent to proving the theorem?

3   What are the basic differences between a direct proof and an indirect proof?

4   Make a study of the truth tables of symbolic logic.

Prove these three theorems.

5   If $a,b \in N$ and $a < b$, then $a^2 < b^2$.

6   If $a,b,c,d \in N$ and $a < b < c$, then $a + d < b + d < c + d$.

7   If $n \in N$, there exists no natural number k such that $k + n = n$.

Use the principle of finite induction to prove these two theorems.

8   If n is a natural number, then $1 \leq n$.

9   The sum of the first n natural numbers is $\dfrac{n(n + 1)}{2}$.

10   Given the definitions $a^1 = a$ and $a^{n+1} = a^n \cdot a$ for all natural numbers a and n, use the principle of finite induction to prove the following properties of exponents.
(a) If a is a natural number, then a^k is a natural number for all numbers k.
(b) $1^k = 1$ for all natural numbers k.
(c) If $a,b,n \in N$, then $(ab)^n = a^n b^n$.
(d) If $a,m,n \in N$, then $a^m \cdot a^n = a^{m+n}$.
(e) If $a,m,n \in N$, then $(a^m)^n = a^{mn}$.

THE DOMAIN OF INTEGERS

GUIDELINES FOR CAREFUL STUDY

The word "structure" is a name which is ascribed to a very important concept in mathematics. You should take stock at this point to see whether or not you have clearly in mind just what is meant by the structure of a number system. You should also have in mind a clear concept of the natural number system. This chapter will point out some shortcomings of this particular system and make extensions necessary to remove some of them. It will be very important for you to acquire a competent understanding of just what these shortcomings are and what are the legitimate procedures that must be followed in making the desirable extensions.

The following questions will be very helpful as guidelines for the careful study of Chap. 4:

1   What is meant by a number system?
2   What are the basic postulates of the natural number system?
3   What are the definitions of addition and multiplication? Why are they called binary operations?
4   Do you have the properties of natural numbers in mind for ready and effective use?
5   Why should the property of closure be given so much emphasis in the study of number systems?
6   What is the definition of subtraction?
7   When is d said to be the difference of n subtracted from m?
8   Why is subtraction said to be inversely related to addition?
9   What is an integer?
10   What is meant by "additive identity" and "additive inverse"?
11   When two integers m and n are found by valid reasoning to be so related that $m + n = m$, what can be said about the integer n? Why?
12   When two integers k and l are found by valid reasoning to be so related that $k + l = 0$, what can be said about k and l? Why?
13   What is the full significance to be attached to the terms "digit value" and "place value" when applied to the digital symbols used in any numeral?
14   Why is it that a numeral such as 10,423 has no meaning unless the base of the numeration system is known?
15   What is the definition of division?
16   Is the set of integers closed under subtraction? Under division?
17   What is the division algorithm?
18   What new forms are introduced in this chapter for stating the principle of trichotomy?
19   Do you have a clear understanding of the extensions of the definitions of addition and multiplication from natural numbers to integers?
20   In making these extensions what basic principles are followed?
21   How may multiplication be used as a technique for converting a numeral

oriented in a system with a specified base into an equivalent numeral in some different base orientation?

22 In what way is the set of postulates of the natural number system modified to obtain the set of postulates of the domain of integers?

23 Do you understand clearly why these modifications are made?

INTRODUCTION

In the preceding chapter the binary operations of addition and multiplication were defined for natural numbers. Certain properties which intuition and experience have identified as basically characteristic of these operations were then accepted and listed as the postulates of the natural number system. From these properties other properties were derived by means of the techniques of logical deduction. Still other derivable properties characteristic of natural numbers under these two operations were indicated by exercises and suggestions. There is a great deal of operational flexibility and effectiveness possible within this familiar system. However, it has definite restrictions and limitations which necessitate important extensions in order to make possible the further study of number systems that are significant in the structure of elementary mathematics. The consequences of these extensions constitute the subject matter of this chapter and Chaps. 6 and 7.

4-1 THE NEED FOR EXTENSION

It should be rather evident that without closure as a characteristic property of a given operation, the use of any set of numbers with that particular operation would be quite restricted. Recall that for addition the closure postulate (N-1) states that if a and b are any two natural numbers such that $a + b = c$, then c is also a natural number. Contemplation and use of this property leads very naturally to the question: Is it true that for any two natural numbers a and b there exists a natural number a such that $a + d = b$? One counterexample is sufficient to establish the fact that the answer to this question is No. If a is the natural number 8 and b is the

Inadequacy of the natural number system natural number 3, there exists no natural number d such that $8 + d = 3$. The difficulty occurs only when a is greater than or equal to b ($a \geq b$), for when $a < b$ there does exist, by Definition 3-3, a natural number d such that $a + d = b$. When the natural number d does exist, it is called the *difference* between b and a in accordance with this definition:

Difference and subtraction **Definition 4-1** For any two numbers a and b, the *difference* between b and a is said to be d ($d = b - a$) if and only if $a + d = b$. The process of finding the difference between two numbers is called *subtraction*.

While this definition is stated and holds for any numbers, whether they are natural numbers or not, at present it has significance only if a and b are natural numbers and $a < b$. It is this same situation which exists in the early primary grades, where the number system used is essentially the natural number system, and the process of subtraction is limited to the subtraction of a smaller number from a larger number. The restricted context within which the definition of subtraction has significance for natural num-

Importance of order relation
bers emphasizes the importance of the order relation as a fundamental characteristic of the natural number system. The symbol $a > b$, also written $a \geq b$, means that the number represented by a may be the same as that represented by b ($a = b$) or a may be larger than b. Sometimes it is more convenient to state the same relation in a negative form by saying a *is not less than* b. The symbol for this statement is $a \not< b$. Similarly, $a \leq b$, or $a \leqq b$, means a *is less than or equal to* b. Equivalently, this may be stated as a *is not greater than* b. The symbol for this is $a \not> b$. Also, $a \gtrless b$ would be read a *is greater or less than* b. This symbol is rarely used. The more frequently occurring form for stating this relation is $a \neq b$, which is read either as a *is not equal to* b or a *and* b *are distinct*.

The closure postulate N-1 also states that the product of two natural numbers a and b is a natural number p. As one applies this property a question may arise analogous to the one raised previously in connection with addition. If the product of two natural numbers a and b is a natural number p, is it true that there exists a natural number q such that $b \times q = a$? Again it is possible to show by counterexample that the answer to this question is No. If $b = 3$ and $a = 8$ there exists no natural number q such that $3 \times q = 8$. However, if a is a multiple of b then the natural number q does exist. For example, if $a = 15$ and $b = 5$, then $q = 3$ since $5 \times 3 = 15$. When the number q does exist it is called the *quotient* of a divided by b in accordance with this definition:

Quotient and division
Definition 4-2 For any two numbers b and a the *quotient* of a divided by b ($a \div b$) is said to be q if and only if $b \times q = a$. The process of finding the quotient is called *division*.

Just as in the case of subtraction, this definition is stated and holds for all numbers, but it is in a very restricted sense that the quotient q exists when a and b are natural numbers. There is, however, a general rule for division which does provide for unrestricted use of the process with natural numbers. This rule, which can be proved as a theorem, is stated here as a postulate. It is called the *division algorithm*.

Division algorithm for natural numbers
Postulate 4-1 The division algorithm For any two natural numbers a and b, and $b < a$, there exists a unique natural number q such that $a = bq$ or $a = bq + r$ where r is a unique natural number less than b ($r < b$).

Exact division

In this algorithm a is called the *dividend;* b, the *divisor;* q, the *quotient;* and r, when it exists, the *remainder.* If $a = bq$ we usually say that the division is *even* or *exact.*

Example

1 In each case state what the expressed difference is, if it exists: (a) $15 - 4$; (b) $2 - 9$. Give reasons for your answers.
(a) $15 - 4 = 11$ since $4 + 11 = 15$.
(b) For natural numbers the difference $2 - 9$ does not exist since there is no natural number n such that $9 + n = 2$.
2 If the division is exact, find the natural number which is the quotient; if the division is not exact, find the natural numbers which are the quotient and the remainder called for in the division algorithm: (a) $35 \div 5$; (b) $45 \div 6$. Give reasons for your answers.
(a) $35 \div 5 = 7$ since $35 = 5 \times 7$.
(b) $45 \div 6$ gives the quotient 7 and the remainder 3 since $45 = 6 \times 7 + 3$.

While Definition 4-2 and Postulate 4-1 provide the guides for division as used in the early elementary grades, they place considerable restrictions on the general use of the operation. In the later grades these restrictions and those on subtraction grow more and more unsatisfactory. In order to give more freedom of interpretation and use not only to addition and multiplication but also to the new operations of subtraction and division, it becomes necessary and desirable to extend our number system so that the new system will have closure under all four operations. In any such extension it is desirable that the postulates of the old system be valid in the new system. Also, the definitions, operations, and derived properties of the new system must include the definitions, operations, and derived properties of the old system as special cases. This, of course, implies that the associative, commutative, and distributive properties must be preserved in the extension of the definitions of addition and multiplication. Furthermore, these extensions must be such that, when the operations are restricted to the natural numbers, the new definitions will conform to those already established.

EXERCISES

In Exercises 1 and 2 state whether or not there exists a natural number as the indicated difference. If it exists state what it is. In each case state the reason to support your answer.

1 (a) $25 - 10$ $\quad\quad$ (b) $16 - 9$ $\quad\quad$ (c) $7 - 9$
 (d) $8 - 11$ $\quad\quad$ (e) $105 - 25$ $\quad\quad$ (f) $302 - 302$

2 (a) $13 - 15$ $\quad\quad$ (b) $1,126 - 142$ $\quad\quad$ (c) $26 - 72$
 (d) $3 - 3$ $\quad\quad\quad$ (e) $440 - 280$ $\quad\quad$ (f) $17 - 16$

In Exercises 3 and 4 state, in each case, whether or not there exists a natural number such that the indicated division is exact. If it is exact find the natural number q of the division algorithm; if not, find the natural numbers q and r. In each case state the reason to support your answer.

3 (a) $45 \div 5$ (b) $70 \div 14$ (c) $26 \div 3$
   (d) $80 \div 80$ (e) $715 \div 30$ (f) $625 \div 25$

4 (a) $56 \div 28$ (b) $1 \div 1$ (c) $326 \div 7$
   (d) $205 \div 13$ (e) $352 \div 15$ (f) $723 \div 3$

5 In what ways do Exercises 1 to 4 indicate the need to extend the number system to include numbers other than natural numbers?

4-2 THE CONCEPT OF INTEGER

The proposed extension of the natural number system is to provide for the two newly defined operations, subtraction and division. This calls for two distinct extensions and a decision as to which operation shall be considered in the first extension. We shall select subtraction and postulate the existence of a new set of numbers, called *integers*, and then proceed to investigate the characteristics of the new number system to which they lead.

Integer

Postulate 4-2 For any two natural numbers a and b there exists a unique number z such that $z + a = a + z = b$. The number z is called an *integer*.

$N \subset Z$

When $a < b$ we know, by Definition 3-3, that the integer z is a natural number. An important implication of this statement is that $N \subset Z$ where $N = \{n \mid n \text{ is a natural number}\}$ and $Z = \{z \mid z \text{ is an integer}\}$. Since the definitions of addition and multiplication for natural numbers remain unchanged, the problem of extension becomes that of interpreting what modifications, if any, of these definitions are necessary to give the operations meaning when applied to those integers which are not natural numbers and yet preserve these properties when the integers are natural numbers. To accomplish this it is sufficient to accept as basic postulates for integers those which state the closure, associative, commutative, and distributive properties for natural numbers, and shape the extended definitions within the content of their implications. Our problem thus has resolved itself into that of determining what these implications are for the integer z in the relation $a + z = b$ of Postulate 4-2 when $a = b$ and when $a > b$.

When $a = b$, whether they are elements of the set Z or of the subset N, the relation $a + z = b$ takes the form $a + z = a$. Under this hypothesis the integer z is such that when it is added to any integer a, the resultant

Additive identity

sum is a. It is then called the *additive identity* and given the symbol 0.

Under these conditions Postulate 4-2 may be restated for this special case of integers as follows:

Zero is an integer

Postulate 4-3 There exists one and only one integer 0, called the *additive identity*, such that $0 + a = a + 0 = a$ for any integer a.

Since, by this postulate, $0 + a = a$ for any natural number a we shall extend the concept of Definition 3-3 to say that $a > 0$ for any natural number a. The natural numbers have now been identified as that proper subset of $Z = \{z \mid z$ is an integer$\}$ for which $z > 0$. In other words, the two phrases "a natural number" and "an integer greater than zero" may be used interchangeably. Postulate N-6 may now be extended to give this new form for stating the principle of trichotomy:

Extension of the principle of trichotomy

For any two elements a and b of Z exactly one of these three relations holds: $a = b$, $a = b + c$ $(a > b)$, or $a + c = b$ $(a < b)$ for some integer $c > 0$.

The set C of cardinal numbers

At this point the set of natural numbers has been extended to the set of cardinal numbers, $C = \{0,1,2,3,4, \ldots\}$.† This extension gives rise to the question: Does there exist an integer z such that $z + a = a + z = 0$? There is evidence, based on experience, that such numbers do exist. For example, if a person deposits \$50 to the credit of his bank account and later withdraws \$50, the net effect on his account is that of having added 0 dollars to it. If b represents the bank account, then $b + [(\$50$ deposited$) + (\$50$ withdrawn$)]$ is the same amount as $b + 0$ dollars, since $b + 0 = b$ whether the symbols represent dollars or not.

If 50 represents the money deposited and ⁻50 (call it *bar-fifty*) the money withdrawn, then $b + (50 + {}^{-}50)$ is the same as $b + 0$. In other words, ⁻50 represents a number such that $50 + {}^{-}50 = 0$. Similarly, if $a + z = 0$, then $z = {}^{-}a$, a number which added to a gives 0 as the sum. If $a = 0$, then $z = 0$ since $0 + 0 = 0$ (why?) and ⁻0 = 0. If $a \in C$ and $a \neq 0$, then $a \in N$ and $a + z \in N$ for z any element of C. Thus, it follows that the extension to the set C still does not provide the needed extension from the set N. The further necessary extension is provided by this version of Postulate 4-2:

Additive inverse

Postulate 4-4 For every integer a there exists one and only one integer, called the *additive inverse* of a and indicated by ⁻a, such that ${}^{-}a + a = a + {}^{-}a = 0$.

---

† Some writers call this the *set of whole numbers* and designate it by the symbol W. This practice has not been followed here due to the possible confusion arising from the traditional association of the concept of whole numbers with those numbers which are not fractions.

The import of this postulate is to extend the set C to the set of integers $Z = \{\ldots, {}^-4, {}^-3, {}^-2, {}^-1, 0, 1, 2, 3, 4, \ldots\}$ which includes not only all natural numbers and the additive identity but also the additive inverses of all natural numbers. The natural numbers have been identified as those integers greater than zero. Since by Postulate 4-4 the additive inverse of the natural number a is ${}^-a$ such that ${}^-a + a = 0$ it follows that ${}^-a < 0$. (Why?) Henceforth integers greater than zero (natural numbers) will be designated *Positive and* as *positive integers*. For example, 3 and $+3$ ("positive 3" or "plus 3") *negative* may be used interchangeably. Normally $+3$ will be used only if there is *integers* a desire to emphasize the "positive integer" concept. Integers less than zero (additive inverses of natural numbers) will be designated as *negative integers*. The symbol ${}^-a$ will be reserved to indicate "the additive inverse of a" and $-a$ will be used to indicate a negative integer. Thus ${}^-3$ is the additive inverse of 3, while -3 is read "negative 3" or "minus 3."

Such symbols as ${}^-3$ will be used only in those cases where the emphasis is on the concept of "additive inverse." While ${}^-3 = -3$, it must be remembered that it is incorrect to conclude that ${}^-a$ is necessarily a negative number. For example, ${}^-(-3)$ represents the additive inverse of negative 3. It is the same as ${}^-({}^-3)$ and has the value 3. Why? Furthermore, ${}^-0 = 0$ (why?), and there is no distinction between 0 and its additive inverse.

Example Now that there exists an additive inverse for each integer a, the cancellation law does not have to be stated as a postulate.

If $z \in Z$ and $z + 4 = -5 + 4$, we may use the additive property of equality to add ${}^-4$ to get

$$(z + 4) + {}^-4 = (-5 + 4) + {}^-4$$

Now, from the associative property of addition, we have

$$z + (4 + {}^-4) = -5 + (4 + {}^-4)$$

Postulate 4-4 gives

$$z + 0 = -5 + 0$$

and, from Postulate 4-3, we have

$$z = -5$$

The set Z of The set Z, now clearly defined as
all integers

$$Z = \{\ldots, -5, -4, -3, -2, -1, 0, 1, 2, 3, 4, 5, \ldots\}$$

is frequently thought of as being partitioned into three proper subsets: the set of all negative integers, $\{0\}$, and the set of all positive integers. Furthermore, the set C may be identified as the set of all nonnegative integers and *$N \subset C \subset Z$* the set N as the set of all positive integers. Thus we have $N \subset C \subset Z$. In briefer form the above roster symbol for the set Z is at times written

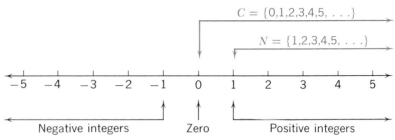

FIGURE 4-1   The set Z of integers

The symbol $\pm$

as $Z = \{0, \pm 1, \pm 2, \pm 3, \pm 4, \pm 5, \ldots\}$, where ± 5, for example, is read "plus or minus 5" and means that both the positive integer $+5$ and the negative integer -5 are included.

The number line for integers

In Fig. 4-1 a one-to-one correspondence between the integers and certain selected points of a number line has been indicated. To accomplish this, first select a reference point and label it 0, then choose a unit length and measure off along the line successive unit lengths starting from 0. By agreement the positive integers will be placed in correspondence with the points to the right of the reference point and the negative integers with those to the left.

Now that the set Z has been fully defined, there are other equivalent forms for stating the principle of trichotomy as extended to integers. If, in the extended form of page 122, one of the integers, say b, is zero, we have this statement:

Other statements of trichotomy

For any integer a exactly one of these relations holds: $a = 0$, $a > 0$, or $a < 0$.

Since the additive inverse of a negative integer is positive, this statement may take the equivalent form:

For any integer a exactly one of these relations holds: $a = 0$, $a > 0$, or $^-a > 0$.

Reference to the number line of Fig. 4-1 provides an easily applied technique for checking to determine which of the order relations holds between two integers a and b. If they correspond to the same point then $a = b$; if a corresponds to a point to the left of the point to which b corresponds, then $a < b$ or $b > a$.

EXERCISES

1 Which of these statements are true and which are false?

(a) $6 \in C$ (b) $0 \in N$ (c) $^-4 \in C$ (d) $^-5 \in Z$
(e) $4 \in N$ (f) $-3 \in N$ (g) $0 \in Z$ (h) $-2 \in C$
(i) $-7 \in Z$ (j) $^-0 \in C$ (k) $^-1 \in N$ (l) $^-0 \in Z$

2 How may the facts that $A \cup \varnothing = A$ and $A \cap \varnothing = \varnothing$ for all sets A be used to support the statement of Postulate 4-3?

3 Explain how the concept of sets may be used to find the answer to these sums.

(a) $4 + 7$ (b) $6 + 0$ (c) $2 + 6$ (d) $0 + 0$

4 Is 0 a positive integer? Explain.
5 Is 0 a negative integer? Explain.
6 If $^-5 = -5$ is a correct statement, why is $^-0 = -0$ incorrect?
7 For $a \in Z$ is it true that ^-a is a negative integer? Why?
8 If $4 + (^-4) = 0$, what is $^-(^-4)$? Why?
9 If $4 + (^-4) = 0$, is it correct to say that $4 + (-4) = 0$? Why?
10 What is $^-(^-n)$?
11 For $a \in Z$ is it true that $^-a \neq 0$? Why?
12 Determine whether each statement is true or false. Give reasons to support your answers.
   (a) ^-c is a negative integer for $c \in C$.
   (b) ^-n is a negative integer for $n \in N$.
13 If $a + 2 = 0$ what is the value of a? Why?
14 If $(-6) + n = 0$ what is the value of n? Why?
15 If $5 + i = 5$ what is the value of i? Why?
16 If $s + (-7) = -7$ what is the value of s? Why?

In Exercises 17 to 22 find n if $n \in N$.

17 $3 + n = 19$ 18 $7 \times n = 14$ 19 $n + 8 = 23$
20 $n \times 5 = 25$ 21 $21 + 79 = n$ 22 $35 \times 72 = n$

23 If k and l are integers such that $k + l = 0$, what can be said about k and l? Explain.
24 If a and b are integers such that $a + b = a$, what can be said about a and b? Explain.
25 If $n \in N$ such that $n + 5 = 7 + 5$, what is the value of n? What property of natural numbers is needed to support your answer?

Use the properties postulated in this section for the set Z to determine a value for the integer a in each of the following exercises. Cite the proper authority for each statement you make.

*26 $a + 2 = 3 + 2$

*27 $\quad ^{-}3 + 4 = a + 4$

*28 $\quad a + (-6) = (-4) + (-6)$

*29 $\quad a + 2 = b + 2, b \in Z$

*30 $\quad (^{-}5) + a = (^{-}5) + b, b \in Z$

*31 $\quad a + (-6) = b + (-6), b \in Z$

*32 $\quad a + c = b + c, b,c \in Z.$

*33 $\quad a + 3 = 9$ *34 $\quad a + 0 = 16$ *35 $\quad 0 + a = 8$

*36 $\quad 10 + a = 25$ *37 $\quad a + 5 = 0$ *38 $\quad 17 + a = 0$

4-3 THE DOMAIN OF INTEGERS

The discussion of the preceding section pointed out that in making the extension from natural numbers to integers, the operations of addition and multiplication retain, by definition, their respective properties for all natural numbers (all integers $z > 0$) and the newly postulated additive identity ($z = 0$). With the postulated existence of an additive inverse for every integer z the cancellation law for addition becomes a derivable property (see the example on page 123) rather than one to be postulated. The cancellation law for multiplication must be retained as a postulate, but with a slight modification necessary for use with integers. (See Sec. 4-7.) The trichotomy principle for natural numbers is retained but permits different forms of statement suitable to varying contexts of use. The principle of finite induction remains but is restricted to natural numbers, or positive integers, and so it cannot be stated as a property for all integers.

With these facts in mind and remembering that the definitions for addition and multiplication still need to be established for all integers $z \leq 0$, we may now list the properties which characterize the number system known as the *domain of integers:*

Given $Z = \{ \ldots -4, -3, -2, -1, 0, 1, 2, 3, 4, \ldots \}$ with the two well-defined operations of addition and multiplication. For a, b, $c \in Z$:

Properties of an integral domain

D-1 Closure   If $a + b = x$ and $a \times b = y$, then x, y are unique elements of Z.

D-2 Associative $a + (b + c) = (a + b) + c;$
$a \times (b \times c) = (a \times b) \times c.$

D-3 Commutative $a + b = b + a; a \times b = b \times a.$

D-4 Distributive $a \times (b + c) = (a \times b) + (a \times c).$

D-5 Additive identity There exists one and only one integer 0, called the *additive identity*, such that $0 + a = a + 0 = a$ for every integer a.

D-6 Additive inverse For every integer a there exists one and only one integer, called the *additive inverse of* a and indicated by ^-a, such that $^-a + a = a + {}^-a = 0$.

D-7 Multiplicative identity There exists one and only one integer $1 \neq 0$, called the *multiplicative identity*, such that $1 \times a = a \times 1 = a$ for every integer a.

D-8 Cancellation If $ac = bc$ and $c \neq 0$, then $a = b$.

From these eight postulates, and the five postulates of equality, all the remaining properties of integers may be derived. Notice that Postulates D-1 to D-4 and D-7 are the same as Postulates N-1 to N-5 for the natural number system. Attention has been called to the fact that D-8, the cancellation property for integers, differs slightly from N-7, the cancellation property for natural numbers. In D-7 there is the assumption that the multiplicative identity differs from the additive identity ($1 \neq 0$). This is necessary to eliminate the possibility that any trivial cases, such as a set consisting of only the element 0, may be considered as an integral domain.

4-4 ADDITION AND SUBTRACTION WITH INTEGERS

The properties of the set Z under addition, the principle of place value, and the table of basic addition facts provide the foundation for performing any computation with integers which involves only addition and subtraction.

Basic addition facts
Each such table not only gives basic addition facts but also basic subtraction facts. For example:

1 From Fig. 4-2 we have not only that $4 + 5 = 9$, but also that $9 - 4 = 5$ and $9 - 5 = 4$.
2 From Fig. 4-3, in base **six, 4 + 5 = 13;** also **13 − 4 = 5** and **13 − 5 = 4.**
3 From Fig. 4-4, in base **three, 1 + 2 = 10;** also **10 − 2 = 1** and **10 − 1 = 2.**

These tables have used Postulate D-5 to extend the addition facts for the set N of natural numbers (see Figs. 3-4 to 3-6) to those for the set C of

+	0	1	2	3	4	5	6	7	8	9
0	0	1	2	3	4	5	6	7	8	9
1	1	2	3	4	5	6	7	8	9	10
2	2	3	4	5	6	7	8	9	10	11
3	3	4	5	6	7	8	9	10	11	12
4	4	5	6	7	8	9	10	11	12	13
5	5	6	7	8	9	10	11	12	13	14
6	6	7	8	9	10	11	12	13	14	15
7	7	8	9	10	11	12	13	14	15	16
8	8	9	10	11	12	13	14	15	16	17
9	9	10	11	12	13	14	15	16	17	18

FIGURE 4-2 Basic addition facts (base ten)

cardinal numbers. Postulate D-6 further extends the process of addition to include finding the sum of an integer and its additive inverse. Furthermore, the sum of any two positive integers (not merely those of the tables) is a positive integer. (Why?)

As an important consequence of Postulates D-5 and D-6 and the associative property of addition, it follows that, for $a,b \in Z$,

$$(b + a) + {}^-a = b + (a + {}^-a) = b + 0 = b$$

and
$$(b + {}^-a) + a = b + ({}^-a + a) = b + 0 = b$$

This argument constitutes the proof of Theorem 4-1.

Theorem 4-1 If $a,b \in Z$, then $(b + a) + {}^-a = b = (b + {}^-a) + a$.

+	0	1	2	3	4	5
0	0	1	2	3	4	5
1	1	2	3	4	5	10
2	2	3	4	5	10	11
3	3	4	5	10	11	12
4	4	5	10	11	12	13
5	5	10	11	12	13	14

FIGURE 4-3 Basic addition facts (base six)

+	0	1	2
0	0	1	2
1	1	2	10
2	2	10	11

FIGURE 4-4   Basic addition facts (base **three**)

Example   For any integer b it is true that

$$(b + 6) + {}^-6 = b + (6 + {}^-6) = b + 0 = b$$

or $$(b + {}^-8) + 8 = b + ({}^-8 + 8) = b + 0 = b$$

Theorem 4-1 can be used as an aid in extending addition to the sum of any two integers, whether positive or negative. Also, it can be used to find an interpretation of the difference between two numbers independent of which is the larger or of whether they are positive or negative.

For $a, b \in Z$, if $a > 0$ and $b > 0$ the sum $a + b > 0$ since the closure property for natural numbers implies that the positive integers are closed under addition. Furthermore, if $a > 0$ and $b > 0$ it follows, by definition, that the negative integers $-a$, $-b$, and $-(a + b)$ are the same as the additive inverses of the corresponding positive integers; that is, $-a = {}^-a$, $-b = {}^-b$, and $-(a + b) = {}^-(a + b)$.

These facts may be used to argue Theorem 4-2.

Sum of negative integers   **Theorem 4-2**   If $a, b \in Z$ with $a > 0$ and $b > 0$, then $(a + b) > 0$ and $(-a) + (-b) = -(a + b)$.

The argument will be presented in parallel columns with the general case on the right and, for comparison purposes, this special case on the left: For the two positive integers 3 and 5, their sum $3 + 5$ is a positive integer and $(-3) + (-5) = -(3 + 5)$.

Hypothesis	*Hypothesis*
3 and 5 are positive integers	$a, b \in Z$, $a > 0$, $b > 0$

Conclusion	*Conclusion*
$3 + 5$ is a positive integer	$a + b \in Z$, $a + b > 0$
$(-3) + (-5) = -(3 + 5)$	$(-a) + (-b) = -(a + b)$

Proof

Case I

3 + 5 is a positive integer   Closure $a + b \in Z$, $a + b > 0$

Case II

1	$(^-3 + 3) + (^-5 + 5) = 0$	D-6; D-5	$(^-a + a) + (^-b + b) = 0$
2	$(^-3 + 3) + (^-5 + 5)$ $= (^-3 + ^-5) + (3 + 5)$	D-2; D-3	$(^-a + a) + (^-b + b)$ $= (^-a + ^-b) + (a + b)$
3	$(^-3 + ^-5) + (3 + 5) = 0$	Steps 1 and 2; substitution	$(^-a + ^-b) + (a + b) = 0$
4	$^-3 + ^-5 \in Z$	Closure	$^-a + ^-b \in Z$
5	$^-3 + ^-5 = ^-(3 + 5)$	D-6	$^-a + ^-b = ^-(a + b)$
6	Since $^-3 = -3$ and $^-5 = -5$ $(-3) + (-5) = -(3 + 5)$	Substitution	Since $^-a = -a$ and $^-b = -b$ $(-a) + (-b) = -(a + b)$

As a result of this theorem the sum of two negative integers can be found by following these two steps:

1   Find the sum of the additive inverses of the two given negative integers. This sum will be positive.

2   The sum of the two negative integers will be the negative integer which is the additive inverse of this sum.

Example   What is the sum $(-8) + (-6)$?
Step 1 $^-(-8) = 8$ $^-(-6) = 6$ $8 + 6 = 14$
Step 2 $^-14 = -14$ $(-8) + (-6) = -14$

There remains the question of how to obtain the sum of one integer and the additive inverse of another integer distinct from it. This problem is taken care of by Theorem 4-3.

Theorem 4-3   If $a, b \in Z$, then $b + ^-a = b - a$.

For the special case we shall use $a = 3$ and $b = 5$.

Hypothesis $3, 5 \in Z$	*Hypothesis* $a, b \in Z$
Conclusion $5 + ^-3 = 5 - 3$	*Conclusion* $b + ^-a = b - a$
Proof	
1 $^-3 \in Z$ D-6	$^-a \in Z$

2 $(5 + {}^-3) \in Z$	Closure	$(b + {}^-a) \in Z$
3 $(5 + {}^-3) + 3 = 5$	Theorem 4-1	$(b + {}^-a) + a = b$
4 $5 + {}^-3 = 5 - 3$	Definition 4-1	$b + {}^-a = b - a$

The special case was selected so that $a < b$. In this case we know that $5 - 3 = 2$; whence $5 + {}^-3 = 2$. Thus, for the general case if $a < b$ we have a similar interpretation for the difference $b - a$. It is the number d such that $a + d = b$. In the argument of the general case no condition was placed on the relation between a and b. There still remains the question of what is the interpretation of $b - a$ when $a > b$. From Properties D-2 to D-4 we have $(b + {}^-a) + (a + {}^-b) = (b + {}^-b) + ({}^-a + a) = 0$. Thus, again from D-6, $(b + {}^-a) = {}^-(a + {}^-b)$. Whence, by substitution, $b - a = -(a - b)$.

Thus, to find $b - a$ when $a > b$, first find the positive integer $a - b$. The negative integer which is ${}^-(a - b)$ will be the difference $b - a$.

Example   If $d = 8 - 15$, find d.
Since $15 - 8 = 7$, it follows that $d = -7$.

Finally, as an immediate consequence of Theorems 4-1 and 4-3, we have

Theorem 4-4   If $a,b \in Z$ then $(a + b) - b = a$ and $(a - b) + b = a$.

Hypothesis $a,b \in Z$

Conclusion $(a + b) - b = a$ and $(a - b) + b = a$

Proof

1	$a + b \in Z$	D-1
2	$(a + b) - b = (a + b) + {}^-b$	Theorem 4-3
3	$(a + b) + {}^-b = a$	Theorem 4-1
4	$(a + b) - b = a$	Substitution rule

A similar argument establishes $(a - b) + b = a$.

Addition and subtraction inversely related   The importance of Theorem 4-4 is that it completes the characterization of addition and subtraction as inversely related operations. Each operation cancels the effect of the other. As a further consequence of the theorem it follows that each operation may be used to check the result found by using the other.

Example $(15 + 6) - 6 = 21 - 6 = 15$
$(15 - 6) + 6 = 9 + 6 = 15$

Theorem 4-4 states that these are the results to be expected. There is no need to carry out the operations as done here. The two examples also illustrate how each operation may be used to check the other. These checks may be extended to successive additions or subtractions.

$$2 + 7 + 5 + 6 = [(2 + 7) + 5] + 6$$
$$= (9 + 5) + 6$$
$$= 14 + 6$$
$$= 20$$
$$[(20 - 6) - 5] - 7 = (14 - 5) - 7$$
$$= 9 - 7$$
$$= 2$$

EXERCISES

In these exercises the tables referred to are those of Figs. 4-2 to 4-4.

1 How does each of the tables illustrate the fact that 0 is the additive identity?

In Exercises 2, 3, and 4 explain how the appropriate table may be used to find the indicated differences.

2 Subtract in base ten.

   (a) $6 - 4$ (b) $18 - 7$ (c) $14 - 9$ (d) $7 - 0$

3 Subtract in base **six.**

   (a) **5 − 2**   (b) **3 − 0**   (c) **12 − 4**   (d) **14 − 5**

4 Subtract in base **three.**

   (a) **2 − 2**   (b) **11 − 2**   (c) **10 − 0**   (d) **10 − 1**

5 Check each answer obtained in Exercises 2 to 4 by addition.
6 How do the tables illustrate the fact that integers are commutative under addition?
7 Select two examples from each table to illustrate the fact that integers are associative under addition.
8 Select an example from each table to show that integers are not commutative under subtraction.
9 Select an example from each table to show that integers are not associative under subtraction.
10 Explain each of these statements:

   (a) $5 - 3 \neq 3 - 5$ (b) $5 + {}^-3 = 5 - 3$
   (c) $5 + {}^-3 = {}^-3 + 5$ (d) $5 + {}^-3 \neq 3 + {}^-5$

11 What properties verify these equalities in base **six?**
   (a) **3 + 4 + 5 + 4 = 11 + 13**

(b) **2 + 3 + 4 = 10 + 3**
(c) **4 + 5 + ⁻4 = 5**
(d) **12 + 3 − 4 = (12 + ⁻4) + 3**
= 4 + 3

12 What is the authority to support this statement in which the numerals are in base **three?**

$$(10 - 2) - 1 \neq 10 - (2 - 1)$$

In Exercises 13 and 14 find the sum of each column of numbers. Check by using subtraction.

13	(a) 4	(b) 3	(c) 14	(d) 32
	5	0	61	10
	6	5	32	4
	8	8	67	18
	7	9	72	45

14 Base **six**   Base **three**

	(a) **4**	(b) **12**	(c) **1**	(d) **21**
	3	**31**	**2**	**11**
	1	**42**	**10**	**10**
	4	**20**	**11**	**20**
	2	**34**	**2**	**1**

In Exercises 15 to 17 find each indicated difference. Check by addition.

15	(a) 2,384	(b) 4,567	(c) 3,021	(d) 5,000
	243	824	1,278	3,452

16 Base **six**

	(a) **2341**	(b) **1034**	(c) **4232**	(d) **5000**
	1221	**212**	**1341**	**3452**

17 Base **three**

	(a) **212**	(b) **1011**	(c) **1111**	(d) **2000**
	11	**210**	**222**	**122**

18 Use the sum 29 + 34 to show how the principle of place value, the properties of addition, and the table of basic addition facts enable one to find the indicated sum.

19 In each case state what postulates or theorems are needed to support the results obtained
(a) 8 + 5 + 6 + 2 + 3 + 4 = (8 + 2) + (6 + 4) + (5 + 3)
= 10 + 10 + 8
= 28

(b) $7 + (6 - 3) + (8 - 7) + (4 + 3)$
$$= 7 + (6 + {}^-3) + (8 + {}^-7) + (4 + 3)$$
$$= (7 + {}^-7) + ({}^-3 + 3) + (6 + 8) + 4$$
$$= 0 + 0 + 14 + 4$$
$$= 18$$

(c) $(-6) + (-7) + (-9) = [(-6) + (-7)] + (-9)$
$$= [-(6 + 7)] + {}^-9$$
$$= -13 + {}^-9$$
$$= -(13 + 9)$$
$$= -22$$

*20 These two exercises are designed to illustrate how the principle of place value and the properties of addition and subtraction are used to find such differences. State the authority for each step indicated.

(a) $56 - 24 = (50 + 6) - (20 + 4)$
$$= (50 + 6) + {}^-(20 + 4)$$
$$= (50 + 6) + ({}^-20 + {}^-4)$$
$$= (50 + {}^-20) + (6 + {}^-4)$$
$$= (50 - 20) + (6 - 4)$$
$$= 30 + 2$$
$$= 32$$

(b) $72 - 48 = (70 + 2) - (40 + 8)$
$$= [60 + (10 + 2)] - (40 + 8)$$
$$= (60 + 12) - (40 + 8)$$
$$= (60 + 12) + {}^-(40 + 8)$$
$$= (60 + 12) + ({}^-40 + {}^-8)$$
$$= (60 + {}^-40) + (12 + {}^-8)$$
$$= (60 - 40) + (12 - 8)$$
$$= 20 + 4$$
$$= 24$$

*21 Complete the argument for the unfinished part of the proof of Theorem 4-4.

*22 Use combinations whose sum is 100 to find these sums. How can this procedure be justified?

(a) 25	(b) 41	(c) 78	(d) 91
32	59	56	83
75	60	34	42
39	80	22	17
68	40	66	9

Find the value of x in each exercise if $x \in Z$.

23 $4 + x = 0$ 24 $x + 8 = 8$
25 $5 + x = 5$ 26 $x + 3 = 0$
27 $-x + 5 = 0$ 28 $-x + 0 = -6$

29 $16 + x = 18$ 30 $x + 7 = 3$

31 $5 - x = 2$ *32 $(-x) + 8 = 5$

*33 $(-4) + x = 9$ *34 $(-x) + (-8) = -14$

*35 $(-x) + (-7) = 4$ *36 $(-7) + x = -2$

*37 $(-16) + (-x) = -20$ *38 $(-4) - x = 0$

*39 $(-13) - x = -15$ *40 $(-24) - x = 13$

*4-5 THE NATURE OF SUBTRACTION

Characteristics of subtraction

Historically there have been four distinct processes which, at one time or another through the years, have been used to characterize the process of subtraction. They are (1) *decomposition*, breaking down a number into component parts; (2) *deduction*, taking away a part of a number and retaining the rest, or remainder; (3) *diminution*, decreasing the size of a number; and (4) *comparison*, finding how much larger or how much smaller one number is than another. While each of these interpretations of subtraction places emphasis on an important computational function of the operation, no one of them lends emphasis to its significance in the basic structure of a number system, namely, as the inverse operation of addition.

Just as in the case of addition, there have been many different names used for what we today call "subtraction." In addition to *subtraction* (to subtract), some of the more important of these terms are *diminution* (to diminish), *extraction* (to extract), *detraction* (to detract), and *subduction* (to subduce or subtray). We get the words *minuend* and *subtrahend* from the Latin phrases *numerus minuendus* (number to be diminished) and *numerus subtrahendus* (number to be subtracted), respectively. Since subtraction is more clearly understood in the context of its definition as the inverse operation of addition, the words "minuend" and "subtrahend" are no longer considered as very significant words even in the vocabulary of elementary mathematics. As a consequence, they are being dropped from current usage. Our terms *difference* and *remainder* seem to date from Latin and English writers of the sixteenth century.†

Symbols used to indicate subtraction

The earliest symbols for subtraction came from the Babylonians and Egyptians. In writing their numerals, the Babylonians made use of the subtractive principle and used the symbol $\triangledown \triangleright$ to indicate subtraction. In the Ahmes papyrus, the symbol $\triangle$, indicating a pair of legs walking away, was used. In the manuscript of the Hindu writer Bhāskara (ca. 1150), we find three forms for indicating which of two numbers is the subtrahend: a circle or dot placed over the number, as in $\overset{\circ}{4}$ or $\overset{\bullet}{4}$, or the enclosure of the number in a circle, $④$. Just as in the case of the signs $\bar{p}$ and $+$, the two signs $\bar{m}$ and $-$ were in sharp competition during the fifteenth and sixteenth

---

† David Eugene Smith, "History of Mathematics," vol. II, pp. 94–97, Ginn and Company, Boston, 1925.

centuries. There are many different hypotheses as to the origin of the minus sign (−), none of which seems supported by very strong evidence. The first occurrence in print of this sign, along with the plus sign, was in Widman's arithmetic published in 1489. In spite of the simplicity of the minus sign as a symbol for subtraction, there have been several variations that were popular for a time. There were three in particular which were used fairly widely: ÷, ÷÷, and the two or three successive dots . . . used by Descartes and others.† Just as the earliest use of the plus sign seems to have been to show excess, so the minus sign seems to have been used first as a label denoting deficiency rather than as a symbol to indicate the process of subtraction.

*4-6 THE SUBTRACTION ALGORITHM

Some writers in the past have claimed to identify nine, twelve, and even thirty distinct operational patterns for subtracting one integer from another. The variations lie in different details of procedure and different techniques for capitalizing on the significance of place value. For convenience we might categorize these variations within these four classifications: *decomposition take-away, decomposition additive, equal additions take-away, equal additions additive.*

The decomposition methods

Both *decomposition* methods are based on the idea of breaking a number down into what might be called its component parts. For example, the colored numerals of Fig. 4-5 show the analysis of the process used in finding the difference between 5,032 and 1,687. We first discover that we cannot subtract 7 ones from 2 ones. We then proceed to use one of the 3 tens and convert it to 10 ones, which are combined with the 2 ones to give 12 ones. Since there are only 2 tens remaining in the minuend and there are 8 tens to be subtracted, it now becomes necessary to use one of the 5 thousands to convert to 10 hundreds, of which one may be converted into

† F. Cajori, "A History of Mathematical Notations," vol. I, pp. 229–236, The Open Court Publishing Company, La Salle, Ill., 1928.

			Step 1	3 tens = 2 tens + 10 ones
			Step 2	10 ones + 2 ones = 12 ones
4	10̸	12	Step 3	12 ones − 7 ones = 5 ones
	2̸	12	Step 4	5 thousands = 4 thousands + 10 hundreds
5̸	0	3̸ 2̸	Step 5	10 hundreds = 9 hundreds + 10 tens
1	6	8̸ 7̸	Step 6	10 tens + 2 tens = 12 tens
3	3	4 5	Step 7	12 tens − 8 tens = 4 tens
			Step 8	9 hundreds − 6 hundreds = 3 hundreds
			Step 9	4 thousands − 1 thousand = 3 thousands

FIGURE 4-5 Decomposition method of subtraction

10 tens to be combined with the 2 tens to give 12 tens. The process of subtraction then becomes:

Decomposition

Take-away	*Additive*
7 from 12 = 5	7 and 5 make 12
8 from 12 = 4	8 and 4 make 12
6 from 9 = 3	6 and 3 make 9
1 from 4 = 3	1 and 3 make 4

Either method of decomposition is at times referred to as a "borrowing method." It is likely that this concept is traceable back to the language of abacal reckoning. In using the abacus, the computer literally takes the subtrahend away from the minuend. Thus whenever a line or column has an insufficient number of beads to make the subtraction possible, beads are borrowed from the next higher column.

The equal additions methods The *equal additions* methods are both based on the fact that addition and subtraction are inverse processes (see Theorem 4-4). For example, in Fig. 4-5, since 7 cannot be subtracted from 2, we add 10 ones to the 2 ones to obtain 12 ones. Since 10 ones have been added, we must subtract 10 ones, or 1 ten. We combine this 1 ten with the 8 tens to be subtracted. This procedure is followed in each step. The subtraction process then becomes:

Equal additions

Take-away	*Additive*
7 from 12 = 5	7 and 5 make 12
9 from 13 = 4	9 and 4 make 13
7 from 10 = 3	7 and 3 make 10
2 from 5 = 3	2 and 3 make 5

Checking subtraction The estimation of a difference is an effective and worthwhile check on the reasonableness of an answer in subtraction. While the check of casting out nines is also adaptable to this process, it is of minor significance because it is just about as simple, and much more effective, to add the difference to the subtrahend to obtain the minuend.

EXERCISES

1 What is meant by saying that addition and subtraction are inverse processes? Illustrate.

2 In what way are the words "decomposition," "deduction," "diminution," and "comparison" descriptive of the process of subtraction?

3 For what purposes do we subtract?

4 Answer the following questions about the set $\{1,0,-1\}$. Give reasons for your answers.

(*a*) Is the set closed under addition?

(*b*) Is the set closed under subtraction?

5 Answer the questions of Exercise 4 about the set $\{a,0,-a\}$ where a is any natural number.

6 Prove that the odd integers are not closed either under addition or under subtraction.

7 Construct a rule for using the process of casting out nines as a check for subtraction.

8 First find each of these sums and then use the process of subtraction to check each sum:

(*a*)	157	(*b*)	2,034	(*c*)	25
	456		9,022		413
	203		5,681		1,062
	821		403		67

9 Use each of the four methods of subtraction described in this book to find each of these differences:

(*a*)	2,654	(*b*)	4,326	(*c*)	5,072
	1,231		2,348		1,986

10 Use the examples of Exercise 9 to point out the significance of place value in the process of subtraction.

11 Check each example of Exercise 9 by casting out nines.

12 Construct addition-subtraction tables for these bases: (*a*) **four,** (*b*) **five,** (*c*) **seven,** (*d*) **eight,** (*e*) **nine.**

13 Find these differences in base **twelve.** Check by addition.

(*a*) **6784 − 2341**   (*b*) **5682 − 1378**   (*c*) **5082 − 2598**
(*d*) **8te4 − 6e95**   (*e*) **t04e − 9372**   (*f*) **8007 − 2ete**

14 Find these differences in base **five.** Check by addition.

(*a*) **432 − 112**   (*b*) **314 − 132**   (*c*) **302 − 204**
(*d*) **400 − 124**   (*e*) **403 − 342**   (*f*) **201 − 13**

15 Find these differences in base **eight.** Check by addition.

(*a*) **742 − 345**   (*b*) **1504 − 672**   (*c*) **0605 − 2716**
(*d*) **4003 − 1276**   (*e*) **7002 − 6724**   (*f*) **5000 − 4765**

16 Find these differences in base **four.** Check by addition.

(*a*) **321 − 112**   (*b*) **211 − 122**   (*c*) **230 − 121**
(*d*) **302 − 32**   (*e*) **113 − 21**   (*f*) **300 − 22**

17   Find these differences in base **seven.** Check by addition.

(a) **4625**   (b) **1546**   (c) **203**
    3413   **614**   **124**

(d) **3060**   (e) **1000**   (f) **4023**
    2504   **625**   **645**

18   Find these differences in base **nine.** Check by addition.

(a) **5876**   (b) **5681**   (c) **2078**
    4653   **4723**   **1325**

(d) **8030**   (e) **7132**   (f) **6001**
    7465   **864**   **3504**

4-7 MULTIPLICATION WITH INTEGERS

As with addition, it is desirable to extend the table of basic multiplication facts from the set N of natural numbers to the set C of cardinal numbers before extending it to the set Z of integers. Such a table will provide for computation with nonnegative integers and will provide a basis for extension to multiplication with negative integers.

From D-5 we know that $b + 0 = b$ for any integer b, and also that if m and n are integers such that $m + n = m$, then n *must be 0.* (Why?) Since $b + 0 = b$ the multiplicative property of equality E-5 yields $a \times (b + 0) = a \times b$. From D-4 it then follows that $a \times (b + 0) = (a \times b) + (a \times 0)$. The substitution rule may be used to obtain

$$(a \times b) + (a \times 0) = a \times b$$

Since a, b, and 0 are integers, $a \times b$ and $a \times 0$ are integers. (Why?) Furthermore, $a \times 0$ is an integer such that, when added to $a \times b$, the sum is $a \times b$. Thus, $a \times 0$ must be the same as the additive identity. Whence we have Theorem 4-5.

$a \times 0 = 0$ **Theorem 4-5**   If $a \in Z$, then $a \times 0 = 0$.

Example

$3 + 0 = 3$	D-5
$5 \times (3 + 0) = 5 \times 3$	E-5
$5 \times (3 + 0) = (5 \times 3) + (5 \times 0)$	D-4
$(5 \times 3) + (5 \times 0) = 5 \times 3$	Substitution rule
$15 + (5 \times 0) = 15$	Multiplication
$5 \times 0 = 0$	D-5

×	0	1	2	3	4	5	6	7	8	9
0	0	0	0	0	0	0	0	0	0	0
1	0	1	2	3	4	5	6	7	8	9
2	0	2	4	6	8	10	12	14	16	18
3	0	3	6	9	12	15	18	21	24	27
4	0	4	8	12	16	20	24	28	32	36
5	0	5	10	15	20	25	30	35	40	45
6	0	6	12	18	24	30	36	42	48	54
7	0	7	14	21	28	35	42	49	56	63
8	0	8	16	24	32	40	48	56	64	72
9	0	9	18	27	36	45	54	63	72	81

FIGURE 4-6 Basic multiplication facts (base ten)

This theorem gives the reason for the condition $c \neq 0$ in the statement of the cancellation property (D-8) for integers. For example, $8 \times 0 = 0$ and $12 \times 0 = 0$, from which we have $8 \times 0 = 12 \times 0$, but $8 \neq 12$.

Basic multiplication facts

The table of multiplication facts can now be constructed in any base for the set C (see Figs. 4-6 to 4-8).

If a and b are two integers such that either is zero, then their product is zero. (Why?) What about the case in which it is known that the product is zero and one of the integers is not zero? What can be said about the other integer? For example, assume $b \neq 0$ and $a \times b = 0$. Since $0 \times b = 0$ (why?), it follows that $a \times b = 0 \times b$ (why?). From this last statement and D-8 it follows that $a = 0$. This completes the proof of Theorem 4-6.

Theorem 4-6 If $a,b \in Z$ and $a \times b = 0$, then either $a = 0$ or $b = 0$.

×	0	1	2	3	4	5
0	0	0	0	0	0	0
1	0	1	2	3	4	5
2	0	2	4	10	12	14
3	0	3	10	13	20	23
4	0	4	12	20	24	32
5	0	5	14	23	32	41

FIGURE 4-7 Basic multiplication facts (base **six**)

×	0	1	2
0	0	0	0
1	0	1	2
2	0	2	11

FIGURE 4-8 Basic multiplication facts (base **three**)

This theorem states that if the product of two integers is zero, then at least one of the integers is zero. This is a situation which occurs frequently, particularly in the solution of equations.

Example Find the integer a such that $3(a - 5) = 0$.

Since $(a - 5)$ is an integer (why?), we have two integers whose product is zero. Furthermore, since $3 \neq 0$, Theorem 4-6 states that $a - 5 = 0$. This implies that $a + {}^-5 = 0$ and, therefore, $a = 5$. (Why?)

If a is replaced by 5 in the given product we have

$$3(5 - 5) = 3(5 + {}^-5) = 3 \times 0 = 0$$

In this case we say that $a = 5$ is the solution of the equation $3(a - 5) = 0$.

Theorems 4-2 and 4-3 provide the needed extension of addition to the set Z of integers. Theorems 4-7 and 4-8, along with Theorem 4-5, will complete the definition of the two basic operations of the domain of integers, addition and multiplication.

$a \times ({}^-b) =$ ${}^-(a \times b)$ **Theorem 4-7** If $a,b \in Z$, then $a \times ({}^-b) = {}^-(a \times b)$.

As a parallel case, the theorem will be argued for the positive integers 3 and 4.

Hypothesis $3,4 \in Z$ *Hypothesis* $a,b \in Z$

Conclusion $3 \times ({}^-4) = {}^-(3 \times 4)$ *Conclusion* $a \times ({}^-b) = {}^-(a \times b)$

Proof

Statements	*Reasons*	*Statements*
1 ${}^-4 \in Z$ such that $4 + ({}^-4) = 0$	Additive inverse	${}^-b \in Z$ such that $b + ({}^-b) = 0$
2 $3 \times [4 + ({}^-4)] = 0$	E-5; Theorem 4-5	$a \times [b + ({}^-b)] = 0$
3 $3 \times [4 + ({}^-4)] =$ $(3 \times 4) + [3 \times ({}^-4)]$	Distributive	$a \times [b + ({}^-b)] =$ $(a \times b) + [a \times ({}^-b)]$

4 $(3 \times 4) +$ Substitution $(a \times b) +$
   $[3 \times (^-4)] = 0$ rule $[a \times (^-b)] = 0$
5 $3 \times ^-4 \in Z$ Closure $a \times (^-b) \in Z$
6 $3 \times (^-4) = ^-(3 \times 4)$ Additive inverse $a \times (^-b) = ^-(a \times b)$

Product of positive and negative integers

This theorem states that for any two integers, the product of one by the additive inverse of the other is the additive inverse of their product. If the two integers are both taken to be positive, as in the argument of the special case of the theorem, then the theorem states that the product of a positive integer by a negative integer is always a negative integer.

$$3 \times (-4) = 3 \times (^-4) = ^-(3 \times 4) = ^-12 = -12$$

If, in step 2 of the argument for Theorem 4-7, $^-3$ and ^-a are used as multipliers instead of 3 and a, we have:

2 $^-3(4 + ^-4) = 0$ E-5; Theorem 4-5 $^-a(b + ^-b) = 0$
3 $(^-3 \times 4) + (^-3 \times ^-4) = 0$ D-4; substitution $(^-a \times b) +$
                                    rule $(^-a \times ^-b) = 0$
4 $^-(3 \times 4) + (^-3 \times ^-4) = 0$ Theorem 4-7; $^-(a \times b) +$
                                    substitution rule $(^-a \times ^-b) = 0$
5 $^-3 \times ^-4 = ^-[^-(3 \times 4)]$ D-6 $^-a \times ^-b =$
                                                           $^-[^-(a \times b)]$
6 $^-3 \times ^-4 = 3 \times 4$ Exercise 10, $^-a \times ^-b = a \times b$
                                    Sec. 4-2

Thus we have Theorem 4-8.

$^-a \times ^-b = a \times b$

Theorem 4-8 If $a,b \in Z$, then $^-a \times ^-b = a \times b$.

Product of two negative integers

This theorem states that for any two integers, the product of their respective additive inverses is the same as the product of the two integers. If the two integers are both taken as positive, as in the argument of the special case, then the theorem states that the product of two negative integers is always positive. $(-3) \times (-4) = ^-3 \times ^-4 = 3 \times 4 = 12$.

4-8 INEQUALITY AND INTEGERS

Since the positive integers have been identified with natural numbers, any inequality relation which holds for natural numbers will remain a true statement if only positive integers are involved. For example, Definition 3-3 applies directly to integers if it is changed to Definition 4-3.

$a < b$ for
$a,b \in Z$ **Definition 4-3** If for $a,b \in Z$ it is true that $a + c = b$ for some positive integer c, then a is said to be *less than* b ($a < b$), or b is *greater than* a ($b > a$).

Theorem 3-3, and any other such property which involves only addition, remains true when stated for integers. This theorem is restated here as Theorem 4-9.

Theorem 4-9 If a, b, and c are integers such that $a < b$, then $a + c < b + c$.

Note that there is no condition that any of the integers be positive, negative, or zero. The reader should adapt the proof of Theorem 3-3 to the proving of Theorem 4-9. (See Exercise 25 at end of this section.)

Example The theorem covers all cases for which $a < b$.

(*a*) Given $3 < 6$, then $3 + 2 < 6 + 2$, or $5 < 8$
$\qquad 0 < 4$, then $0 + 2 < 4 + 2$, or $2 < 6$
$\qquad -4 < -1$, then $-4 + 2 < -1 + 2$, or $-2 < 1$
(*b*) Given $3 < 6$, then $3 - 2 < 6 - 2$ or $1 < 4$

This case follows directly from the theorem since

$$3 - 2 = 3 + {}^{-}2 = 3 + (-2) \qquad \text{and} \qquad 6 - 2 = 6 + {}^{-}2 = 6 + (-2)$$

Similarly
$$\begin{array}{lll} 0 - 2 < 4 - 2 & \text{or} & -2 < 2 \\ -4 - 2 < -1 - 2 & \text{or} & -6 < -3 \end{array}$$

There are problems, however, when extension to integers is attempted with any inequality relations involving multiplication. There are three distinct cases which have to be considered due to the fact that the set Z may be partitioned into the three disjoint sets: the set $\{0\}$, the set of positive integers, and the set of negative integers.

Theorem 3-9 states that *if $a,b,c \in N$ such that $a < b$, then $ac < bc$.* Consider this theorem in the three following cases:

Case I $c = 0$. If $a = 3$ and $b = 8$ then $3 < 8$. From Theorem 4-5 we have $0 \times 3 = 0$ and $0 \times 8 = 0$, and $0 \times 3 = 0 \times 8$. Thus, multiplication of an inequality by zero changes it to an equality.

Case II $c > 0$. In this case the positive integer c may be identified as a natural number, and the theorem remains true as stated here.

Theorem 4-10 If $a,b,c \in Z$ such that $a < b$ and $c > 0$, then $ac < bc$.

Sense of an inequality

Case III $c < 0$. In this case multiplication by c changes the relation "less than" to the relation "greater than." In such cases the *sense of the inequality* is said to be changed. This extension of Theorem 3-9 is expressed in Theorem 4-11.

Theorem 4-11 If $a,b,c \in Z$ such that $a < b$ and $c < 0$, then $ac > bc$.

The proofs of these theorems are left as exercises. (See Exercises 26 and 27 at the end of this section.) However, the example presents the argument in an illustrative special case for each of the theorems as stated for integers.

Example (*Case II.*) Given $a = 3$, $b = 8$, and $c = 4$. In this case $3 < 8$ and $4 > 0$.

1. There exists a positive integer, in this case 5, such that $3 + 5 = 8$. — Definition 4-3
2. $(3 + 5) \times 4 = 8 \times 4$ — E-5
3. $(3 \times 4) + (5 \times 4) = 8 \times 4$ — D-4, substitution rule
4. $5 \times 4 > 0$ — Closure for N
5. $3 \times 4 < 8 \times 4$ — Definition 4-3

Example (*Case III.*) Given $a = 3$, $b = 8$, and $c = -4$. In this case $3 < 8$ and $-4 < 0$.

1. There exists a positive integer, in this case 5, such that $3 + 5 = 8$. — Definition 4-3
2. $(3 + 5) \times (-4) = 8 \times (-4)$ — E-5
3. $[3 \times (-4)] + [5 \times (-4)] = 8 \times (-4)$ — D-4
4. $[3 \times (-4)] + [^-(5 \times 4)] = 8 \times (-4)$ — Theorem 4-7; substitution rule
5. $[3 \times (-4)] + [^-(5 \times 4)] + (5 \times 4)$
   $= [8 \times (-4)] + (5 \times 4)$ — E-4
6. $[3 \times (-4)] + [^-(5 \times 4) + (5 \times 4)]$
   $= [8 \times (-4)] + (5 \times 4)$ — D-2
7. $[3 \times (-4)] + 0 = [8 \times (-4)] + (5 \times 4)$ — D-6; substitution rule
8. $3 \times (-4) = [8 \times (-4)] + (5 \times 4)$ — D-5; substitution rule
9. $3 \times (-4) > 8 \times (-4)$ — Definition 4-3

EXERCISES

1. Cite the authority to support each of these statements.
   (*a*) $3 \times 4 = 4 \times 3$

(b) $(5 \times 12) \times 3 = 5 \times (12 \times 3)$
(c) $8 \times (12 + 4) = (8 \times 12) + (8 \times 4)$
(d) $3 \times (5 \times 1) = 3 \times 5$

2 Find $(^-a) \times b$ in each case.

(a) $a = 2, b = 3$ (b) $a = 5, b = 4$
(c) $a = 6, b = 0$ (d) $a = 7, b = 1$
(e) $a = 1, b = 2$ (f) $a = 0, b = 12$
(g) $a = -1, b = 3$ (h) $a = -2, b = -3$
(i) $a = -5, b = -7$

3 Find $(^-a) \times (^-b)$ in each case.

(a) $a = 3, b = 5$ (b) $a = 1, b = 4$
(c) $a = 8, b = -1$ (d) $a = 7, b = 4$
(e) $a = 1, b = 1$ (f) $a = -2, b = 3$
(g) $a = -4, b = -5$ (h) $a = 8, b = -3$
(i) $a = -1, b = -1$

4 Find these products.

(a) $2 \times 3 \times 4$ (b) $(-3) \times 6 \times 2$
(c) $(-2) \times (-4) \times 5$ (d) $(-1) \times 4 \times (-6)$
(e) $8 \times (-3) \times (-5)$ (f) $(-4) \times (-8) \times (-6)$

5 Find these products in two ways: without using the distributive property and using the distributive property.

(a) $2(4 + 6)$ (b) $(3 + 7)5$ (c) $6[7 + (-3)]$
(d) $-4(3 + 2)$ (e) $[8 + (-6)]7$ (f) $-4[5 + (-2)]$
(g) $6[8 + (-10)]$ (h) $[7 + (-16)]3$ (i) $10[(-8) + (-2)]$
(j) $-6[(-3) + (-5)]$ (k) $12(8 - 3)$ (l) $-4(7 - 5)$
(m) $[(-4) - 2]3$ (n) $8[6 - (-4)]$ (o) $-2[(-6) - (-3)]$

6 Explain how each of these results is obtained.
(a) $(a + b)(x + y) = ax + ay + bx + by$
(b) $(a + b)(x - y) = a(x - y) + b(x - y)$
(c) $(a + 3)(b + 5) = ab + 3b + 5a + 15$
*(d) $(x + 2)(x + 1) = x^2 + 3x + 2$
*(e) $(y - 5)(y - 3) = y^2 - 8y + 15$
*(f) $(x + 8)(x - 5) = x^2 + 3x - 40$

7 Find the value of x in each case, and verify your answer.

(a) $4(x - 2) = 0$ (b) $^-3(x + 5) = 0$
(c) $-6(2x - 8) = 0$ (d) $-5(-3x - 9) = 0$

8 Arrange this set of integers in order from left to right with the smallest on the left.

$$\{8, -4, 0, 1, -6, -10, 7, 9, 3, -1\}$$

9 Construct a number line for integers and place each integer of Exercise 8 in its proper position on the line.

10 Construct multiplication tables in bases **two, nine,** and **twelve.**

In Exercises 11 to 16 use the commutative property to check each product obtained.

11 Find these products in base ten.

    (*a*) 2,504   (*b*) 1,234   (*c*) 8,256
         368   506   675

    (*d*) 5,026   (*e*) 14,856   (*f*) 28,609
       4,208   2,907   32,048

12 Find these products in base **six.**

    (*a*) **245**   (*b*) **503**   (*c*) **342**
       34   **21**   **54**

    (*d*) **435**   (*e*) **4004**   (*f*) **3124**
      241   **413**   **503**

13 Find these products in base **three.**

    (*a*) **12**   (*b*) **102**   (*c*) **112**
      21   **12**   **102**

    (*d*) **2112**   (*e*) **1201**   (*f*) **2212**
      201   **2102**   **2120**

14 Find these products in base **two.**

    (*a*) **11**   (*b*) **1101**   (*c*) **11011**
      10   **111**   **1110**

15 Find these products in base **nine.**

    (*a*) **38**   (*b*) **87**   (*c*) **54**
      42   **26**   **37**

    (*d*) **326**   (*e*) **708**   (*f*) **714**
      48   **214**   **408**

    (*g*) **6104**   (*h*) **7258**   (*i*) **8702**
     672   **3840**   **3056**

16 Find these products in base **twelve.**

(a) **39**
        46
        ‾‾

(b) **1t**
        e4
        ‾‾

(c) **te**
        24
        ‾‾

(d) **208**
        76
        ‾‾

(e) **5t4**
        et
        ‾‾

(f) **tte**
        et9
        ‾‾

(g) **2147**
        378
        ‾‾

(h) **6ete**
        20et
        ‾‾

(i) **ette**
        tete
        ‾‾

Follow these instructions for Exercises 17 to 19: (a) Estimate each product. (b) Compute each product. (c) Check each product by casting out nines in base ten; **fives** in base **six;** and **e**'s in base **twelve.**

17   Base ten

(a) 7,526
        3,426
        ‾‾

(b) 8,231
        5,042
        ‾‾

(c) 7,103
        4,219
        ‾‾

18   Base **six**

(a) **4213**
        2013
        ‾‾

(b) **5421**
        3134
        ‾‾

(c) **2144**
        5423
        ‾‾

19   Base **twelve**

(a) **2356**
        4589
        ‾‾

(b) **et03**
        2846
        ‾‾

(c) **te48**
        e6te
        ‾‾

20   Find these products in base ten. Check your answers.

(a) -24
        32
        ‾‾

(b) 423
        -46
        ‾‾

(c) -203
        -406
        ‾‾

(d) 4,721
        -326
        ‾‾

(e) $-5,402$
        $-2,336$
        ‾‾

(f) $-7,064$
        205
        ‾‾

21   Argue the truth of each of these statements.
        (a) If $x + 3 > 0$, then $x > -3$.
        (b) If $x - 5 < 0$, then $x < 5$.
        (c) If $4 - x > 0$, then $x < 4$.
        (d) If $-x + 7 < 0$, then $x > 7$.
        (e) If $-x - 6 > 0$, then $x < -6$.
        (f) If $-12 - x < 0$, then $x > -12$.

*22   Prove each of these statements.
        (a) The even integers are closed under multiplication.
        (b) The odd integers are closed under multiplication.

*23 Given the set $S = \{0, 1, \perp, \triangle, \square\}$ with the two operations $*$ and $\#$ defined on S. Assume these statements as true and illustrate each.

(a) S is commutative under each operation.

(b) S is associative under each operation.

(c) $*$ is distributive over $\#$.

*24 Prove that $^-a = (-1) \times a$ for every integer a.

*25 Prove Theorem 4-9.

*26 Prove Theorem 4-10.

*27 Prove Theorem 4-11.

*28 Prove this theorem: If $a, b \in Z$ such that $a \times b \neq 0$, then $a \neq 0$ and $b \neq 0$.

*29 If $a, b \in Z$ such that $a < b$, does it follow that $a^2 < b^2$? Why?

*30 Given the integers a, b, c, d, e, and f with $a < b$, $c < d$, $e > 0$, and $f > 0$. Prove that $ae + cf < be + df$.

4-9 CONVERSION TO DIFFERENT BASES BY MULTIPLICATION

Equivalent numerals in different bases

In Sec. 1-12 the observation was made that the symbol representing the number of letters in the English alphabet is 26—ten, **22—twelve,** or **101—five.** These are equivalent numerals all representing the same number. This fact suggests that it is possible to change from a numeral in a given base to an equivalent numeral in some other base. There are different methods for carrying out such a conversion. In order to make use of its important informational value, multiplication will be used at this time.

Example Use multiplication to change **22—twelve** to its equivalent numeral in base ten.

The symbol 22 means $2(10) + 2(1)$ in any base. Since we are given that the symbol is in base **twelve,** $2(10) + 2(1)$ means 2 twelves and 2 ones. When this is translated into base ten it becomes $2(12) + 2(1)$. The arrow ($\rightarrow$) of Fig. 4-9 means "translates into." Thus the equivalent numeral is 26-ten. Notice that in the figure the computation is done in base ten, because at the arrow the thinking was translated from base **twelve** to base ten.

Twelve	Ten
$22 =$	
$2(10) + 2(1) \longrightarrow$	$2(12) + 2(1)$
	$= 24 + 2$
	$= 26$

FIGURE 4-9 Conversion from base **twelve** to base ten: **10—twelve** = 1 twelve = 1 ten + 2 ones = 12

Twelve	Five
22 =	
2(10) + 2(1) ──→	**2(22) + 2(1)**
	= 44 + 2
	= 101

FIGURE 4-10   Conversion from base **twelve** to base **five: 10—twelve = 1 twelve =** 2 fives + 2 ones = **22—five**

Example   Use multiplication to change **22—twelve** to its equivalent numeral in base **five.** In Fig. 4-10 the multiplication and addition are all done in base **five.** Note that **22—twelve** is *not* the same as **22—five. 22—twelve** converts into **101—five.**

Example   Use multiplication to change the numeral 268-ten to its equivalent numeral in base **six.** Before we can make this conversion it is necessary to be able to multiply in base **six.** Figure 4-7 gives all the basic multiplication facts for base **six.**

Note: The digits 6-ten and 8-ten become two-figure numerals in base **six,** since they are each larger than five and smaller than thirty-six (see Fig. 4-11).

The most difficult part of the computation is finding the product $2(14^2)$. It is necessary to compute in base **six.**

$$14 \times 14 = 14 \times (10 + 4)$$
$$= (14 \times 10) + (14 \times 4) \qquad \text{Why?}$$
$$= (14 \times 10) + [(10 + 4) \times 4] \qquad \text{Why?}$$
$$= (14 \times 10) + (10 \times 4) + (4 \times 4) \qquad \text{Why?}$$
$$= (14 \times 10) + (40 + 24)$$
$$= 140 + 104$$
$$= 244$$
$$2 \times (14^2) = 2 \times 244 = 532\text{—six}$$

Ten	Six
6 ──→	10
8 ──→	12
10 ──→	14
268	
= $2(10^2)$ + 6(10) + 8(1) ──→	$2(14^2)$ + 10(14) + 12(1)
	$2(14^2)$ = 532
	10(14) = 140
	12 = 12
268 ──→	1124

FIGURE 4-11   Conversion from base ten to base **six**

The addition is then carried out in base **six.** 268—ten converts into **1124—six.**

It should be obvious that any such conversion can be checked by multiplication. In this example the check is to convert **1124—six** to its equivalent numeral in base ten.

$$\begin{aligned}
\textbf{1124} &= \textbf{1(10}^3\textbf{)} + \textbf{1(10}^2\textbf{)} + \textbf{2(10}^1\textbf{)} + \textbf{4(1)—six} \\
&= 1(6^3) + 1(6^2) + 2(6^1) + 4(1)\text{—ten} \\
&= 216 + 36 + 12 + 4\text{—ten} \\
&= 268\text{—ten}
\end{aligned}$$

EXERCISES

1 What is incorrect in each of these numerals?

   (*a*) **2641—five**   (*b*) 3e4t—ten   (*c*) **5069—nine**

2 Why is it necessary to use the word and not a symbol when identifying the base of a numeral system if the discussion involves two or more different bases?

3 By grouping letters of the English alphabet determine the numeral in bases **four** and **nine** to be used in telling how many letters there are in the alphabet.

4 Check each of the numerals in Exercise 3 by using multiplication to convert 26—ten to the equivalent numeral in each base.

5 Each of the given numerals is written in base ten. Convert each one to an equivalent numeral in base **two:** (*a*) 2; (*b*) 4; (*c*) 8; (*d*) 16; (*e*) 32; (*f*) 5; (*g*) 6; (*h*) 7.

6 Use Exercise 5 as an aid in explaining each of these conversions:
   (*a*) 10—ten = **1010—two;**
   (*b*) 100—ten = $(\textbf{1010})^{10}$**—two**
   (*c*) 1000—ten = $(\textbf{1010})^{11}$**—two**

7 What change should be made in a numeral to indicate that the number it represents has been multiplied by the base of the system? By any power of the base?

8 A test frequently used in base ten to determine whether a number is even or not is to see if the ones digit of the representing numeral is divisible by two. Can this test be used in these bases: (*a*) **two;** (*b*) **three;** (*c*) **four;** (*d*) **five;** (*e*) **six;** (*f*) **seven;** (*g*) **eight;** (*h*) **nine;** (*i*) **twelve?** Why or why not?

In Exercises 9 to 13 use multiplication to convert each given numeral to an equivalent numeral in each specified new base. Also use multiplication in each check.

9 5266—ten: (*a*) **twelve;** (*b*) **eight;** (*c*) **six**
   Check each new numeral by converting it to base ten.

10 **7320—eight:** (*a*) ten; (*b*) **five;** (*c*) **twelve**
    Check each new numeral by converting it to base **eight.**

11   **etet—twelve:** (a) ten; (b) **nine;** (c) **four**
Check each new numeral by converting it to base **twelve.**

12   **111011000—two:** (a) **five;** (b) ten; (c) **twelve**
Check each new numeral by converting it to base **two.**

13   **1748—nine:** (a) **five;** (b) **twelve;** (c) **six**
Check by converting the given numeral and each new numeral to an equivalent numeral in base ten.

4-10   MORE ABOUT FACTORIZATION

In Sec. 3-17 the concepts of prime number and composite number were defined. These definitions were within the restrictions of the natural number system. With the extension to integers these concepts do not change, but it does become desirable to revise the definitions to adjust to the inclusion of zero and the negative integers. For example, if the integer p is a prime it may have as factors not only p and 1 but also $-p$ and -1.

Prime number   **Definition 4-4**   An integer $p > 1$ is a *prime number* if it is divisible only by p, $-p$, 1, or -1.

Composite number   **Definition 4-5**   An integer $c > 1$ is a *composite number* if it has divisors other than c, $-c$, 1, or -1. More simply, an integer $c > 1$ is *composite* if it is not prime.

There are occasions when it is desirable to extend the discussion of division and factorization to negative integers. This can be done by recalling that if $a > 0$ then $-a < 0$ and $-a = (-1) \times a$. (See Exercise 24, Sec. 4-8.) When this extension is made, Postulates 3-1 to 3-3, as well as other properties, still apply.

A similar modification of the division algorithm is desirable.

Division algorithm   **Postulate 4-5   The division algorithm**   For any two integers a and b, with $b > 0$, there exist unique integers q and r such that

$$a = bq + r \qquad \text{where } 0 \leq r < b\dagger$$

In the statement of the division algorithm the condition on b is that it be positive ($b > 0$). As previously noted, this condition does not rule out application to negative integers; it merely makes provision for statement of the algorithm in a form which is more effective in application. The significant

---

† The symbol $0 \leq r < b$ means that r may have the value 0 but, if not, it is a positive integer less than b. It is read "0 is less than or equal to r which is less than b," or "r is greater than or equal to 0 but less than b."

condition is that the divisor cannot be zero ($b \neq 0$). Why rule out the possibility that the divisor might be zero? The answer to this question may be found in Definition 4-2 where it is stated that $q = a \div b$ if and only if $q \times b = a$. If $b = 0$, then $a = 0$ since $q \times 0 = 0$ for any integer q. So, if the divisor is 0, the dividend must also be 0 and the quotient can be any *Division by* integer. There would exist *no unique* value for q. For this reason *division* *zero undefined* *by zero is undefined.* Furthermore, where $r \neq 0$ it may seem inconsistent with Definition 4-2 to call q the quotient of a divided by b. However, it seems unnecessarily meticulous to assign q one label when $r = 0$ and another when $r \neq 0$. We shall call q the quotient in both cases and simply state that, when $r = 0$, the quotient q exists and the division is *exact*, that is, *b* *b divides a* *divides a*, or *a is divisible by b*; when $r \neq 0$ there exist both a quotient q and a remainder r. In such cases the only possible remainders are 1, 2, 3, . . . , $b - 1$. For example, when any integer is divided by 5, if the division is not exact, then the only possible remainders are 1, 2, 3, and 4.

An important application of the division algorithm is to provide a means for testing a given positive integer to determine whether or not it is a prime number. The test consists of trying consecutive primes as divisors.

Test for **Example**   Determine whether these integers are prime or not: (a) 105;
primeness (b) 169; (c) 191.

(a)   2$\lfloor$105   3$\lfloor$105
           52 r 1   35 r 0

The second trial reveals that $105 = 3 \times 35$, and so it is not a prime.

(b)   2$\lfloor$169   3$\lfloor$169   5$\lfloor$169
           84 r 1   56 r 1   33 r 4

        7$\lfloor$169   11$\lfloor$169   13$\lfloor$169
           24 r 1   15 r 4   13 r 0

The last trial reveals that $169 = 13 \times 13$, so that it is not a prime number.
In such a test it is not necessary to try any composite numbers as divisors. For example, since $6 = 2 \times 3$ and neither 2 nor 3 is a divisor of 169, it follows that 6 cannot divide 169. Why? Notice that in the successive tests made, as the divisor increases the quotient decreases until they become equal.

(c)   2$\lfloor$191   3$\lfloor$191   5$\lfloor$191
           95 r 1   63 r 2   38 r 1

        7$\lfloor$101   11$\lfloor$191   13$\lfloor$191
           27 r 2   17 r 4   14 r 9

There are no exact divisions and the next consecutive prime, 17, is larger than the last quotient obtained. From this it follows that 191 has no prime factors and, hence, is itself a prime number.

Prime divisors

   The test of the previous example reveals that 3 and 35 are both factors, or divisors, of 105. A continuation of the process would reveal that $35 = 5 \times 7$. Thus the *prime divisors, prime factors*, of 105 are 3, 5, and 7. Since $3 \times 5 \times 7 = 105$, it should be clear that still other divisors of 105 are $15 = 3 \times 5$ and $21 = 3 \times 7$. From these facts it follows that 1, 3, 5, 7, 15, 21, 35, and 105 are all of the divisors (factors)† of 105. The integers 1, 3, 5, 7, 15, 21, and 35 are *proper divisors*.

Proper divisors

Definition 4-6 A *proper divisor* of an integer a is an integer $b \geq 1$ and less than a ($1 \leq b < a$) which is a divisor of a.

Example Since $286 = 2 \times 11 \times 13$, the divisors of 286 are 1, 2, 11, 13, 22, 26, 143, and 286; the prime divisors are 2, 11, and 13; and the proper divisors are 1, 2, 11, 13, 22, 26, and 143.

EXERCISES

For each integer list: (*a*) all divisors, (*b*) all prime divisors, and (*c*) all proper divisors.

1 6	2 169	3 33	4 12
5 81	6 36	7 49	8 110
9 210	10 125	11 64	12 500

*13 p^2 where p is a prime
*14 p^3 where p is a prime
*15 p^4 where p is a prime
*16 Use the concept of proper divisor to rewrite the definition of a prime number.

In Exercises 17 to 24 test each integer to determine whether or not it is a prime number.

17 83	18 133	19 269	20 389
*21 529	*22 617	*23 883	*24 2,197

*25 What are the twin primes between 300 and 350? (See Exercise 4, Sec. 3-17.)
*26 Of the integers less than 100 there are seven sets of three consecutive composite numbers, seven sets of five, and only one of seven. Identify each of these 15 sets of composite numbers.

It has been conjectured that every even integer greater than 4 can be expressed as the sum of two (not necessarily distinct) odd primes. (For

† It should be remembered that the words "divisor" and "factor" are used to mean "positive divisor" and "positive factor." It has been pointed out previously that all of this discussion can be extended to negative integers by merely considering their additive inverses which are positive.

example, $6 = 3 + 3$; $10 = 3 + 7$ or $5 + 5$.) In Exercises 27 to 34 write each integer as the sum of two odd primes in as many ways as you can.

27	8	28	14	29	22	30	36
*31	34	*32	46	*33	122	*34	202

*4-11 THE NATURE OF DIVISION

Uses of division

Definition 4-2 gives explicit emphasis to the important inverse relation between division and multiplication. Furthermore it implies the no less important functions of division as an operation which can be used to find how many times one integer is contained as a factor in another integer; to find a number which is contained in the dividend as many times as unity is contained in the divisor; to separate a large group into a number of groups of equal size; to find how many smaller groups of equal size are contained in a larger group; to find how many times a smaller number may be subtracted from a larger number; or to find a number which has the same relation to unity that the dividend has to the divisor.

Terms used in division

The terms *dividend* and *divisor* come from the Latin expressions *numerus dividendus* (number to be divided) and *numerus divisor* (dividing or divider number). As in the cases of addition, subtraction, and multiplication, the word *numerus* gradually disappeared through technical usage and translation. In early practice the word "result" was used to indicate the answer to the actual division process.† Other terms used at one time or another are "product," "part," "exiens," "outcome," and "quotient," with "quotient" finally winning general approval. The term "remainder," used to designate that part left over in cases of uneven division, has likewise won general approval over such words as *numerus residuus* and *residuus*.‡

The symbol ÷

The symbol ÷ was first used as a symbol to indicate subtraction.§ The early symbols for division were varied just as were the symbols for multiplication. Abbreviations, words, literal symbols, a straight horizontal or slanting line, one dot, a colon, one or two parentheses, as in 6)30 or 6)30(, and variations of the form $6\overline{)30}$, have been some of the more popular symbols used to indicate division. Of these we have retained the following: the symbol ÷, the colon, both the horizontal and the slanting line, the parentheses, and variations of the form $\underline{}\overline{}$. The colon is used primarily to emphasize the ratio concept of division, and the horizontal or slanting line to emphasize the fractional aspect.

The division process, historically, has been the most difficult of the elementary computational processes. From the struggle for efficiency in operation our present algorithm has evolved as the one which seems to

---

† David Eugene Smith, *op. cit.*, p. 131.
‡ *Ibid.*, p. 132.
§ F. Cajori, *op. cit.*, pp. 240–244.

combine most effectively the benefits of place value with simplicity in recording the results of the incidental operations necessary to carry out the division.

The simplest of all methods of historical significance is the one we use today as a shortcut in cases of division with one-digit and simple two-digit divisors (formerly referred to as short division). The basic techniques of this method have remained the same, and there has been only slight change in form. The oldest form of division is probably that of the Egyptians in which they used an adaptation of duplation and mediation for multiplication. Another method of interest with a limited amount of merit is the *repiego* method. As in the case of multiplication, this method employs the principle of decomposition of one of the numbers—in this case the divisor—into simple factors. For example, the quotient of 2,212 ÷ 28 can be obtained very simply by thinking of 28 as 7 × 4 and then considering either the continued division of (2,212 ÷ 7) ÷ 4 or that of (2,212 ÷ 4) ÷ 7. The merit of this method lies in the possibility of reducing to simple oral division a problem which otherwise would require the recording of incidental steps in arriving at the desired result.

Historians have been unable to fix an exact date for the origin of the algorithm we use today. The first appearance in printed form seems to have been in an arithmetic, by Calandri, published in 1491.† Two forms bearing strong resemblance to it are illustrated in Fig. 4-12. Form A was used in the fourteenth century by Maximus Planudes. Form B, which dates to the fifteenth century, is known as the *a danda* (by giving) method. It derives

† David Eugene Smith, *op. cit.*, p. 141.

Form *A*   Form *B*
                        (*a danda*)   FIGURE 4-12 Early variations of modern method

```
          137
  75 | 10,282
       2 78
        532
          7
```

FIGURE 4-13 South American method

its name from the fact that at each step, an additional digit from the dividend is "given" to the remainder to derive a new partial dividend for the succeeding step. This form is definitely recognizable as a forerunner of the form we use today, the only difference being that the remainder at each step is copied before the new digit is "given" to it. An abbreviated form of the *a danda* method is illustrated in Fig. 4-13. The abbreviation is due to the fact that all partial products are omitted. This modification, which requires somewhat more mental effort than the form we use, seems to be in current use in some South American countries.

These forms are but a selected few of the many different patterns which have been of historical significance in the struggle to overcome the difficulties incident to the implication of the division algorithm. In fact, an examination of arithmetic texts in current use in the United States will reveal variations, each carrying some argument in its favor over the others in use.

4-12 THE EUCLIDEAN ALGORITHM

Greatest common divisor (g.c.d.)

As a consequence of Definition 4-2 any factor of an integer is recognized as a divisor of the integer. Thus the concept of *greatest common divisor* (*g.c.d.*) is synonymous with that of greatest common factor (g.c.f.). The procedure used in Sec. 3-17 to find the greatest common factor of two numbers a and b consisted in breaking each number down into its prime factors and then selecting all those factors which were common to both a and b. The product of these common factors then was the sought-for g.c.f. For example, the g.c.f. of 390 and 462 is 6 since $390 = 2 \times 3 \times 5 \times 13$ and $462 = 2 \times 3 \times 7 \times 11$. Thus $6 = 2 \times 3$ is the g.c.f. This scheme works very well for small numbers whose prime factors are readily recognizable, but not so well in other cases.

Example Find the g.c.f. of 2,958 and 3,162.

The primes 2 and 3 are rather easily recognizable as factors of each of the given integers.

$$2,958 = 2 \times 3 \times 493$$
$$3,162 = 2 \times 3 \times 527$$

There is no simple way of determining whether 493 and 527 are prime numbers or, if not prime, whether they have any common factors. Actually they are factorable and each has 17 as a factor. Thus the complete factorization of the two numbers is

$$2,958 = 2 \times 3 \times 17 \times 29$$
$$3,162 = 2 \times 3 \times 17 \times 31$$

From Eratosthenes's sieve (Fig. 3-18) 17, 29, and 31 are seen to be primes. Therefore the g.c.f. of 2,958 and 3,162 $= 2 \times 3 \times 17 = 102$.

In cases such as that of this example where prime factorization is difficult, the division algorithm offers a simple means of finding the g.c.f. of any two integers. The process used is known as the *euclidean algorithm*, which we shall postulate. Since division, rather than factorization, is the process used, the result is called the greatest common divisor (g.c.d.) rather than the greatest common factor (g.c.f.).

The euclidean algorithm

Postulate 4-6   The euclidean algorithm   To find the greatest common divisor of two positive integers, divide the larger integer by the smaller one. Use the remainder of this division as the new divisor and the former divisor as the new dividend. Continue this process of dividing the old divisor by the new remainder until the remainder is zero. The last divisor used is the g.c.d. of the given integers.

The division algorithm is the authority for this process, which is illustrated in Fig. 4-14, where we find the g.c.d. of the two integers 2,958 and 3,162 of the previous example. In the first step, for example, $3,162 = 1 \times 2,958 + 204$. Use of the distributive property will establish that any common divisor of 204 and 2,958 will also be a divisor of 3,162. Therefore, the g.c.d. of 204 and 2,958 will also be the g.c.d. of 2,958 and 3,162. In the second step $2,958 = 14 \times 204 + 102$. The same reasoning shows that the g.c.d.

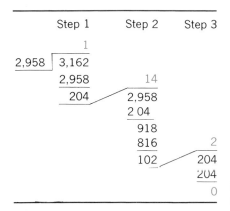

FIGURE 4-14   The euclidean algorithm

of 102 and 2,958 is also the g.c.d. of 204 and 2,958, which has already been established as the g.c.d. of 2,958 and 3,162. The remainder in the next division is 0. Therefore, 204 = 2 × 102. Thus 102 is the g.c.d. of the given numbers 2.958 and 3,162.

This process may be extended to finding the g.c.d. of any set of three or more integers. The process merely consists of applying the euclidean algorithm to pairs of integers until the desired g.c.d. has been determined.

Example   What is the g.c.d. of 2,958; 3,162; 11,169; and 13,617?

In the example of Fig. 4-14 the g.c.d. of 2,958 and 3,162 was found to be 102.

The same process will reveal that 153 is the g.c.d. of 11,169 and 13,617.

Thus, 102 contains *all* divisors common to the first two numbers and 153 contains *all* divisors common to the second pair. Any divisor common to all four numbers is, therefore, contained in the g.c.d. of 102 and 153. A third use of the euclidean algorithm produces 51 as the g.c.d. of all four integers.

Lowest common multiple (l.c.m.)   Factorization used in finding the lowest common multiple (l.c.m.) of two numbers, discussed in Sec. 3-17, has the same limitations as in finding the greatest common factor. For large numbers the process can be simplified by multiplying one of the numbers by the quotient obtained by dividing the other number by the g.c.d. of the two numbers. If we represent the two numbers by a and b, the g.c.d. by d, and the l.c.m. by l, then

$$l = a \times \frac{b}{d}$$

or

$$l = \frac{a}{d} \times b$$

where $\frac{b}{d}$ means $b \div d$.

Example   Find the l.c.m. of 2,958 and 3,162.

Previously the factored forms of these two numbers were displayed as

$$2{,}958 = 2 \times 3 \times 17 \times 29$$
$$3{,}162 = 2 \times 3 \times 17 \times 31$$

If we use l to represent the l.c.m., we have from these factored forms

$$l = 2 \times 3 \times 17 \times 29 \times 31$$

Thus, we may write l in either of these forms

$$l = 2{,}958 \times 31$$
$$l = 3{,}162 \times 29$$

or

These are exactly the values obtained from the above formulas.

$$l = 2{,}958 \times \frac{3{,}162}{102}$$

$$l = \frac{2{,}958}{102} \times 3{,}162$$

EXERCISES

1 When and for what purposes do we divide?
2 Why is division by zero an undefined process?
3 How are multiplication and division related?
4 Why can the multiplication tables of this chapter be used also as tables of the basic division facts?
5 Use the division algorithm as a guide in constructing a rule for checking the result of any division.
6 Carry out these divisions and then use the rule of Exercise 5 to check each one.

(a) $35\overline{)2{,}467}$ (b) $117\overline{)24{,}531}$ (c) $403\overline{)71{,}245}$

7 Construct a scheme for checking division by casting out nines. (*Hint:* Use Exercise 5 as a guide.)
8 Check each of the divisions of Exercise 6 by casting out nines.
9 How can division be checked by using subtraction?
10 Carry out these divisions and then check each one by subtraction.

(a) $52\overline{)164}$ (b) $315\overline{)1{,}375}$ (c) $508\overline{)1{,}339}$

11 Use the *repiego* method to find each of these quotients orally:

(a) $1{,}248 \div 24$ (b) $2{,}205 \div 21$ (c) $3{,}735 \div 45$

12 Use these examples to illustrate the importance of place value in division:

(a) $6{,}798 \div 33$ (b) $9{,}672 \div 18$ (c) $41{,}935 \div 321$

13 Express the results of Exercise 12 in the form of the division algorithm.
14 Break each of the given integers into its prime factors and then find the greatest common divisor and the least common multiple in each example.

(a) 20, 35, 75   (b) 42, 56, 70
(c) 66, 150, 210, 300   (d) 132, 165, 231, 297

15 Use the euclidean algorithm to find the greatest common divisor and the least common multiple of each pair of integers.

(a) 1,785, 2,590   (b) 3,927, 7,161
(c) 9,666, 206,745   (d) 22,134, 32,538

16 Use the euclidean algorithm to find the greatest common divisor of each set of integers.

(*a*) {814, 1,295, 1,517}

(*b*) {29,172, 30,420, 185,484}

(*c*) {2,565, 7,315, 11,115, 21,615}

(*d*) {8,283, 9,789, 21,335, 38,905}

17 How can the euclidean algorithm be used to determine whether or not two integers are relatively prime?

18 In each case find the quotient and remainder in base **twelve.**

(*a*) **724 ÷ 42**	(*b*) **690 ÷ 16**	(*c*) **812 ÷ 38**
(*d*) **6028 ÷ 3t**	(*e*) **t4e7 ÷ e3**	(*f*) **253t01 ÷ 2te**

19 In each case find the quotient and remainder in base **eight.**

(*a*) **3456 ÷ 21**	(*b*) **7065 ÷ 57**	(*c*) **32646 ÷ 52**
(*d*) **16735 ÷ 213**	(*e*) **70204 ÷ 316**	(*f*) **200074 ÷ 76**

*20 In each case find the greatest common divisor of the numbers written in base **five.**

(*a*) **3031; 4441**

(*b*) **13230; 43412; 100240; 222220**

*21 In each case find the greatest common divisor of the numbers written in base **nine.**

(*a*) **4251; 11282**

(*b*) **17863; 27264; 48010**

*22 (*a*) Find the least common multiple of **3031—five** and **4441—five.**

(*b*) Find the least common multiple of **4251—nine** and **11282—nine.**

4-13 CONVERSION TO DIFFERENT BASES BY DIVISION

In Sec. 4-9 attention was given to the technique of converting, by multiplication, from numerals in one base to equivalent numerals in another base. The division algorithm provides a means for making similar conversions by using division as the technique.

First, recall that in base **b,** the digital symbols used in writing any numeral are **0, 1, . . . , b** − 1. Second, as a consequence of the division algorithm, the same numerals represent the only possible remainders when any integer is divided by the integer *b*. These two facts combine to give us the technique for conversions to a new base by division, which is illustrated by the following example.

Example Use division to carry out these three conversions:

1 Convert **11004—five** to its equivalent numeral in base ten.

2 Convert the resulting numeral in base ten to its equivalent numeral in base **twelve.**

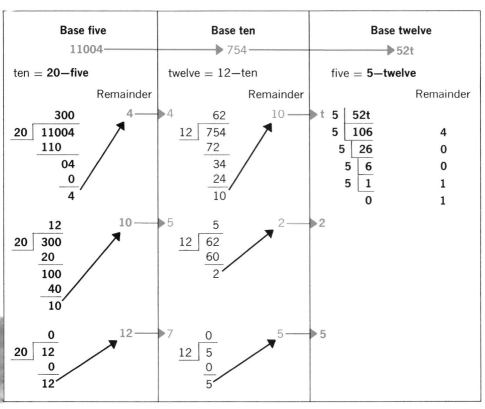

FIGURE 4-15   Conversion of numerals to different bases. In each column the computation is carried out in the base indicated at the top of the column. The arrows which connect two columns are to be read "transforms into." For example, in the conversion from base **five** to base ten we have **4—five** transforms into 4—ten; **10—five** transforms into 5—ten; and **12—five** transforms into 7—ten.

3   Check the entire process by converting the resultant numeral in base **twelve** to its equivalent numeral in base **five.**

Each of these conversions is carried out in Fig. 4-15. The computations are all in the base in which the numeral is given. In the left column, the question is basically: How many tens are there in **11004?** Since the numeral is written in base **five** it is necessary to divide in this base. Hence the divisor, ten, is written as **20,** since ten is 2 fives + 0 ones. The division shows that in **11004** there are **300** tens with a remainder of **4.** This remainder thus becomes the digital symbol in ones position in the numeral in base ten. The division is continued to determine how many tens there are in **300—five.** There are **12** tens with a remainder of **10.** Since the computation is in base **five, 10** means "1 five." Therefore **10—five** > 5—ten, which becomes the digital symbol in tens position in the numeral in base ten. Similarly, **12—five** → 7—ten, which is the digital symbol in hundreds position in the numeral in base ten.

Observation of the process reveals that the division is carried out until the final quotient figure is zero, each remainder being recorded in the notation of the new base. Then the numeral in the new base is obtained by reading the remainders in the reverse order to that in which they were obtained.

This same process is repeated in each of the two remaining bases, ten and **twelve.**

Since one can convert from any given base to any other base either by multiplication or by division, either process may be used to check the other. Also, as in the example, any conversion may be checked by the same process as that used in the conversion.

Example Use division to convert 2,158—ten into its equivalent numeral in base **twelve.** Check the work by using multiplication to convert the resultant numeral into its equivalent numeral in base ten.

The necessary computation for this example is outlined in Fig. 4-16. The reader should make sure that the details of each operation are understood

Base ten		Base **twelve**
Remainder	Digit symbol	

Step 1   12 ⌐2,158 — — — — — — — — — — — The symbol in
Step 2   12 ⌐179   10 ——————————→ t   base **twelve** is
Step 3   12 ⌐14   11 ——————————→ e   **12et;** read "one-
Step 4   12 ⌐1   2 ——————————→ 2   two-e-te"
            0   1 ——————————→ 1

2,158—(ten) ——————————→ **12et**—(twelve)

Check

Base ten	Base **twelve**

12et means:
Step 5 $1(12^3) + 2(12^2) + 11(12) + 10 = $ ←—— $1(10^3) + 2(10^2) + e(10) + t$
Step 6 $\begin{cases} 1(1,728) + 2(144) + 11(12) + 10 = \\ 1,728 + 288 + 132 + 10 = 2,158 \end{cases}$
            2,158—(ten) ←———— **12et** = (twelve)

FIGURE 4-16 Conversion from base ten to base **twelve.**
Step 1: 2,158 ÷ 12 = 179 with remainder of 10 which transforms into **t—twelve**
Step 2: 179 ÷ 12 = 14 with remainder 11 → **e—twelve**
Step 3: 14 ÷ 12 = 1 with remainder 2 → **2—twelve**
Step 4: 1 ÷ 12 = 0 with remainder 1 → **1—twelve**
Step 5: To check—**12et—twelve** → $1(12^3) + 2(12^2) + 10$—ten
Step 6: To get the numeral in base ten the indicated operations are performed in base ten.

and can be supplied. It should be recognized that there are other combinations of the two processes by which such a conversion and its check can be accomplished.

EXERCISES

1   Use multiplication for each of these conversions.
   (*a*) **6207—eight** to an equivalent numeral in base **six**
   (*b*) **6207—eight** to an equivalent numeral in base **twelve**
   (*c*) Check by converting from base **twelve** to base **six**
2   Consider the numeral **tete—twelve.** Use division to convert this numeral to an equivalent numeral in base ten. Check by using multiplication to convert back to base **twelve.**
3   Use division for each of these conversions. Check each conversion by multiplication.
   (*a*) **106324—seven** to base **twelve**
   (*b*) **5264—seven** to base ten; base **five**
   (*c*) **30414—eight** to base **seven**
   (*d*) **13330—six** to base **four;** base **nine**
4   Make each of these conversions by division. Check each conversion by using division to convert the new numeral back to the old base.
   (*a*) 3043—ten to base **twelve**
   (*b*) **5122—twelve** to base ten
   (*c*) **5122—twelve** to base **five**
   (*d*) **5122—eight** to base **twelve**
   (*e*) **ette—twelve** to base **six**
   (*f*) **ette—twelve** to base ten
   (*g*) 463—ten to base **two**
   (*h*) **342—five** to base **two**

4-14 REVIEW EXERCISES

1–23   Write a carefully constructed answer to each of the guideline questions for this chapter.
24   What are the definitions of the operations of subtraction and division?
25   What is the chief reason for extending the natural number system to the domain of integers?
26   What are the basic differences in the postulates of the two systems?
27   How may set concepts be used to support the statement that $7 + 0 = 7$?

28 Is this a true statement: ^-i is negative if $i \in Z$?

29 What is $^-(^-i)$ where $i \in Z$?

30 What is $^-(-2)$?

31 What is the value of i in each of these statements?

(a) $i + 3 = 3$ (b) $i + 2 = 0$

(c) $i + (-5) = 0$ (d) $i + 6 = 9$

(e) $i + 4 = 2$ (f) $i + {}^-4 = 2$

(g) $i - 8 = 0$ (h) $i - 5 = 2$

(i) $i - 6 = 8$ (j) $i - (^-3) = 4$

32 Use the discussion of the paragraph immediately following Theorem 4-3 to prove that $5 - 8 = -(8 - 5)$.

33 Find these sums.

(a) $(-2) + (-5) + (-7) + (-15)$

(b) $5 + (-3) + 6 + (-4)$

(c) $(-18) + (-4) + 15 + 6$

34 Find these sums and check by subtraction.

(a) $8 + 7 + 6 + 9 + 2 + 12$

(b) $23 + 15 + 14 + 32 + 41$

(c) $3 + 16 + 8 + 20 + 32$

35 Find these sums and check by casting out nines.

(a) 1,423	(b) 90,234	(c) 51,672
8,076	5,102	438
3,405	16,708	2,406
7,236	80,350	80,076
8,091	19,876	19,043

36 Find these differences and check by addition.

(a) 80,764	(b) 52,461	(c) 70,020
78,923	10,738	59,318

37 Find these differences in base **eight** and check by addition.

(a) **7215**	(b) **6054**	(c) **40071**
4102	**5763**	**12063**

38 Use these differences to illustrate how place value and the properties of integers are used: (a) $74 - 52$; (b) $61 - 48$.

39 What properties of the domain of integers justify the subtraction and addition of 10 in the equal additions method of subtraction (page 137)?

40 Follow the argument for Theorem 4-5 to prove that $26 \times 0 = 0$.

41 Construct a number line with these integers placed in proper order on it: $\{8, -6, 0, 1, 7, -3, -7, 4, 5, -1\}$.

42 Find these products. Use the commutative property to provide a check.

(a) 726
503

(b) −28
39

(c) 372
−514

(d) −416
−728

(e) 3,026
670

(f) 7,234
−1,009

43 Find each of these products in each indicated base. Use the commutative property to provide a check.

(a) **Eight**
723
415

(b) **Nine**
−208
567

(c) **Two**
−10111
−1101

(d) **Twelve**
8t05
−e1t

(e) **Five**
−10234
−2103

(f) **Six**
4531
2045

*44 Given the integers a, b, and c and the two operations * and # defined as follows:

$$a * b = 3a + b \qquad a \# b = 3 \cdot a \cdot b$$

where + and · mean ordinary addition and multiplication for integers.
(a) Is * associative or commutative?
(b) Is # associative or commutative?
(c) Is # distributive over *?

45 Use multiplication to carry out each of the indicated conversions. Check by using division.
(a) **1101440—five** to base **twelve**
(b) **11011001—two** to base ten
(c) **200021—four** to base **nine**

46 Convert the numeral **26351—seven** to an equivalent numeral in base **twelve.** Check by converting each numeral to its equivalent numeral in base ten.

47 (a) What is 0 ÷ 4? Explain.
(b) What is 4 ÷ 0? Explain.

48 For each integer, list all divisors, all prime divisors, and all proper divisors. (a) 14; (b) 27; (c) 157; *(d) 38,346

49 Determine which of these integers are prime and which are composite.

(a) 781
*(e) 593

(b) 127
*(f) 1,669

(c) 2,747
*(g) 37,789

(d) 233
*(h) 1,951

*50 For each integer of Exercise 49, list all divisors, all prime divisors, and all proper divisors.

51 Find the greatest common divisor and the least common multiple of each set of integers.

(*a*) {10,778; 37,089}

(*b*) {22,355; 121,506}

52 Find the greatest common divisor of these sets of integers.

(*a*) {69,954; 28,558; 51,483}

(*b*) {17,862; 151,827; 7,786; 58,395}

*INVITATIONS TO EXTENDED STUDY

1 Prove this theorem: If $a,b,c \in Z$, then $a + (b - c) = (a + b) - c$.

2 Prove this theorem: If $a,b,c \in Z$, then $a - (b - c) = (a - b) + c$.

3 Why are 5 and 2 the only prime numbers whose difference is 3?

4 List all the prime numbers less than 1,000.

5 What are the twin primes less than 1,000? (See Exercise 4, Sec. 3-17.)

6 How many distinct factors will the square of a prime number have? The cube of a prime? Why?

7 How many distinct factors will a number have if it is the product of three distinct prime numbers?

8 A *perfect number* is a number which is the sum of all its factors except the number itself. For example, 6 is a perfect number since $6 = 1 + 2 + 3$. A formula for finding perfect numbers is $2^{n-1} (2^n - 1)$ where n is a positive integer and $2^n - 1$ is a prime number. Find two more perfect numbers and check them by finding the sum of their factors.

9 Show that, if the formula $2^n - 1$ is to give a prime number, n must be a prime.

10 If the factors of 220, other than the number itself, are added, the sum is 284. Similarly, if the factors of 284, other than the number itself, are added, the sum is 220. For this reason these two numbers are called *amicable numbers*. Verify the above statement concerning the sums of the factors of these two numbers.

11 Verify that 1,184 and 1,210 is another pair of amicable numbers.

12 The greatest common divisor of 3,450 and 5,775 is 75, symbolically (5,775; 3,450) = 75. It can be shown that 75 can be expressed in the form $75 = 3 \cdot 5,775 - 5 \cdot 3,450$ or, in equivalent form, $(5,775; 3,450) = 3 \cdot (5,775) + (-5) \cdot 3,450$. Verify this last statement. Now use the euclidean algorithm to find (5,775; 3,450) and show how the integral multipliers

3 and -5 can be computed from the results obtained in the steps of the algorithm.

This exercise illustrates a very important property of numbers, namely, if d is the greatest common divisor of the positive integers a and b, then there exist integers p and q such that $d = pa + qb$.

13   Find (63,973; 6,105) and determine the integers p and q such that (63,973; 6,105) $= p \cdot 63,973 + q \cdot 6,105$.

MODULAR ARITHMETIC

The process of factoring and the operation of division, as defined for natural numbers and integers, make it desirable to have an effective pattern for ready recognition of prime factors and divisors. Such a pattern may be derived from an important relation which exists between subsets of the set Z of integers. Carl Friedrich Gauss (1777–1855), who first discovered this relation in 1801, called it a *congruence relation*. This discovery led mathematicians to turn a great deal of their attention to a study of the basic relations existing between numbers and between sets of numbers. One such development has been given the name of *modular arithmetic*. In this chapter a brief look is taken at this study of the relations between sets of integers and at some of the details pertinent to the problems of factoring and division.

The following guidelines can be helpful in the study of Chap. 5.

1 What is a prime number?
2 What is a composite number?
3 What is the fundamental theorem of arithmetic?
4 What is meant by a well-defined binary operation?
5 What is meant by closure of a set under an operation?
6 What is the definition of division?
7 What is the division algorithm?
8 When is a division said to be exact?
9 What is the definition of a group? An abelian group?
10 What is meant when one integer is said to be congruent to another integer modulo m?
11 What is meant by addition modulo m?
12 What is meant by multiplication modulo m?
13 What are the various divisibility tests given in this chapter?
14 How may the congruence relation be used to devise other such tests?
15 How may tests for divisibility by prime numbers be used to develop tests for divisibility by composite numbers?

INTRODUCTION

In the two previous chapters the emphasis has been on the structure of two important number systems of elementary mathematics, the natural number system and the domain of integers. These two systems are called
Two-operation two-operation systems since there are two operations, addition and multi-
systems plication, fundamental to the use of number within each system.

The congruence relation will be used as a means for developing a form of *modular arithmetic* to serve as a basis for defining finite number systems of both the one-operation and two-operation types. These will be in contrast to the infinite number systems of Chap. 3 (The Natural Number System) and Chap. 4 (The Domain of Integers), whose basic sets were the infinite sets N and Z, respectively. Furthermore, the relation will be used as an effective aid in developing certain important divisibility tests.

5-1 THE CONCEPT OF GROUP

Among the simplest of all mathematical systems is the *group*, which consists of a set of elements and one well-defined operation which has certain characteristic properties in accordance with this definition.

Definition of a group

Definition 5-1 A *group* consists of a set of elements $G = \{a,b,c, \ldots\}$ and one well-defined operation (*) such that the set satisfies the following postulates under the given operation:†

G-I Closure If $a * b = x$, then x is in G.

G-2 Associative $a * (b * c) = (a * b) * c$.

G-3 Identity There exists in G a unique element i, called the *identity* element, such that $i * a = a * i = a$ for each a in G.

G-4 Inverse For each element a in G there exists in G a unique element a^{-1}, called the *inverse of a*, such that $a^{-1} * a = a * a^{-1} = i$.

If, in addition to these four properties, the elements of the set G also satisfy the commutative property under the operation (*), the group is said to be *Abelian group* a *commutative*, or an *abelian*, *group*.

G-5 Commutative $a * b = b * a$.

Example Given the set $G = \{a,b\}$ and the operation (*) defined by the table of Fig. 5-1, show that the system is a group.
The table is read: $a * a = a$; $a * b = b$; $b * a = b$; and $b * b = a$.

| An examination of the table reveals that the set has closure under the

---

† The symbol *, call it "star," is used to represent an unspecified operation. For specific types of groups the operation * will be translated into the appropriate operational symbol. For example, for an *additive group* * will be interpreted as +, the sign for addition.

$*$	a	b
a	a	b
b	b	a

FIGURE 5-1 Operational table for $G = \{a,b\}$

given operation since a and b are the only elements seen in the table, and they are both elements of the set.

2 One example will be given of the associative property:

$$a * (b * a) = a * b = b \qquad \text{from Fig. 5-1}$$
$$(a * b) * a = b * a = b$$

It thus follows by substitution that $a * (b * a) = (a * b) * a$.

To establish this property for the system would call for the verification of all eight such possible groupings. Why eight?

3 From the first row we see that $a * a = a$ and $a * b = b$, which exhibits a as the identity element. This is verified by column 1.

4 Since a is the identity element, G-4 states that the only criterion needed to establish an element k as the inverse of element l is that $k * l = l * k = a$. The table reveals that $a * a = a$ and $b * b = a$, so that each element is its own inverse.

5 The system is, in fact, an abelian group since $a * b = b * a$.

Example Does the set Z of integers with addition form a group? (The operation $*$ has been defined to be addition, $+$.)

1 By D-1 (page 126) the set Z is closed under addition, so that G-1 is satisfied.

2 By D-2 the set is associative under addition, so that G-2 is satisfied.

3 By D-5 there exists an additive identity, so that G-3 is satisfied.

4 By D-6 for each integer there is an additive inverse, so that G-4 is satisfied.

The set Z of integers with addition does form a group since all four group properties are satisfied. Furthermore, this group is abelian since by D-3 the set Z is commutative under addition, and so G-5 is satisfied.

Example Does the set Z of integers with multiplication form a group? (The operation $*$ has been defined to be multiplication, $\times$.)

D-1 and D-2 state that the set Z is closed and is associative under multiplication. D-7 states that there exists a multiplicative identity, but there does not exist in Z a multiplicative inverse for each element of Z. Although G-1, G-2, and G-3 are satisfied, G-4 is not. It therefore follows that the set Z with multiplication is not a group.

*	0	1
0	0	1
1	1	0

FIGURE 5-2 Operation table for $S = \{0,1\}$

#	0	1
0	0	0
1	0	1

FIGURE 5-3 Operation table for $S = \{0,1\}$

Each of the two previous examples deals with an infinite set of elements. The next two examples will provide a similar illustration, one system a group and the other not, using a finite set of elements.

Example Does the set $S = \{0,1\}$, with the operation (*) as defined by the table in Fig. 5-2, form a group?

From a comparison of S with the set G of the first example in this section, and of the two corresponding operation tables, it should be evident that the examples are the same except that a has been replaced by 0 and b by 1. Just as G with the operation * is a finite group with two elements, so the set S with the operation * is a finite group of two elements.

Example Does the set $S = \{0,1\}$, with the operation # (sharp) as defined in the table of Fig. 5-3, define a group?

Examination of the table will reveal that the system has closure, is both associative and commutative. Line 2 of the table reveals that 1 is the identity and that 1 is its own inverse element. There is no inverse element for 0 since $0 * 0 = 0 * 1 = 0$. The system, therefore, is not a group.

EXERCISES

In all the exercises of this section the associative property holds for each stated operation. Which of the systems are groups? Which are abelian groups?

1 The set $\{0,1,2\}$ with the operation * defined by the table

*	0	1	2
0	0	1	2
1	1	2	0
2	2	0	1

EXERCISE 1

2   The set {0,1,2} with the operation # defined by the table

#	0	1	2
0	0	0	0
1	0	1	2
3	0	2	1

EXERCISE 2

3   The set {0,1,2,3} with the operation * defined by the table

*	0	1	2	3
0	0	1	2	3
1	1	2	3	0
2	2	3	0	1
3	3	0	1	2

EXERCISE 3

4   The set N of natural numbers with addition
5   The set N of natural numbers with multiplication. (Does each element have an inverse? For example, does there exist $n \in N$ such that $n \times 2 = 2 \times n = 1$?)
6   The set C of cardinal numbers with addition
7   The set C of cardinal numbers with multiplication
8   The set of all even integers with addition
9   The set of all even integers with multiplication
10 The set of all odd integers with addition
11 The set of all odd integers with multiplication
*12 The set {0,1,2,3} with the operation # defined by the table

#	0	1	2	3
0	0	0	0	0
1	0	1	2	3
2	0	2	0	2
3	0	3	2	1

EXERCISE 12

*13 The set {a,b,c} with the operation * defined by the table

*	a	b	c
a	a	b	c
b	b	a	d
c	c	b	a

EXERCISE 13

*14   The set $\{a,b,c,d\}$ with the operation * defined by the table

*	a	b	c	d
a	a	b	c	d
b	b	d	a	c
c	c	a	d	b
d	d	c	b	a

EXERCISE 14

5-2 CALENDAR ARITHMETIC

Suppose you were asked the question: What day of the week will it be 90 days from today? Whatever pattern of thinking you might use in arriving at an answer to the question, it would in one way or another involve determining the number of weeks in 90 days and the number of extra days. The simplest way to determine these facts is to divide 90 by 7. The quotient will indicate the number of weeks and the remainder will be the number of extra days: $90 \div 7 = 12$ with 6 as a remainder. There are 12 weeks and 6 days in 90 days. After this has been determined the 12 may be discarded, since any number of weeks from a given day in the week is that same day. All that is necessary to determine the answer to the question is to use the remainder 6 and count that number of days from the given day. For example, assume the "today" of the question to be Wednesday. Twelve weeks from Wednesday will be Wednesday, and 6 days more will be Tuesday of the next week. So if the original question had been "What day of the week will it be 90 days from Wednesday?" the answer would be "Tuesday."

It is to be noticed that the answer would be the same if the 90 days were changed to 20, 41, 76, 370, or any other number of days which yields the remainder 6 when divided by 7. ($20 = 2 \times 7 + 6$; $41 = 5 \times 7 + 6$; $76 = 10 \times 7 + 6$; $370 = 52 \times 7 + 6$.) In other words, for this computation we are not concerned with the quotient in the division by 7, but are interested only in the remainder.

This problem can be simplified still further in the following manner:

1   Associate each day of the week with a number in this manner: Sunday 0, Monday 1, Tuesday 2, Wednesday 3, Thursday 4, Friday 5, and Saturday 6. Note that the numbers used are the only possible remainders when 7 is used as a divisor.

2   Since Wednesday is the day of the week involved in the question, we take its associated number 3 and add it to 90. ($90 + 3 = 93$.)

3   Divide 93 by 7 and discard the quotient, and the remainder will be the number associated with the day which is 90 days from Wednesday. $93 \div 7 = 13$ with a remainder of 2. This is the number associated with Tuesday, the day already determined as the answer to the question.

Example

1 What day of the week will it be 21 days from Friday? The number associated with Friday is 5. Then $21 + 5 = 26$; $26 \div 7 = 3$ with a remainder of 5, so 21 days from Friday is Friday. This, of course, checks with the fact that 21 days is the same as 3 weeks. The division shows this since $21 \div 7 = 3$ with 0 as a remainder.

2 What day of the week will it be 38 days from Sunday? The number associated with Sunday is 0. Then $38 + 0 = 38$, and $38 \div 7$ leaves a remainder of 3. Since $0 + 3 = 3$, Wednesday, the day associated with 3, is the day of the week 38 days from Sunday.

The fact that the remainder of the division is the only result of concern in this type of arithmetic affords a means for further simplification of the problem. This simplification is the result of a notation which discards the quotient and indicates only the remainder of a particular division.

From the division algorithm (page 151) we know that for any given integers a and b with $b > 0$, there exist integers q and r such that $a = bq + r$ where $0 \leq r < b$. This fact implies that $a - r = bq$ (Definition 4-1). In other words, the difference $a - r$ is exactly divisible by b. The congruence relation provides a convenient method for stating such results. The divisor
Modulus is called the *modulus*, usually abbreviated to *mod*, and the symbol $\equiv$, which is read "is congruent to," replaces the equality symbol, $=$. In other words, $a \equiv r \pmod{b}$, which is read "a is congruent to r mod b," means that the difference $a - r$ is exactly divided by b. What does $a \equiv 0 \pmod{b}$ mean?

For comparison, the numbers mentioned previously in this section are listed in the symbolism of the division algorithm and also in the new symbolism of congruency.

$$90 = 12 \times 7 + 6 \qquad 90 \equiv 6 \pmod{7}$$
$$20 = 2 \times 7 + 6 \qquad 20 \equiv 6 \pmod{7}$$
$$41 = 5 \times 7 + 6 \qquad 41 \equiv 6 \pmod{7}$$
$$76 = 10 \times 7 + 6 \qquad 76 \equiv 6 \pmod{7}$$
$$370 = 52 \times 7 + 6 \qquad 370 \equiv 6 \pmod{7}$$
$$93 = 13 \times 7 + 2 \qquad 93 \equiv 2 \pmod{7}$$
$$21 = 3 \times 7 + 0 \qquad 21 \equiv 0 \pmod{7}$$
$$26 = 3 \times 7 + 5 \qquad 26 \equiv 5 \pmod{7}$$
$$38 = 5 \times 7 + 3 \qquad 38 \equiv 3 \pmod{7}$$

These are merely examples of the symbolism and the type of computation
Modular in a very convenient and useful form of arithmetic known as *modular*
arithmetic *arithmetic*. In this type of arithmetic the computations of the previous examples would be carried out in a manner illustrated in this example.

Example What day of the week will be (1) 90 days from Wednesday, (2) 21 days from Friday, (3) 38 days from Sunday?

1 $90 \equiv 6$ (mod 7). 3 is the number associated with Wednesday. $6 + 3 = 9$. $9 \equiv 2$ (mod 7). If one prefers, the addition may be performed first. $90 + 3 = 93$. $93 \equiv 2$ (mod 7). 2 is the number associated with Tuesday, and so Tuesday is the day of the week 90 days from Wednesday.

2 $21 \equiv 0$ (mod 7). 5 is associated with Friday. $0 + 5 = 5$, and $5 \equiv 5$ (mod 7). Or, $21 + 5 \equiv 26$. $26 \equiv 5$ (mod 7). Thus Friday is the day of the week 21 days from Friday.

3 $38 \equiv 3$ (mod 7). 0 is associated with Sunday. $3 + 0 = 3$. $3 \equiv 3$ (mod 7). 3 is the number associated with Wednesday, and so Wednesday is the day of the week 38 days from Sunday.

5-3 CONGRUENCE MODULO m

An equivalent concept of the congruence of two integers is that given by the following definition.

Congruent modulo m **Definition 5-2** Two integers a and b are said to be *congruent modulo m* if and only if they have the same remainder upon division by the integer m.†

Properties of congruence There are three theorems which are of fundamental importance to the brief development of modular arithmetic which is to be pursued here. They will be stated without proof. They might be proved as exercises by the interested student. (See Exercises 15, 16, 18, and 19, page 181.)

These theorems are stated for integers, and so all symbols occurring are used to represent integers.

Theorem 5-1 $a \equiv b$ (mod m) if and only if $a - b$ is divisible by m.

Corollary $a \equiv b$ (mod m) if and only if $a = km + b$.

It follows as an immediate consequence of this corollary that the concept of congruence of the previous section is equivalent to that expressed in Definition 5-2. (See Exercise 17, page 181.)

Theorem 5-2 If $a \equiv b$ (mod m) and $c \equiv d$ (mod m), then $a + c \equiv b + d$ (mod m).

†This definition was given in 1801 by Gauss (1777–1855). See E. T. Bell, "The Development of Mathematics," pp. 175–176, McGraw-Hill Book Company, New York, 1940.

Theorem 5-3 If $a \equiv b \pmod m$ and $c \equiv d \pmod m$, then $ac \equiv bd \pmod m$.

Example

1 $34 = 5 \cdot 6 + 4$ and $49 = 5 \cdot 9 + 4$
2 $16 = 5 \cdot 3 + 1$ and $36 = 5 \cdot 7 + 1$
Whence $34 \equiv 49 \pmod 5$ and $16 \equiv 36 \pmod 5$.
From (1) and (2) and the addition properties of integers we have

$$34 + 16 \qquad\qquad 49 + 36$$
$$= (5 \cdot 6 + 4) + (5 \cdot 3 + 1) \qquad = (5 \cdot 9 + 4) + (5 \cdot 7 + 1)$$
$$= (5 \cdot 6 + 5 \cdot 3) + (4 + 1) \qquad = (5 \cdot 9 + 5 \cdot 7) + (4 + 1)$$
$$= 5 \cdot 9 + 5 \qquad\qquad = 5 \cdot 16 + 5$$

Since $34 + 16$ and $49 + 36$ have the same remainder upon division by 5 it follows that $34 + 16 \equiv 49 + 36 \pmod 5$.

In similar manner using the properties of integers under multiplication and addition it can be shown that

$$34 \cdot 16 \equiv 49 \cdot 36 \pmod 5 \qquad \text{(see Exercise 5, page 181)}$$

In modular arithmetic there are interesting examples of finite groups. In such an arithmetic with m as a modulus the only elements to be considered are the possible remainders in any division involving an integral dividend and the positive integer m as the divisor, namely, the integers 0, 1, 2, 3, . . . , $m - 1$. Before considering any such systems it is necessary to define the operations to be used.

Addition modulo m

Definition 5-3 Addition modulo m means that ordinary addition is used, but the results are all reduced to the remainders upon division by m.

Multiplication modulo m

Definition 5-4 Multiplication modulo m means that ordinary multiplication is used, but the results are all reduced to the remainders upon division by m.

Example Find the sum and product of the integers 5 and 6 (mod 7).

$$5 + 6 = 11 \equiv 4 \pmod 7$$
$$5 \times 6 = 30 \equiv 2 \pmod 7$$

Example Show that the set of integers $\{0,1,2,3,4\}$ with addition mod 5 is a group.

Before proceeding with the test, be sure that you understand how the addition table is derived. As an example, in the row labeled 4 at the left

+	0	1	2	3	4
0	0	1	2	3	4
1	1	2	3	4	0
2	2	3	4	0	1
3	3	4	0	1	2
4	4	0	1	2	3

FIGURE 5-4 Addition modulo 5

×	1	2	3
1	1	2	3
2	2	0	2
3	3	2	1

FIGURE 5-5 Multiplication modulo 4

these sums are recorded in order from left to right (see Fig. 5-4): $4 + 0 = 4 \equiv 4$; $4 + 1 = 5 \equiv 0$; $4 + 2 = 6 \equiv 1$; $4 + 3 = 7 \equiv 2$; $4 + 4 = 8 \equiv 3$. (All congruencies are mode 5.)

1 *Closure* Only the elements of the set, i.e., the integers 0, 1, 2, 3, 4 are seen in the table.
2 *Associative* The addition involves only integers, and modular addition of integers is associative.
3 *Identity* The identity element is 0. This can be seen to be true by looking at the sums in the row or column headed by 0.
4 *Inverse* The identity element 0 is seen once and only once in each row and column. This means that each element has an additive inverse. For example, 1 is the inverse of 4 and 4 is the inverse of 1 since $1 + 4 = 4 + 1 = 0$.

Is this group an abelian group? Why?

Example Show that the set of integers $\{1,2,3\}$ with multiplication modulo 4 does not form a group. The multiplication table is given in Fig. 5-5.
This set does not have closure under the given operation, because $2 \times 2 = 4 \equiv 0$ (mod 4), and 0 is not in the given set. Hence the set is not a group.
Can you determine another reason why this system is not a group?

EXERCISES

1 How does the table of Fig. 5-4 differ from the table of basic addition facts for integers in base **five?** Explain the differences.

2   How does the table of Fig. 5-5 differ from the table of basic multiplication facts for natural numbers in base **four?** Explain the differences.

3   The calendar date for any selected day of the week will occur one or two days later in the week next year. What is the explanation of this fact?

4   A person desiring to take out a 90-day loan prefers that the payment come due on a Monday. On what day of the week should he execute the loan?

5   Use the pattern of the example following Theorem 5-3 to show that $34 \times 16 \equiv 49 \times 36$ (mod 5).

6   Show that: (a) $58 + 90 \equiv 37 + 13$ (mod 7)
           (b) $58 \times 90 \equiv 37 \times 13$ (mod 7)

In Exercises 7 to 14 which systems are groups? Are they also abelian? You are to assume the associative property.

7   The set $\{0, 1\}$ with multiplication modulo 2. Compare with the example of Fig. 5-3.

8   The set $\{0,1\}$ with addition modulo 2. Compare with the example of Fig. 5-2.

9   The set $\{1\}$ with multiplication modulo 2.

10   The set $\{0,1,2,3,4,5\}$ with addition modulo 6.

11   The set $\{1,2,3,4,5\}$ with multiplication modulo 6.

12   The set $\{0,1,2,3,4,5,6\}$ with addition modulo 7.

13   The set $\{0,1,2,3,4,5,6\}$ with multiplication modulo 7.

14   The set $\{1,2,3,4,5,6\}$ with multiplication modulo 7.

*15   Prove Theorem 5-1.

*16   Prove the corollary to Theorem 5-1.

*17   Prove that the concept of congruence of Sec. 5-2 is equivalent to that stated in Definition 5-2.

*18   Prove Theorem 5-2.

*19   Prove Theorem 5-3.

*20   Prove: If $a \equiv b$ (mod m), then $ac \equiv bc$ (mod m). (*Hint:* Use Theorem 5-1.)

*21   Give one counterexample to show that the fact that $ac \equiv bc$ (mod m) does not necessarily imply that $a \equiv b$ (mod m).

5-4   TESTS FOR DIVISIBILITY

In the reduction of fractions to lower terms, in the finding of greatest common divisors of two or more numbers, and in many other forms of division exercises, it is very desirable to be able to test large numbers for divisibility by smaller numbers. The techniques of congruence afford very simple means for testing any number for divisibility by 2, 3, 4, 5, 6, 8, 9, 11, and composite numbers of which these numbers are the only factors.

Recall that any integer may be written in the form

$$a(10^n) + b(10^{n-1}) + \cdots + c(10) + d$$

where the coefficients represented by $a, b, \ldots, c, d$ are one-digit, non-negative integers with $a \neq 0$, and n is any positive integer. This fact and Theorems 5-1 to 5-3 provide the means for establishing the tests. In each case a specific number will be used for convenience of illustration; the reader should make sure that the general application is understood.

Divisibility by 2 Consider the number 3,472. This number may be written: $3(10^3) + 4(10^2) + 7(10) + 2(1)$. Since $10 \equiv 0 \pmod 2$, it follows, by Theorem 5-3, that $10^2 \equiv 0$ and $10^3 \equiv 0$. Again by Theorem 5-3, $3(10^3) \equiv 1 \times 0 = 0$, $4(10^2) \equiv 0 \times 0 = 0$, $7(10) \equiv 1 \times 0 = 0 \pmod 2$. By Theorem 5-2, since $2 \equiv 0$ and $1 \equiv 1 \pmod 2$, $3,472 = 3(10^3) + 4(10^2) + 7(10) + 2 \cdot 1 \equiv 0 + 0 + 0 + 0 = 0 \pmod 2$. Since $3,472 \equiv 0 \pmod 2$, it follows from Theorem 5-1 that 3,472 is divisible by 2.

An analysis of this example will reveal that, since $10^n \equiv 0 \pmod 2$ for any positive integer n, the test for divisibility by 2 is:

Divisibility by 2 An integer is divisible by 2 if and only if the digit in the ones place is divisible by 2.

Divisibility by 4 Since $10 \equiv 2$ and $10^2 \equiv 0 \pmod 4$, it follows that $10^n \equiv 0 \pmod 4$ if $n \geq 2$. These facts may be used to give two tests for divisibility by 4.

Divisibility by 4 1 An integer is divisible by 4 if and only if twice the tens digit plus the ones digit is a number which is divisible by 4.

2 An integer is divisible by 4 if and only if the last two digits form a number divisible by 4.

Example Test 3,472 for divisibility by 4.

1 $3,472 = 3(10^3) + 4(10^2) + 7(10) + 2(1)$
$\equiv (3 \times 0) + (4 \times 0) + (7 \times 2) + (2 \times 1) \qquad \text{mod } 4$
$= 14 + 2 = 16 \equiv 0 \qquad \text{mod } 4$

(Only powers of 10 have been reduced in order to illustrate the statement of test 1.) Therefore 3,472 is divisible by 4.

2 $3,472 = 3,400 + 72 = 34(100) + 72$. $100 = 10^2 \equiv 0 \pmod 4$. This means that any number of hundreds is divisible by 4. Therefore, in 3,472, we only need to examine 72. Since $72 = 18 \times 4$ it follows that 3,472 is divisible by 4.

Divisibility by 8   Argument similar to that used for 4 will give two tests for divisibility by 8.

Divisibility by 8

1   An integer is divisible by 8 if and only if 4 times the hundreds digit + 2 times the tens digit + the ones digit is a number divisible by 8.

2   An integer is divisible by 8 if and only if the last three digits form a number which is divisible by 8.

The reader should derive these two tests. (See Exercise 2, page 186.)

Divisibility by powers of 5

Divisibility by 5 and powers of 5   Since $10 = 2 \times 5$, the tests for divisibility by 5 and powers of 5 (25, 125, 625, and so on) are similar to those for 2 and powers of 2 (4, 8, 16, and so on). (See Exercises 2 to 7, page 186.)

Divisibility by 3   Since $10 \equiv 1 \pmod 3$, it follows that $10^n \equiv 1 \pmod 3$. Thus if Theorems 5-2 and 5-3, with $m = 3$, are applied to any number, the result will be the sum of the digits in the number. For example, $3{,}472 = 3(10^3) + 4(10^2) + 7(10) + 2(1)$. If we reduce *only* the powers of 10, we then have

$$3{,}472 = 3(10^3) + 4(10^2) + 7(10) + 2(1)$$
$$\equiv (3 \times 1) + (4 \times 1) + (7 \times 1) + (2 \times 1) \qquad \mod 3$$
$$= 3 + 4 + 7 + 2$$

Thus, the question of whether or not the number 3,472 is divisible by 3 has been reduced to the question of whether or not the sum of the digits gives a number which is divisible by 3. The sum in this case is 16 and $16 \equiv 1 \pmod 3$. Therefore, 3,472 is *not* divisible by 3. In fact, $3{,}472 = 1{,}157 \cdot 3 + 1$ and the 1 of the test is exhibited as the remainder of the division.

Theorem 5-2 provides a shortcut to this addition by making it possible to drop any digit, or sum of digits, which is congruent to $0 \pmod 3$, so we have the same result as before. The sum of the digits $\equiv 1 \pmod 3$.

Divisibility by 3

An integer is divisible by 3 if and only if the sum of its digits is a number divisible by 3.

Divisibility by 9   Since $10 \equiv 1 \pmod 9$, the test for divisibility by 9 is similar to that for 3.

Divisibility by 9

An integer is divisible by 9 if and only if the sum of its digits is a number divisible by 9.

The same shortcut as that used in the test for 3 can be used in the test for 9. Of course 9 replaces 3.

This is the property needed to validate the checks of addition and multiplication by casting out nines presented on pages 85 and 96. There are three principles to be established.

Principle I   The excess of nines in any given integer is the excess in the sum of its digits.

The truth of this statement follows immediately from the fact that $10^n \equiv 1$ (mod 9) where n is any positive integer. Furthermore, this excess is, of course, the remainder obtained upon dividing the given number by 9.

Example   The excess of nines in 7,642 is 1, since

$$7,642 = 7(10^3) + 6(10^2) + 4(10) + 2(1)$$
$$\equiv (7 \cdot 1) + (6 \cdot 1) + (4 \cdot 1) + (2 \cdot 1) \qquad \text{mod } 9$$
$$= 7 + 6 + 4 + 2 \equiv 1 \qquad \text{mod } 9$$
$$7,642 = 849 \cdot 9 + 1$$

Principle II   The excess of nines in the sum of two integers is the same as the excess in the sum of their respective excesses.

Proof   By the division algorithm the integers l and m may be written in the form $l = 9 \cdot r + a$ and $m = 9 \cdot s + b$ where a and b are the excesses of nines in l and m, respectively. Hence $l \equiv a$ and $m \equiv b$ (mod 9), and by Theorem 5-2

$$l + m \equiv a + b \qquad \text{mod } 9$$

This statement is equivalent to the statement of Principle II. The principle can be extended to apply to the sum of any number of integers.

Principle III   The excess of nines in the product of two integers is equal to the excess in the product of their respective excesses.

Proof   From the argument for Principle II, $l \equiv a$ and $m \equiv b$ (mod 9). By Theorem 5-3 it then follows that $lm \equiv ab$ (mod 9). Thus the excess of nines in the product lm is exhibited as being equal to the excess in ab. This principle can be extended to apply to the product of any number of integers.

Divisibility by 6   Any number which is divisible by 6 must also be divisible by both 2 and 3. Conversely, any number which is divisible by both 2 and

3 is divisible by 6. For these reasons the test for divisibility by 6 is a combination of the tests for 2 and 3.

Example Is the integer 346,752 divisible by 6?
Since the ones digit is divisible by 2, the number is divisible by 2.
The sum of the digits is $3 + 4 + 6 + 7 + 5 + 2 = 27 \equiv 0$ (mod 3). Therefore 346,752 is divisible by 6.
The shortcut suggested in Theorem 5-2 can be applied very effectively in this case by noting that $3 + 6 = 9 \equiv 0; 4 + 5 = 9 \equiv 0; 7 + 2 = 9 \equiv 0$ (mod 3).

Divisibility by 11

Divisibility by 11 Since $10 = 0 \cdot 11 + 10$ or $10 = 1 \cdot 11 + (-1)$, we may use either of the congruences $10 \equiv 10$ (mod 11) or $10 \equiv -1$ (mod 11). From these facts we have the following table of values mod 11: $1 \equiv 1, 10 \equiv -1; 10^2 = 10 \times 10 \equiv (-1) \times (-1) = 1; 10^3 = 10 \times 10^2 \equiv (-1) \times (1) = -1; 10^4 = 10 \times (10^3) \equiv (-1) \times (-1) = 1;$ and so on with alternating values of 1 and -1. From this we derive the test for divisibility by 11, which is best stated in three steps.
For any given integer:

1 Add the digits in the odd places, starting with the ones digit. Call this sum S.

2 Add the digits in the even places, starting with the tens digit. Call this sum T.

3 Find the difference between S and T. The original number will be divisible by 11 if and only if this difference is divisible by 11.

Example Test the number 382,349 for divisibility by 11.

$$S = 9 + 3 + 8 = 20$$
$$T = 4 + 2 + 3 = 9$$
$$S - T = 20 - 9 = 11 \equiv 0 \qquad \text{mod } 11$$

Therefore, 382,349 is divisible by 11.

Divisibility tests for composite numbers

The test for divisibility by 6 serves as an example for determining tests for divisibility by simple composite numbers. The number must be broken down into factors which are relatively prime. For example, the test for divisibility by 12 would resolve into the tests for divisibility for 3 and 4, since $12 = 3 \times 4$. It is to be noted, however, that 12 may also be factored as 2×6, but tests for 2 and 6 would not suffice, since any number which is divisible by 6 is also divisible by 2. The number 6 itself will serve as a counterexample since it is divisible by both 6 and 2, but it is not divisible by 12. The number 18 is another number divisible by both 6 and 2, but not by 12. What is a test for divisibility by 18?

No test for divisibility by 7 has been derived. It is possible to follow the procedures used above, or use other techniques, to derive such a test.† No such test is a simple one, and because of this fact it is usually simpler to proceed to divide by 7 rather than apply the test. It is also possible to derive tests for other prime numbers such as 13, 17, 19, but the difficulty of the test supersedes the practicality of its use. In fact, the size of the dividend enters into the determination of the practicality of the use of any of the tests devised.

EXERCISES

1   Test each of these numbers for divisibility by 12: (a) 27,624; (b) 75,942; (c) 257,816.
2   Derive the two tests for divisibility by 8.
3   Derive two tests for divisibility by 16.
4   Derive the test for divisibility by 5.
5   Derive one test for divisibility by 25.
6   Derive one test for divisibility by 125.
7   Derive one test for divisibility by 625.
8   What are tests for divisibility by 15, 18, 22, 24?
9   Why are there not two distinct tests for divisibility by 25, 125, and 625, as there are for 4, 8, and 16?

Test each of the numbers in Exercises 10 to 18 for divisibility by 2, 3, 4, 5, 6, 8, 9, 11, 12, 15, 18, 22, 24.

10	55,692	11	4,740	12	180,840
13	36,036	14	79,460	15	1,665,158
16	213,240	17	60,720	18	3,526,740

19   Derive tests for divisibility by **2** and **4** in base **five.**
20   Derive tests for divisibility by **2, 3, 4,** and **5** in base **six.**
21   Derive tests for divisibility by **2, 3, 4, 6,** and **8** in base **nine.**

5-5   REVIEW EXERCISES

1–15   Write a carefully constructed answer to each of the guideline questions for this chapter.
16   What is the congruence relation?
*17   Prove that the congruence relation is an equivalence relation. (See Definition 2-3.)

† For example, see E. Rebecca Matthews, A Simple "7" Divisibility Rule, *Mathematics Teacher*, **62**:461–464 (1969); Calvin T. Long, A Simpler "7" Divisibility Rule, *Mathematics Teacher*, **64**:473–475 (1971); and Frank Smith, Divisibility Rules for the First Fifteen Primes, *Arithmetic Teacher*, **18**:85–87 (1971).

18 Write in roster form the set whose elements are those positive integers less than 50 which are congruent to zero for each given integer as a modulus: (*a*) 2; (*b*) 3; (*c*) 4; (*d*) 5; (*e*) 6; (*f*) 9; (*g*) 10.

19 Which sets of Exercise 18 are proper subsets of other sets of the same exercise?

20 Use the relations between the moduli of Exercise 18 to explain the set relations as stated in Exercise 19.

21 Use two methods to express the results of each of these operations as nonnegative integers less than the modulus: *First*, carry out the operation with numbers as given and then reduce using the stated modulus; *second*, reduce each number, using the modulus, and then complete the operation, making a second reduction if necessary.

(*a*) 15 + 32 (mod 3)   (*b*) 28 + 96 (mod 2)
(*c*) 74 + 23 (mod 5)   (*d*) 63 + 91 (mod 6)
(*e*) 84 − 21 (mod 5)   (*f*) 112 − 56 (mod 2)
(*g*) 17 × 84 (mod 7)   (*h*) 37 × 49 (mod 9)
(*i*) 19 × (18 + 15) (mod 7)   (*j*) 19 + (18 + 32) (mod 5)
(*k*) 26 × 72 × 31 (mod 5)   (*l*) 32 × 89 × 71 (mod 4)
(*m*) 10^2 (mod 2)   (*n*) 10^3 (mod 2)
(*o*) $8(10^4)$ (mod 5)   (*p*) $7(10^5)$ (mod 5)
(*q*) $6(10^2)$ (mod 9)   (*r*) $5(10^7)$ (mod 3)
(*s*) $2(10^8)$ (mod 3)   (*t*) $7(10^6)$ (mod 9)

In Exercises 22 to 26 you are given that the associative property holds for each stated operation.

22 Does the set {0,1,2,3,4,5,6,7} with addition modulo 8 form a group?

23 Does the set {0,1,2,3,4,5,6,7} with multiplication modulo 8 form a group?

24 Does the set {0} with addition form a group?

*25 Does the set {0,1,2,3,4} with addition and multiplication modulo 5 form an integral domain? (See Sec. 4-3. Assume the distributive property.)

*26 Does the set {0,1,2,3} with addition and multiplication modulo 4 form an integral domain? (Assume the distributive property.)

*27 Why will a set of nonzero integers, with a composite number as modulus, not form a group?

28 Construct tests for divisibility by each of these numbers: (*a*) 32; (*b*) 33; (*c*) 36; (*d*) 44.

*29 Construct tests for divisibility by **2, 3,** and **6** in base **seven.**

*30 Construct tests for divisibility by **2** and **3** in base **four.**

*31 Construct tests for divisibility by **2, 4,** and **7** in base **eight.**

*INVITATIONS TO EXTENDED STUDY

1 For the set of integers {0,1,2,3,4,5,6,7,8,9,10}, construct the addition table for addition modulo 11.

2 For the set of integers {1,2,3,4,5,6,7,8,9,10}, construct the multiplication table for multiplication modulo 11.

3 For the set of integers {0,1,2,3,4,5,6,7,8,9,10,11}, construct the addition table for addition modulo 12.

4 For the set of integers {1,2,3,4,5,6,7,8,9,10,11}, construct the multiplication table for multiplication modulo 12.

5 Which of the systems of Exercises 1 to 4 are groups and which are not?

6 For the set of integers **{0,1,2,3,4}**, construct the addition and multiplication tables modulo **11** in base **five.**

7 For the set of integers **{1,2,3,4}**, construct the addition and multiplication tables modulo **12** in base **five.**

8 Which of the systems of Exercises 6 and 7 are groups and which are not?

9 Use the above exercises as a guide in arguing this theorem: Any set of integers with multiplication modulo m is not a group if m is a composite number.

10 Construct a test for divisibility by 7. ($Hint:$ $1 \equiv 1$, $2 \equiv 2$, $3 \equiv 3$, $4 \equiv -3$, $5 \equiv -2$, $6 \equiv -1$, all mod 7.)

For the remaining exercises, use the elements of the set G in Definition 5-1 to prove the theorems.

11 If $a,b,c \in G$ and $a * c = b * c$, then $a = b$.

12 If there exists any element u in G such that $u * a = a * u = a$ for $a \in G$, then $u = i$.

13 If for any given $a \in G$ there exists another element $x \in G$ such that $x * a = a * x = i$, then $x = a^{-1}$.

14 For any a in G it is true that $(a^{-1})^{-1} = a$.

THE FIELD OF
RATIONAL NUMBERS

GUIDELINES FOR CAREFUL STUDY

With the domain of integers the structure of the number system of elementary mathematics is only partially complete. Extensions of the foundation structure of the natural number system were made possible through the introduction of the concepts of additive identity and additive inverse. This led to the identification of the additive identity as the integer zero, the natural numbers as positive integers, and their additive inverses as negative integers. With the introduction of subtraction as the inverse operation to addition a complete facility of operational procedure was established with respect to addition and subtraction. Although the definition of division has meaning when applied either to natural numbers or to integers and provides greater latitude in the use of multiplication, neither system is closed under this new operation. It is the purpose of Chap. 6 to extend the domain of significance for both multiplication and division by introducing the concept of a multiplicative inverse. This enables us to enlarge the number system to include the field of rational numbers which is characterized by closure under each of the four operations: addition, subtraction, multiplication, and division.

The following questions will be very helpful as guidelines for the careful study of Chap. 6.

1   What is meant by a number system?

2   What are the basic postulates of the natural number system? Of the domain of integers? Of a group?

3   What are the basic definitions of addition, subtraction, multiplication, and division? Why is division by zero undefined?

4   Why are addition and subtraction said to be inverse operations?

5   Do you have sufficient command of the properties, concepts, and theorems of Chaps. 3 and 4 to be able to use them efficiently?

6   What is the division algorithm, and how does the notion of place value contribute to a better understanding of the algorithm and greater facility in its use?

7   Why are multiplication and division said to be inverse operations?

8   What is a rational number?

9   What is meant by the multiplicative inverse of a nonzero number?

10   When two rational numbers a and b, by any process of valid reasoning, are found to be so related that $a \times b = 1$, what can be said about the two numbers?

11   What are the definitions of equality, greater than, less than, addition, subtraction, and multiplication as applied to rational numbers?

12   Why are addition, subtraction, multiplication, and division referred to as rational operations or field operations?

13   Can you illustrate the fact that integers are neither associative nor commutative under subtraction?

14   Can you argue the truth of this statement: In the domain of integers multiplication is distributive over subtraction?

15 It is true that for $a,b,c \in Q$ and $c \neq 0$, division is distributive over both addition and subtraction in the form $(a + b) \div c$ but not in the form $c \div (a + b)$. Can you illustrate the truth of this statement?

16 What is a number field?

17 What are some of the basic distinctions in the use of a common fraction as the representation of a rational number and as a ratio?

18 Do you have a clear understanding of the terminology of both common fractions and decimal fractions?

19 How does the use of decimal fractions simplify the manipulation of fractions?

20 Why can the set of all decimal fractions be considered as a proper subset of all fractions? What are the implications of this fact?

21 Can you analyze any common fraction and tell into what kind of decimal fraction it will convert?

22 Can you convert any repeating infinite decimal or any finite decimal into its equivalent common fraction in its lowest terms?

23 Can you operate efficiently using common and decimal fractions?

24 Can you use scientific notation when computing with very large or very small numbers?

25 Do you understand the basic principles of rounding numbers?

26 Why can the set of all percents be considered as a proper subset of the set of all decimal fractions? What implications does this relation have toward the study of the topic of percentage?

27 Can you use the percentage formula intelligently as an aid in the analysis and solution of percentage problems?

28 What are some of the precautions that must be observed in using percents?

INTRODUCTION

The domain of integers provides for much less restricted operational procedures than does the system of natural numbers. In spite of this fact there still remain limitations on the computational efficiency desirable for an adequate number system. In the present chapter a more liberal interpretation of the closure property for multiplication will lead to a number system which will provide for the removal of some of these remaining limitations on computation.

6-1 THE CONCEPT OF RATIONAL NUMBER

In order to have more freedom of interpretation and use both for multiplication and for division, it becomes necessary to make a further extension of our number system. This extension is to be made not only so that the set of all integers will be a subset of the extended set of numbers but also so that in the new system we shall have closure not only under multiplication and division but also under addition and subtraction. This, of course, means that the definitions of the four operations will have to be extended to apply to the new numbers. Furthermore, these extensions will have to be such that, when the operations are restricted to the subset of integers or even to the subset of natural numbers, the new definitions will conform to those already established.

We shall postulate the existence of these new numbers and then proceed to investigate the significant characteristics of the new number system to which they lead.

Postulate 6-1   For any two integers a and b ($b \neq 0$) there exists a number q such that $q \times b = a$. The number q is called a *rational number*.

We may now combine the concepts of Definition 4-2 and Postulate 6-1 to get a clearer idea of just what is meant by a rational number. In the definition q is identified as the quotient of $a \div b$. Another accepted, and very useful, form for expressing this concept is the fraction $\frac{a}{b}$, or a/b. Postulate 6-1 restricts the existence of such quotients to only those cases in which $b \neq 0$. Why is this restriction necessary? This fraction is recognizable as the means of expressing a quotient between the integers a and b. Thus we might formulate the explicit definition of a rational number as follows:

Rational number   **Definition 6-1**   A *rational number* is a number which can be expressed in the form of the quotient of two integers $\frac{a}{b}$ where $b \neq 0$.

Note that the definition does not state that the number *must* be, but only that it *can* be, expressed as the quotient of two integers. A specific illustration is the fact that the integers are rational numbers since each integer can be expressed in the form of the quotient of two integers. For example,

$$3 = \frac{3}{1} \qquad -5 = \frac{-5}{1} \qquad 1{,}000 = \frac{1{,}000}{1} \qquad -600 = \frac{-600}{1}$$

Thus the set of integers and the set of natural numbers are proper subsets of the set of rational numbers. This statement will become even more significant later. When the different operations have been defined for rational numbers, it will be seen that in any calculation using these processes, the integer a and the rational number $a/1$ may be used interchangeably, causing no effect on the results of the computation.

*Integers a
proper subset
of the rational
numbers*

It has been established that the set of natural numbers is a proper subset of the set of integers, which is a proper subset of the set of rational numbers. From this it follows that whatever might be the definitions of addition, subtraction, and multiplication for the set of rational numbers, they must be such that they conform to the definitions and properties already established for integers (D-1 to D-8, pages 126 and 127) and natural numbers (N-1 to N-8, page 97). The properties of equality (E-1 to E-5, page 83) also apply. As an immediate consequence we have that all postulates previously assumed and all theorems previously proved may be restated for rational numbers except in those cases where the intrinsic nature of any hypothesis is such that it restricts the conclusions either to integers or to natural numbers. An illustration of such a situation is the postulate of finite induction (N-8). The nature of this postulate is such that it relates *only* to natural numbers. Thus, in the domain of integers or the field of rational numbers, it applies only to the subset of positive integers. Space limitations prevent formal restatement of these postulates and theorems in the specific context of rational numbers. They will be used, however, in appropriate places as established authority for deductions made, leaving to the reader the substitution of "rational numbers" for "natural numbers" or "integers" as the case may be.

Equality　**Definition 6-2**　Two rational numbers $\frac{a}{b}$ and $\frac{c}{d}$ are said to be equal if and only if $ad = bc$.

Example　The rational numbers $\frac{3}{4}$ and $\frac{15}{20}$ are equal since $3 \times 20 = 60$ and $4 \times 15 = 60$.

The rational numbers $\frac{3}{4}$ and $\frac{5}{7}$ are not equal, since $3 \times 7 = 21$ and $4 \times 5 = 20$, and $21 \neq 20$.

The equality of rational numbers involves a concept of equality which is basically different from that used in dealing with natural numbers and integers. For two natural numbers or two integers, $a = b$ carries the meaning "*a* is the same as *b*." For two rational numbers this is not the case. For example, $\frac{3}{4}$ and $\frac{15}{20}$ are two distinct rational numbers. Thus $\frac{3}{4} = \frac{15}{20}$ cannot be read "$\frac{3}{4}$ is the same as $\frac{15}{20}$," implying an identity of symbols as in the use of integers. However, Definition 6-2 states that the two rational numbers are equal since the integer 3×20 is the same as the integer 4×15. Thus, for rational numbers, equality carries the meaning of *equal in value*, or *equivalence*, and not that of identity of symbols.

*Identity
relation*

*Equivalence
classes*

The relation of equality which exists between natural numbers or between integers is sometimes called an *identity relation*. While the equality of rational numbers is not an identity relation it can be shown to be an equivalence relation. (See Exercises 8 to 10 at the end of this section.) Under this concept of equality rational numbers may be separated into *equivalence classes*, that is, sets of rational numbers which are equal in value, or which satisfy the equality relation of Definition 6-2, such as the infinite sets

$$\left\{ \cdots, \frac{-5}{-10}, \frac{-7}{-14}, \frac{1}{2}, \frac{2}{4}, \frac{3}{6}, \frac{4}{8}, \cdots \right\}$$

$$\left\{ \cdots, \frac{-18}{-24}, \frac{-15}{-20}, \frac{3}{4}, \frac{6}{8}, \frac{9}{12}, \frac{12}{16}, \frac{15}{20}, \cdots \right\}$$

Thus $\frac{3}{4}$ and $\frac{15}{20}$ belong to the same equivalence class. Similarly, any other rational number of this set will meet the condition of Definition 6-2 when compared either with $\frac{3}{4}$ or $\frac{15}{20}$.

Since the elements of any equivalence class may be used interchangeably in any computation, the element of lowest terms is most generally used as a symbol to represent the class. For example,

$$\frac{1}{2} = \left\{ \cdots, \frac{-5}{-10}, \frac{-7}{-14}, \frac{1}{2}, \frac{2}{4}, \frac{3}{6}, \frac{4}{8}, \cdots \right\}$$

$$\frac{3}{4} = \left\{ \cdots, \frac{-18}{-24}, \frac{-15}{-20}, \frac{3}{4}, \frac{6}{8}, \frac{9}{12}, \frac{12}{16}, \frac{15}{20}, \cdots \right\}$$

Any other element of the class could be used just as well as an identifying symbol. Thus a rational number may be considered as defined by its equivalence class. In this sense equality of rational numbers can be interpreted as an identity relation. $\frac{3}{4} = \frac{15}{20}$ may be read "$\frac{3}{4}$ is the same as $\frac{15}{20}$," meaning that the equivalence class represented by $\frac{3}{4}$ is the same as the equivalence class represented by $\frac{15}{20}$. For this reason no interpretation difficulties result from the use of the same symbol for equality whether we are dealing with natural numbers, integers, or rational numbers.

EXERCISES

1 Which pairs of rational numbers are equivalent?

(a) $\frac{1}{7}$ and $\frac{3}{21}$ (b) $\frac{2}{2}$ and $\frac{1}{1}$ (c) $\frac{-6}{5}$ and $\frac{9}{-5}$

(d) $\frac{3}{2}$ and $\frac{18}{12}$ (e) $\frac{0}{6}$ and $\frac{0}{1}$ (f) $\frac{-8}{3}$ and $\frac{8}{-3}$

(g) $\frac{0}{3}$ and $\frac{1}{3}$ (h) $\frac{0}{-3}$ and $\frac{0}{7}$ (i) $\frac{-6}{-7}$ and $\frac{6}{7}$

(j) $\frac{-3}{5}$ and $\frac{-9}{15}$ (k) $\frac{-12}{-15}$ and $\frac{4}{5}$ (l) $\frac{-5}{-5}$ and $\frac{1}{1}$

(m) $\frac{9}{2}$ and $\frac{10 \times 9}{10 \times 2}$ (n) $\frac{4}{13}$ and $\frac{-5 \times 4}{-5 \times 13}$ (o) $\frac{5}{7}$ and $\frac{4 \times 5}{5 \times 7}$

(p) $\frac{-3}{11}$ and $\frac{6 \times {}^-3}{6 \times 11}$ (q) $\frac{15}{14}$ and $\frac{6 \times 15}{14 \times 6}$ (r) $\frac{7}{3}$ and $\frac{7}{3} \times \frac{5}{6}$

(s) $\frac{10}{12}$ and $\frac{5}{6}$ (t) $\frac{9}{6}$ and $\frac{3}{4}$ (u) $\frac{8}{9}$ and $\frac{15}{15} \times \frac{8}{9}$

2 Show that $\dfrac{-8}{-16}$ is an element of the equivalence class $\dfrac{1}{2}$.

3 Show that $\dfrac{-3}{-4}$ is an element of the equivalence class $\dfrac{3}{4}$.

4 (a) Show that neither $\dfrac{3}{-4}$ nor $\dfrac{-3}{4}$ is an element of the equivalence class $\dfrac{3}{4}$.

   (b) Are $\dfrac{3}{-4}$ and $\dfrac{-3}{4}$ members of the same equivalence class?

5 (a) Select 10 more elements of the equivalence class $\frac{3}{4}$.
   (b) Select 10 more elements of the equivalence class $\frac{1}{2}$.

6 Select 15 elements for each equivalence class represented by these symbols.

   (a) $\dfrac{2}{3}$ (b) $\dfrac{5}{8}$ (c) $\dfrac{7}{2}$ (d) $\dfrac{10}{12}$ (e) $\dfrac{8}{1}$

   (f) $\dfrac{0}{2}$ (g) $\dfrac{27}{9}$ (h) $\dfrac{-4}{5}$ (i) $\dfrac{0}{-7}$ (j) $\dfrac{-3}{1}$

7 Use the rational numbers $\frac{1}{3}$, $\frac{3}{9}$, and $\frac{6}{18}$ to illustrate that the relation specified in Definition 6-2 is an equivalence relation.

* 8 Prove that the relation given in Definition 6-2 is an equivalence relation.

* 9 Prove that $\dfrac{-a}{-b}$ and $\dfrac{a}{b}$, for $b \neq 0$, belong to the same equivalence class. In particular, for $a > 0$ and $b > 0$, prove that $\dfrac{-a}{-b}$ and $\dfrac{a}{b}$ belong to the same equivalence class.

*10 Prove that $\dfrac{-a}{b}$ and $\dfrac{a}{-b}$, for $b \neq 0$, belong to the same equivalence class. In particular, for $a > 0$ and $b > 0$, prove that $\dfrac{-a}{b}$ and $\dfrac{a}{-b}$ belong to the same equivalence class.

Positive and negative rational numbers

If a,b are positive integers, rational numbers of the form $\dfrac{a}{b}$ or $\dfrac{-a}{-b}$ are defined to be *positive*, and those of the form $\dfrac{-a}{b}$ or $\dfrac{a}{-b}$ are defined to be *negative*. Thus, rational numbers, like integers, are partitioned into the three disjoint sets of positive numbers, negative numbers, and zero. Furthermore, since $\dfrac{a}{b} = \dfrac{-a}{-b}$ and $\dfrac{-a}{b} = \dfrac{a}{-b}$, it is the convention with rational numbers to consider the divisor as positive.

6-2 USING RATIONAL NUMBERS

As in the case of natural numbers and integers, addition and multiplication are the two basic operations to be defined. Subtraction and division then will be defined as the respective inversely related operations.

Addition   **Definition 6-3**   The sum of the rational numbers $\dfrac{a}{b}$ and $\dfrac{c}{d}$ is $\dfrac{ad + bc}{bd}$.

$\left(\dfrac{a}{b} + \dfrac{c}{d} = \dfrac{ad + bc}{bd} \right)$.

This definition guarantees uniqueness and closure since $ad + bc$ and bd are integers. Why? Also $bd \neq 0$. Why?

Addition is commutative. By definition, $\dfrac{c}{d} + \dfrac{a}{b} = \dfrac{cb + da}{db}$. Since a, b, c, and d are integers, it follows that $cb + da = bc + ad = ad + bc$, and $db = bd$. Why? Therefore

$$\frac{c}{d} + \frac{a}{b} = \frac{a}{b} + \frac{c}{d} \qquad \text{Why?}$$

While it is a bit longer, it is not a very involved argument to show that the associative property holds under this definition of addition. (See the exercises at the end of this section.)

There are occasions when the sum of two rational numbers is found erroneously by the incorrect formula $\dfrac{a}{b} + \dfrac{c}{d} = \dfrac{a + c}{b + d}$. It is interesting to note that even with this incorrect definition, addition of rational numbers would have the properties of closure, associativity, and commutativity. The definition, however, does not meet two other very important restrictions on the definitions of the operations to be applied to any set of numbers.

1   *The definition of an operation must result in a well-defined operation.* As an illustration that this incorrect definition does not meet this condition, consider the two sums $\frac{2}{3} + \frac{5}{7}$ and $\frac{4}{6} + \frac{5}{7}$. Since $\frac{2}{3} = \frac{4}{6}$ it follows that, if the operation of addition is well defined, the two sums must be rational numbers from the same equivalence class. This is the case when Definition 6-3 is used.

$$\frac{2}{3} + \frac{5}{7} = \frac{2 \cdot 7 + 3 \cdot 5}{3 \cdot 7} = \frac{14 + 15}{21} = \frac{29}{21}$$

$$\frac{4}{6} + \frac{5}{7} = \frac{4 \cdot 7 + 6 \cdot 5}{6 \cdot 7} = \frac{28 + 30}{42} = \frac{58}{42}$$

and $\frac{29}{21} = \frac{58}{42}$ since $29 \times 42 = 21 \times 58$. However, by the incorrect definition we get

$$\frac{2}{3} + \frac{5}{7} = \frac{2 + 5}{3 + 7} = \frac{7}{10} \quad \text{and} \quad \frac{4}{6} + \frac{5}{7} = \frac{4 + 5}{6 + 7} = \frac{9}{13} \qquad \frac{7}{10} \neq \frac{9}{13}$$

since $7 \times 13 \neq 10 \times 9$.

2   *The definition of any given operation must not introduce inconsistencies in working with numbers.* The correct definition of addition applied to the rational numbers $\dfrac{a}{1} + \dfrac{c}{1}$ gives for the sum $\dfrac{a + c}{1}$. This is in agreement with the result expected for the sum of the integers a and c.

The incorrect definition being discussed gives the sum $\frac{a}{1} + \frac{c}{1} = \frac{a+c}{1+1} = \frac{a+c}{2}$, thus introducing an inconsistency which cannot be tolerated.

Multiplication   **Definition 6-4**   The product of the rational numbers $\frac{a}{b}$ and $\frac{c}{d}$ is $\frac{ac}{bd}$. $\left(\frac{a}{b} \times \frac{c}{d} = \frac{ac}{bd}\right)$.

From this definition it is not too difficult to prove that rational numbers have the properties of uniqueness, closure, associativity, and commutativity under multiplication, and also that multiplication is distributive over addition. (See the exercises at the end of this section.) Furthermore, $\frac{a}{1} \times \frac{c}{1} = \frac{a \times c}{1}$. This result shows that the one-to-one correspondence $a \leftrightarrow a/1$, where $a \in Z$, is preserved under multiplication. This fact and the corresponding one for addition complete the statement of the

$Z \subset Q$ previous section that the set Z of integers is a proper subset of the set Q where

$$Q = \left\{ q \mid q = \frac{a}{b}, \ a,b \in Z, \ b \neq 0 \right\}$$

In other words, in any computation involving integers and rational numbers any integer a may be replaced by any element of the equivalence class $a/1$.

From Definition 6-4 it follows that $\frac{a}{b} \times \frac{1}{1} = \frac{a \times 1}{b \times 1} = \frac{a}{b}$. Therefore,

Multiplicative identity $\frac{1}{1}$ is the *multiplicative identity* for rational numbers. It is usually symbolized by 1. Also, since for any nonzero integer d, $\frac{d}{d} = \frac{1}{1}$, it follows that any such rational number belongs to the equivalence class $\frac{1}{1}$ and may be used as a replacement for the multiplicative identity in any computation. Also

$$\frac{a}{b} + \frac{0}{1} = \frac{(a \times 1) + (b \times 0)}{b \times 1} = \frac{a+0}{b} = \frac{a}{b}$$

and

$$\frac{a}{b} + \frac{0}{d} = \frac{(a \times d) + (b \times 0)}{b \times d} = \frac{ad+0}{bd} = \frac{ad}{bd} = \frac{a}{b} \times \frac{d}{d} = \frac{a}{b}$$

Since $\frac{0}{1}$ and $\frac{0}{d}$ belong to the same equivalence class they may be used inter-

Additive identity   changeably as the *additive identity* for rational numbers. It is usually symbolized by 0. Note that the two identities are distinct elements. ($1 \neq 0$.)

Subtraction   **Definition 6-5**   *Subtraction* of two rational numbers is the inverse operation to the addition of two rational numbers. $\frac{a}{b} - \frac{c}{d} = \frac{e}{f}$ if and only if $\frac{a}{b} = \frac{c}{d} + \frac{e}{f}$.

Additive
inverse
By analogy with integers, $^-\left(\dfrac{c}{d}\right)$ is the symbol for the *additive inverse* of the rational number $\dfrac{c}{d}$. Using Definition 6-3 and the properties of integers, we have

$$\frac{c}{d} + \frac{^-c}{d} = \frac{(c \times d) + (d \times\, ^-c)}{d \times d}$$

$$= \frac{(c \times d) +\, ^-(c \times d)}{d \times d}$$

$$= \frac{0}{d \times d} = 0$$

Thus $^-\left(\dfrac{c}{d}\right) = \dfrac{^-c}{d}$.

It can be proved, just as in the case of integers (Theorem 4-3), that $\dfrac{a}{b} - \dfrac{c}{d} = \dfrac{a}{b} +\, ^-\left(\dfrac{c}{d}\right)$, or, better for computation purposes, $\dfrac{a}{b} - \dfrac{c}{d} = \dfrac{a}{b} + \dfrac{^-c}{d}$. From this it follows that the difference obtained by subtracting $\dfrac{c}{d}$ from $\dfrac{a}{b}$ may be symbolized by $\dfrac{a}{b} - \dfrac{c}{d} = \dfrac{ad - bc}{bd}$.

Example Find these two differences: (a) $\dfrac{17}{6} - \dfrac{5}{4}$; (b) $\dfrac{3}{4} - \dfrac{-2}{3}$.

(a)

$$\frac{17}{6} - \frac{5}{4} = \frac{(17 \times 4) - (6 \times 5)}{6 \times 4}$$

$$= \frac{68 - 30}{24}$$

$$= \frac{38}{24} = \frac{19 \times 2}{12 \times 2}$$

$$= \frac{19}{12}$$

(b)

$$\frac{3}{4} - \frac{-2}{3} = \frac{(3 \times 3) - [4 \times (-2)]}{4 \times 3}$$

$$= \frac{9 - (-8)}{4 \times 3}$$

$$= \frac{9 +\, ^-(-8)}{12}$$

$$= \frac{9 + 8}{12}$$

$$= \frac{17}{12}$$

We are now in a position to examine a bit more carefully the process of division and the rational number q, presented in Definition 6-1 and

Postulate 6-1. From the postulate there exists a rational number q such that $q \times b = a$ where b and a are integers with $b \neq 0$. From the definition this number is the quotient of $a \div b$. Also recall that another symbol introduced to represent this quotient is the fraction a/b. We shall use $a \div b$ and a/b interchangeably. There are two distinct cases to consider.

Case I $a = b$ Under this hypothesis $q = b/b$, or the multiplicative identity for rational numbers. Furthermore, under this hypothesis $q \times b = a$ becomes $q \times b = b$, and by D-7 (page 127) $q = 1$, the multiplicative identity for integers. Note that this identifies the integer 1 as the rational number b/b, where b is any nonzero integer. This further supports the use of 1 as the symbol for the multiplicative identity for rational numbers.

Case II $a \neq b$ Here there are two hypotheses on a which we must examine.

Hypothesis 1 $a = 1$ Under this condition $q \times b = a$ becomes $q \times b = 1$. Such familiar considerations as "two halves make a whole" ($2 \times \frac{1}{2} = 1$), "three thirds make a whole" ($3 \times \frac{1}{3} = 1$), and "four fourths make a whole" ($4 \times \frac{1}{4} = 1$) furnish a strong intuitive background for the acceptance of the following special version of Postulate 6-1.

Multiplicative inverse

Postulate 6-2 For every rational number $b \neq 0$ there exists a rational number $\frac{1}{b}$ such that $\frac{1}{b} \times b = b \times \frac{1}{b} = 1$. The number $\frac{1}{b}$ is called the *reciprocal* of b, or the *multiplicative inverse* of b. Another symbol frequently used instead of $\frac{1}{b}$ is b^{-1}.

Notice, incidentally, that in this case $q = 1/b$ is indeed the quotient of $1 \div b$, as stated in Definition 6-1, since $q \times b = 1/b \times b = 1$. Furthermore, $1/b \neq 0$. For, if $1/b = 0$ then $b \times 1/b = b \times 0 = 0$ by Theorem 4-5. This is impossible since $b \times 1/b = 1$ by Postulate 6-2.

Hypothesis 2 $a \neq 1$ Under this condition we can appeal to the properties of multiplication to assist us in getting an interpretation of $q = a \div b$.

Division

Theorem 6-1 If a and b are two integers such that $b \neq 0$, then $a \times \frac{1}{b} = a \div b$.

Hypothesis a, b are integers; $b \neq 0$.

Conclusion $a \times 1/b = a \div b$

Since a and b are integers and $b \neq 0$, we know that a, b, $1/b$, and a/b are rational numbers. This means that we can use the properties of integers or rational numbers in dealing with a and b and the properties of rational numbers with a/b and $1/b$.

We are to prove that the product of the integer a by the multiplicative inverse of the nonzero integer b is the rational number a/b, or the quotient of $a \div b$. From Definition 6-1 we know that any number q can be the quotient of $a \div b$ if and only if $q \times b = a$. This means that if we are to be able to prove that $a \times 1/b$ is the quotient $a \div b$, we must be able to establish that $(a \times 1/b) \times b = a$.

The development of the proof will follow the parallel-column pattern used in previous theorems with the special case on the left and the general case on the right. For this special case we shall use the integers 5 and 3 and seek to show that $5 \times \frac{1}{3} = 5 \div 3$.

Proof

Statements	*Reasons*	*Statements*
The rational number $\frac{1}{3}$ exists.	Postulate 6-2	The rational number $\frac{1}{b}$ exists.
$5 \times \frac{1}{3} = \frac{5}{1} \times \frac{1}{3} \in Q$	Closure	$a \times \frac{1}{b} = \frac{a}{1} \times \frac{1}{b} \in Q$
$\left(5 \times \frac{1}{3}\right) \times 3 =$ $5 \times \left(\frac{1}{3} \times 3\right)$	Associative property	$\left(a \times \frac{1}{b}\right) \times b =$ $a \times \left(\frac{1}{b} \times b\right)$
$\left(5 \times \frac{1}{3}\right) \times 3 = 5 \times 1$	Multiplicative inverse and substitution	$\left(a \times \frac{1}{b}\right) \times b = a \times 1$
$\left(5 \times \frac{1}{3}\right) \times 3 = 5$	Multiplicative identity	$\left(a \times \frac{1}{b}\right) \times b = a$
Since $5 \times \frac{1}{3}$ is a rational number q such that $q \times 3 = 5$, it follows that $q =$ $5 \div 3$, or $5 \times \frac{1}{3} =$ $5 \div 3$.	Definition 6-1	Since $a \times \frac{1}{b}$ is a rational number q such that $q \times b = a$, it follows that $q =$ $a \div b$, or $a \times \frac{1}{b} =$ $a \div b$.

Corollary If a, b, and c are any three integers such that $c \neq 0$, then $(a + b) \div c = (a \div c) + (b \div c)$.

Theorem 6-2 is proved as an important illustration of the possibility of extending to rational numbers theorems previously proved for integers. It is to be noted that the only essential difference in the statements and proofs of Theorems 6-1 and 6-2 is the replacement of "integers" by "rational numbers."

Theorem 6-2 If r and s are any two rational numbers such that $s \neq 0$, then $r \times \dfrac{1}{s} = r \div s$.

Hypothesis r and s are rational numbers; $s \neq 0$.

Conclusion $r \times 1/s = r \div s$

Proof

Statements	*Reasons*
The rational number $\dfrac{1}{s}$ exists.	Postulate 6-2
$r \times \dfrac{1}{s}$ is a rational number.	Closure
$\left(r \times \dfrac{1}{s}\right) \times s = r \times \left(\dfrac{1}{s} \times s\right)$	Associative property
$\left(r \times \dfrac{1}{s}\right) \times s = r \times 1$	Multiplicative inverse and substitution
$\left(r \times \dfrac{1}{s}\right) \times s = r$	Multiplicative identity
Since $\left(r \times \dfrac{1}{s}\right)$ is a rational number q	Definition 6-1

such that $q \times s = r$, it follows that

$r \times \dfrac{1}{s} = r \div s$.

These two theorems are indeed the authority for the familiar rule of elementary arithmetic: In division you invert the divisor and multiply.

Just as subtraction was defined as the inverse operation of addition, now division is defined as the inverse operation of multiplication. If a/b and c/d are any two rational numbers such that $c/d \neq 0$, then

$$\left(\frac{a}{b} \times \frac{c}{d}\right) \div \frac{c}{d} = \frac{a}{b} \quad \text{and} \quad \left(\frac{a}{b} \div \frac{c}{d}\right) \times \frac{c}{d} = \frac{a}{b}$$

The proof of these facts is left as an exercise (see Exercise 26, page 206). How do the facts presented here emphasize the fact that the multiplication tables you constructed in previous chapters can be used as tables giving the basic division facts?

Example

(a) Find this quotient: $\dfrac{5}{9} \div 3$.

$$\frac{5}{9} \div 3 = \frac{5}{9} \div \frac{3}{1} = \frac{5}{9} \times \frac{1}{3} = \frac{5 \times 1}{9 \times 3} = \frac{5}{27}$$

(b) Find this quotient: $\dfrac{8}{17} \div \dfrac{4}{3}$.

$$\frac{8}{17} \div \frac{4}{3} = \frac{8}{17} \times \frac{3}{4} = \frac{8 \times 3}{17 \times 4}$$

$$= \frac{(2 \times 4) \times 3}{17 \times 4}$$

$$= \frac{(2 \times 3) \times 4}{17 \times 4}$$

$$= \frac{6}{17}$$

(c) Use two methods to find the quotient $\left(\dfrac{7}{8} + \dfrac{5}{3}\right) \div \dfrac{5}{12}$.

(1)
$$\frac{7}{8} + \frac{5}{3} = \frac{(7 \times 3) + (8 \times 5)}{8 \times 3}$$

$$= \frac{21 + 40}{24}$$

$$= \frac{61}{24}$$

$$\left(\frac{7}{8} + \frac{5}{3}\right) \div \frac{5}{12} = \frac{61}{24} \div \frac{5}{12}$$

$$= \frac{61}{24} \times \frac{12}{5}$$

$$= \frac{61 \times 12}{(2 \times 12) \times 5} = \frac{61 \times 12}{(2 \times 5) \times 12}$$

$$= \frac{61}{10}$$

(2)
$$\left(\frac{7}{8} + \frac{5}{3}\right) \div \frac{5}{12} = \left(\frac{7}{8} \div \frac{5}{12}\right) + \left(\frac{5}{3} \div \frac{5}{12}\right)$$

$$= \left(\frac{7}{8} \times \frac{12}{5}\right) + \left(\frac{5}{3} \times \frac{12}{5}\right)$$

$$= \frac{7 \times (3 \times 4)}{(2 \times 4) \times 5} + \frac{5 \times (4 \times 3)}{3 \times 5}$$

$$= \frac{(7 \times 3) \times 4}{(2 \times 5) \times 4} + \frac{4 \times (3 \times 5)}{1 \times (3 \times 5)}$$

$$= \frac{21}{10} + \frac{4}{1}$$

$$= \frac{(21 \times 1) + (10 \times 4)}{10 \times 1}$$

$$= \frac{21 + 40}{10}$$

$$= \frac{61}{10}$$

EXERCISES

1 Select 15 elements of each equivalence class: (a) $\frac{0}{1}$; (b) $\frac{1}{1}$.

2 Prove that the additive and multiplicative identities for rational numbers are distinct.

3 Find these sums.

(a) $\frac{2}{3} + \frac{5}{7}$ (b) $\frac{1}{2} + \frac{1}{8}$ (c) $\frac{9}{11} + \frac{5}{6}$

(d) $\frac{15}{4} + \frac{8}{7}$ (e) $\frac{5}{2} + \frac{9}{5}$ (f) $\frac{11}{4} + \frac{7}{9}$

(g) $\frac{5}{6} + \frac{-3}{4}$ (h) $\frac{-7}{9} + \frac{-3}{5}$ (i) $\frac{-11}{3} + \frac{7}{2}$

4 Find these differences.

(a) $\frac{8}{9} - \frac{2}{5}$ (b) $\frac{7}{12} - \frac{8}{11}$ (c) $\frac{-15}{4} - \frac{7}{12}$

(d) $\frac{21}{9} - \frac{8}{5}$ (e) $\frac{4}{15} - \frac{-3}{8}$ (f) $\frac{-7}{9} - \frac{-2}{5}$

5 Show that these sums are equal.

(a) $\left(\frac{5}{6} + \frac{2}{3}\right) + \frac{1}{7} = \frac{5}{6} + \left(\frac{2}{3} + \frac{1}{7}\right)$

(b) $\left(\frac{7}{8} + \frac{-5}{4}\right) + \frac{3}{2} = \frac{7}{8} + \left(\frac{-5}{4} + \frac{3}{2}\right)$

(c) $\left(\frac{9}{11} + \frac{-8}{15}\right) + \frac{-8}{3} = \frac{9}{11} + \left(\frac{-8}{15} + \frac{-8}{3}\right)$

6   What property of rational numbers is illustrated by the examples of Exercise 5?

7   (a) Is this a true equality?

$$\left(\frac{8}{9} - \frac{3}{2}\right) - \frac{5}{6} = \frac{8}{9} - \left(\frac{3}{2} - \frac{5}{6}\right)$$

   (b) What does this example prove about rational numbers under subtraction?

8   Find these products.

   (a) $\frac{3}{5} \times \frac{7}{8}$ (b) $\frac{5}{11} \times \frac{6}{6}$ (c) $\frac{7}{2} \times \frac{3}{14}$

   (d) $\frac{-2}{3} \times \frac{8}{11}$ (e) $\frac{15}{4} \times \frac{-3}{7}$ (f) $\frac{-6}{7} \times \frac{-4}{5}$

   (g) $\frac{18}{5} \times \frac{-7}{7}$ (h) $\frac{-9}{4} \times \frac{-5}{5}$ (i) $\frac{14}{3} \times \frac{9}{2}$

9   Show that these products are equal.

   (a) $\left(\frac{3}{5} \times \frac{4}{7}\right) \times \frac{8}{11} = \frac{3}{5} \times \left(\frac{4}{7} \times \frac{8}{11}\right)$

   (b) $\left(\frac{7}{3} \times \frac{11}{4}\right) \times \frac{2}{5} = \frac{7}{3} \times \left(\frac{11}{4} \times \frac{2}{5}\right)$

10   What property of rational numbers is illustrated by the examples of Exercise 9?

11   Find these quotients.

   (a) $\frac{8}{9} \div 4$ (b) $\frac{7}{3} \div 5$ (c) $\frac{6}{9} \div 2$

   (d) $\frac{7}{8} \div \frac{2}{3}$ (e) $\frac{18}{5} \div \frac{9}{5}$ (f) $\frac{-7}{3} \div \frac{2}{3}$

   (g) $\frac{18}{35} \div \frac{-9}{7}$ (h) $\frac{7}{13} \div \frac{6}{6}$ (i) $\frac{-6}{5} \div \frac{7}{7}$

12   (a) Is this a true equality?

$$\left(\frac{8}{15} \div \frac{2}{3}\right) \div \frac{4}{5} = \frac{8}{15} \div \left(\frac{2}{3} \div \frac{4}{5}\right)$$

   (b) What does this example prove about rational numbers and division?

13   Show that these are true equations.

   (a) $\frac{2}{3} \times \left(\frac{3}{4} + \frac{5}{7}\right) = \left(\frac{2}{3} \times \frac{3}{4}\right) + \left(\frac{2}{3} \times \frac{5}{7}\right)$

   (b) $\frac{-3}{7} \times \left(\frac{8}{5} + \frac{9}{11}\right) = \left(\frac{-3}{7} \times \frac{8}{5}\right) + \left(\frac{-3}{7} \times \frac{9}{11}\right)$

   (c) $\left(\frac{7}{2} + \frac{3}{5}\right) \times \frac{6}{11} = \left(\frac{7}{2} \times \frac{6}{11}\right) + \left(\frac{3}{5} \times \frac{6}{11}\right)$

   (d) $\left(\frac{8}{5} - \frac{2}{3}\right) \times \frac{7}{4} = \left(\frac{8}{5} \times \frac{7}{4}\right) - \left(\frac{2}{3} \times \frac{7}{4}\right)$

(e) $\dfrac{11}{2} \times \left(\dfrac{7}{9} - \dfrac{5}{4}\right) = \left(\dfrac{11}{2} \times \dfrac{7}{9}\right) - \left(\dfrac{11}{2} \times \dfrac{5}{4}\right)$

14 What properties of rational numbers are illustrated by the examples of Exercise 13?

15 Show that these are true equalities.

(a) $\left(\dfrac{5}{9} + \dfrac{3}{4}\right) \div \dfrac{2}{3} = \left(\dfrac{5}{9} \div \dfrac{2}{3}\right) + \left(\dfrac{3}{4} \div \dfrac{2}{3}\right)$

(b) $\left(\dfrac{-7}{8} + \dfrac{11}{2}\right) \div \dfrac{-3}{5} = \left(\dfrac{-7}{8} \div \dfrac{-3}{5}\right) + \left(\dfrac{11}{2} \div \dfrac{-3}{5}\right)$

(c) $\left(\dfrac{12}{7} - \dfrac{5}{3}\right) \div \dfrac{-4}{11} = \left(\dfrac{12}{7} \div \dfrac{-4}{11}\right) - \left(\dfrac{5}{3} \div \dfrac{-4}{11}\right)$

16 What properties of rational numbers are illustrated by the examples of Exercise 15?

17 Prove that for any two nonzero integers, the product of their respective multiplicative inverses is equal to the multiplicative inverse of their product.

In Exercises 18 to 21 construct your own examples to illustrate each stated property of rational numbers.

18 Rational numbers are closed under addition, subtraction, multiplication, and division.

19 Rational numbers are associative under addition and multiplication.

20 Rational numbers are commutative under addition and multiplication.

21 For rational numbers multiplication is distributive over both addition and subtraction.

22 Prove that if $a \neq 0$ then $\dfrac{b}{a}$ is the multiplicative inverse of the rational number $\dfrac{a}{b}$.

23 Prove that the set Q of rational numbers, with addition and multiplication as defined, is an integral domain.

24 If r is a rational number, then r^2 and r^3 are rational numbers.

25 Use $\tfrac{2}{3}$ and $\tfrac{5}{7}$ to illustrate the two statements

$$\left(\dfrac{a}{b} \times \dfrac{c}{d}\right) \div \dfrac{c}{d} = \dfrac{a}{b} \quad \text{and} \quad \left(\dfrac{a}{b} \div \dfrac{c}{d}\right) \times \dfrac{c}{d} = \dfrac{a}{b}$$

26 Prove the general statement: If $\dfrac{a}{b}$ and $\dfrac{c}{d}$ are any two rational numbers with $\dfrac{c}{d} \neq 0$, then

$$\left(\dfrac{a}{b} \times \dfrac{c}{d}\right) \div \dfrac{c}{d} = \dfrac{a}{b} \quad \text{and} \quad \left(\dfrac{a}{b} \div \dfrac{c}{d}\right) \times \dfrac{c}{d} = \dfrac{a}{b}$$

27 Given q, r positive integers with $q > r$. Which of these three statements is the true statement (a) if $p > 0$; (b) if $p < 0$?

$$\dfrac{p}{q} = \dfrac{p}{r} \qquad \dfrac{p}{q} > \dfrac{p}{r} \qquad \dfrac{p}{q} < \dfrac{p}{r}$$

*28–31   For $\dfrac{a}{b}, \dfrac{c}{d}, \dfrac{e}{f} \in Q$ prove the statements of Exercises 18 to 21.

In Exercises 32 to 35 write the roster symbol for the set S.

*32 $S = \left\{ \dfrac{a}{b} \,\middle|\, a,b \in Z, \; a \geq -4, \; b > 0, \; a + b = 6 \right\}$

*33 $S = \left\{ \dfrac{a}{b} \,\middle|\, a,b \in Z, \; 0 < b < 10, \; a \times b = 0 \right\}$

*34 $S = \left\{ \dfrac{a}{b} \,\middle|\, a,b \in Z, \; b > 0, \; a > 0, \; a \times b = 1 \right\}$

*35 $S = \left\{ \dfrac{a}{b} \,\middle|\, a,b \in Z, \; b > 1, \; a \times b = a + b \right\}$

*36   Are the sets of Exercises 34 and 35 equal sets?

Use a symbol other than the given set selector to indicate the sets of Exercises 37 to 39.

*37 $R = \left\{ \dfrac{c}{d} \,\middle|\, c,d \in Z, \; d > 0, \; c < 0, \; c \times d > 0 \right\}$

*38 $T = \left\{ \dfrac{x}{y} \,\middle|\, x,y \in Z, \; y > 0, \; x < 0, \; x - y > 0 \right\}$

*39 $I = \left\{ \dfrac{e}{f} \,\middle|\, e,f \in Z, \; f > 0, \; \dfrac{e}{f} = \dfrac{1}{1} \right\}$

6-3 DEFINITION OF A NUMBER FIELD

With the definition of division as the inverse operation to multiplication and the extension of the number system to include rational numbers, a number system has been obtained which retains all the operations and properties of an integral domain and which also has the property of closure under the newly defined operation of division with nonzero divisors. Such a number system is called a *number field* in accordance with this definition:

Definition 6-6   A *number field* is a set F of numbers, with two well-defined binary operations, addition and multiplication, which satisfy the following postulates:

Field properties   For a, b, c, in F:

F-1   Closure   If $a + b = x$ and $a \times b = y$, then x and y are unique elements of F.

F-2   Associative $(a + b) + c = a + (b + c)$; $a \times (b \times c) = (a \times b) \times c$.

F-3 **Commutative** $a + b = b + a; a \times b = b \times a.$

F-4 **Distributive** $a \times (b + c) = (a \times b) + (a \times c)$

F-5 **Additive identity** There exists a number 0 in F, called the *additive identity*, such that $a + 0 = a$ for every number a in F.

F-6 **Additive inverse** For every a in F there exists a number ^-a in F such that $a + (^-a) = 0$. The number ^-a is called the *additive inverse* of a.

F-7 **Multiplicative identity** There exists a number $1 \neq 0$ in F, called the *multiplicative identity*, such that $a \times 1 = a$ for every number a in F.

F-8 **Multiplicative inverse** For every nonzero number a in F there exists a number $\dfrac{1}{a}$, or a^{-1}, in F such that $a \times \dfrac{1}{a} = 1$. The number $\dfrac{1}{a}$ is called the *multiplicative inverse* of a.

Note that the uniqueness and commutative characteristics for the identities and inverses are not stated in this definition. The proofs that these elements do have these properties are left for Exercises 3 to 5 at the end of this section. While the same statements could have been proved for integers, it seemed best at that time to state them among the postulates defining the system.

Justify this definition of a field:

A number field is an integral domain in which each nonzero element has a multiplicative inverse.

Just as in the case of the integers, the rational numbers can be arranged in order of increasing size. This ordering of the rational numbers is, in fact, dependent upon the ordering of the integers.

Definition 6-7 If $bd > 0$, the rational number $\dfrac{a}{b}$ is said to be greater

Rational numbers are ordered

than $\dfrac{c}{d}$, $\dfrac{a}{b} > \dfrac{c}{d}$, if and only if $ad > bc$, or $ad - bc > 0$. Similarly, for $bd > 0$, $\dfrac{a}{b} < \dfrac{c}{d}$ if and only if $ad < bc$, or $bc - ad > 0$.

In this definition the condition $bd > 0$ simply requires that both divisors have the same sign. Previously, attention has been called to the fact that it is the convention to consider them as both positive, unless there is a real reason for not doing so. When this convention is being followed, of course the condition $bd > 0$ is superfluous.

Example

1 $\frac{3}{4} < \frac{5}{6}$ since $3 \cdot 6 < 4 \cdot 5$. This last statement is true because $20 - 18 = 2$, which is a positive integer.

2 $-\frac{8}{7} < -\frac{2}{3}$ since $(-8) \cdot 3 < 7(-2)$. This last statement is true because $(-14) - (-24) = (-14) + 24 = 10$, which is a positive integer.

The positive rational numbers also have in common with the positive integers the important properties expressed in the following postulates:

Properties of positive rational numbers

1 *Addition* The sum of two positive rational numbers is a positive rational number.

2 *Multiplication* The product of two positive rational numbers is a positive rational number.

3 *Trichotomy* For a given rational number q, one and only one of the following properties holds: q is positive, $q = 0$, or ^-q is positive.

It thus follows, just as in the case of integers, that a positive rational number q can be identified by the relation $q > 0$. Similarly, if q is such that ^-q is positive, we say that q is a negative number and identify it by one or the other of the two equivalent relations $^-q > 0$ or $q < 0$.

Since the set of rational numbers has the property of closure under the operations of addition, subtraction, multiplication, and division by nonzero divisors, these operations are at times referred to as the *four rational*

Field operations

operations. They are also called the *four field operations*, since any set of elements of a field for which these operations are well defined has the same closure properties. This is equivalent to saying that in a field there always exists a solution for the general linear equation in one unknown.

$$ax + b = c \qquad a \neq 0$$

If a, b, and c are numbers of a given number field F, then x is also in F.

The field properties (pages 207 and 208), the two additional properties of addition and multiplication, and the property of trichotomy (above) constitute

Ordered field

the postulates of an *ordered field*. The field of rational numbers and the field of real numbers (see Chap. 7) are the ordered fields that shape the structure of elementary algebra and are thus significant in the study of mathematics.

EXERCISES

1 Compare the field postulates with those for an integral domain (Sec. 4-3) to answer this question: If a set of elements with addition and multiplication is an integral domain, can one say that it is therefore a field?

2   Argue the truth of this statement: A number field is an integral domain in which each nonzero element has a multiplicative inverse.

3   Prove that both the additive and multiplicative identities are commutative with any element of the field.

4   Prove that the additive inverse of any given field element is commutative under addition with that element.

5   Prove that the multiplicative inverse of any nonzero field element is commutative under multiplication with that element.

*6   Given the set $S = \{0,1\}$ with addition and multiplication defined by the accompanying tables. Develop the complete argument to establish the fact that this system is a field.

+	0	1
0	0	1
1	1	0

×	0	1
0	0	0
1	0	1

EXERCISE 6

*7   In each case you are given that the associative and distributive properties hold. Which systems are fields and which are not: (*a*) $S = \{0,1,2\}$; (*b*) $S = \{0,1,2,3\}$; (*c*) $S = \{0,1\}$; (*d*) the set of all nonnegative rational numbers with addition and multiplication?

+	0	1	2
0	0	1	2
1	1	2	0
2	2	0	1

×	0	1	2
0	0	0	0
1	0	1	2
2	0	2	1

+	0	1	2	3
0	0	1	2	3
1	1	2	3	0
2	2	3	0	1
3	3	0	1	2

×	0	1	2	3
0	0	0	0	0
1	0	1	2	3
2	0	2	0	2
3	0	3	2	1

+	0	1
0	0	1
1	1	2

×	0	1
0	0	0
1	0	1

EXERCISE 7

6-4 RATIONAL NUMBERS AS FRACTIONS

One of the very important interpretations of rational numbers is that of a fraction expressing the quotient of two integers $a \div b = a/b \ (b \neq 0)$. This is by no means, however, the complete significance of the concept of fraction.

Early in man's dealing with the quantitative demands of his environment he came in contact with concepts which the integers could not describe adequately. Such concepts as a broken spear, a part of a day, one shepherd's flock of sheep which was not an exact multiple of another shepherd's flock, or a certain distance which was more than 1 day's journey but not as much as 2 days' journey led to the necessity of finding a way to give expression to parts of a whole or parts of a group. The concept of equal parts was not necessarily incorporated in this early idea of fraction but quite possibly was a much later refinement. Our word "fraction" is a derivative of the Latin word *frangere* (to break). Thus, etymologically at least, we continue the tradition that the idea implied by the use of the word "fraction" has to do with a "broken unit."

Through the ages the concept of fraction and the problems connected with its proper use have presented several confusing paradoxes. It has been a concept which (1) represents parts of a unit or is an entity within itself, (2) is a quotient or is not a quotient, (3) expresses a relationship between two numbers or merely expresses a partitioning, and (4) has many of the characteristics of integers yet came into being because of their inadequacy. While the desire for clarification in computation has not removed all the confusing contrasts, the cumbersome symbolism, and the involved computational processes associated with the unit fractions of the form $\frac{1}{2}$, $\frac{1}{15}$, used by the early Egyptians,† it has led to the more precise interpretations and more facile operational procedures of the present.

6-5 GENERAL FRACTIONS

Through their interest in astronomy the Babylonians and Greeks made use of sexagesimal fractions (fractions with denominators expressed in terms of powers of 60). The Greeks also made extensive use of such fractions in their study of geometry. Today we use sexagesimal fractions in our time units. One minute $= \frac{1}{60}$ of an hour (originally *pars minuta prima*), and one second $= \frac{1}{60}$ of a minute, or $\frac{1}{(60)^2} = \frac{1}{3,600}$ of an hour (*pars minuta secunda*).

While, in addition to the sexagesimal fractions, the Greeks developed and used a system of fractions much like the fractions we use, it was the

---

† Howard Eves, "An Introduction to the History of Mathematics," 3rd. ed. pp. 38–39, 44, Holt, Rinehart and Winston, Inc., New York, 1969; and David Eugene Smith, "History of Mathematics," vol. II., pp. 210–211, Ginn and Company, Boston, 1925.

$$
\begin{array}{r}
23 \\
8\ \overline{\smash{)}189} \\
\underline{16} \\
29 \\
\underline{24} \\
5
\end{array}
$$

FIGURE 6-1

Hindus—the originators of our number system along with place value and the use of 10 as a base—who developed a pattern for the general fraction which seemed to capture all the specifics of the various other forms in the generality of its notation. Difficulties of printing caused variations in the mechanics of form which produced, for example, such symbols as $\frac{2}{3}$, $\frac{2}{3}$, 2 : 3, 2/3, and $\frac{2}{3}$ as ways of writing two-thirds. The last three are used today as accepted variations of the symbol.

Fraction as a rational number

When a fraction is used to represent the number of equal parts into which a whole has been divided, one of the equal parts into which a group has been divided, or a quotient of one number divided by another, it is used as a number. It is indeed a rational number for which the four field operations of addition, subtraction, multiplication, and division are well-defined processes.

There is a basic error sometimes made in the extension of division which permits the occurrence of rational numbers (fractions) in carrying out the operation. The error results from a false interpretation of the division algorithm. In a problem where the quotient of one integer divided by another integer is sought, the division algorithm ensures the existence of an *integral* remainder which may be greater than or equal to 0. At times one will find in elementary texts instructions worded somewhat after this pattern: Find the quotient $8\overline{)189}$ and express the remainder as a fraction. In the division of Fig. 6-1 the answer expected for the result, under these instructions, is $23\frac{5}{8}$. The fraction $\frac{5}{8}$ is supposed to be "the remainder expressed as a fraction." It is not. The remainder is the integer 5. It can be expressed in fraction form as $\frac{5}{1}$, or some element of its equivalence class such as $\frac{10}{2}$

Complete quotient

or $\frac{15}{3}$, but not as $\frac{5}{8}$. The number $23\frac{5}{8}$ is the *complete quotient* obtained by dividing 189 by 8 since $8 \times 23\frac{5}{8} = 189$.

6-6 FRACTION AS A RATIO

Official Greek mathematics did not contain fractions. According to Plato, "the experts in this study" could readily accept the concept of visible things being divisible (or broken), but not mathematical units. Thus, while merchants might operate with fractions, mathematicians made use of *ratios of integers*.† This confusing dichotomy in the use of the fraction symbol

†B. L. Van der Waerden, "Science Awakening," p. 49, Erven P. Noordhoff, Ltd., Groningen, Netherlands, 1954.

continues into the modern age. Confusion can be expected when the same symbol is used to represent a number, as in a partitioning, or to represent not a number but merely a relationship between two numbers. The symbol $\frac{3}{4}$ can be used, for example, to represent 3 of the equal fourths of a unit or merely to say that one group of objects when compared with another group of objects is found to be so related to it that for each 3 objects in the one group there are 4 in the other. The symbol $\frac{3}{4}$ thus may represent a number and be used as such (as in the first part of the illustration), or not represent a number at all but be merely a convenient symbol for record-

Fraction as a ratio

ing the quantitative specifics of a relationship. As a ratio, $\frac{3}{4}$ means that the relationship under consideration is such that for each 3 of a group there are 4 of another. It is frequently written as 3:4 and read "three to four." Thus, if there are 7 marbles of exactly the same size and smoothness in a bag and 3 of them are red while 4 are blue, then the chances are 3:4 that a blindfolded person in lifting a single marble from the bag would select a red one. This means that there are 3 chances out of 7 of the marble's being red and 4 chances out of 7 of its being blue.

Proportion

A *proportion* is an expression of equality between two ratios which are equal. Thus, "3 is to 4 as 75 is to 100" ($\frac{3}{4} = \frac{75}{100}$) states a true proportion, since, when 75 is compared with 100, it is found that the ratio is the same as 3 to 4 because there are 3 twenty-fives in 75 and 4 twenty-fives in 100. This latter ratio is frequently read as 75 *per* 100, or 75 percent. Herein lies the justification for the use of the terminology *rate percent*. It means *ratio to one hundred*. Similarly, it is a true proportion to state that the ratio of 3 to 4 is the same as that of the fractional part ($\frac{3}{4}$) to the whole, or $\frac{3}{4}$:1. Just as the ratio of 2:1 may be associated with the numerical concept of 2, so the ratio of $\frac{3}{4}$:1 (three-fourths to one) may be associated with the numerical concept of $\frac{3}{4}$ (three-fourths). If two groups are so related that the ratio of group A to group B is 2:1, we can say that the size of group A is two times that of group B. Similarly, if the ratio is $\frac{3}{4}$:1, we can say that the size of group A is three-fourths the size of B. It is in this concept that we find justification for use of the same symbolism to represent "a fractional part" or "a ratio."

6-7 THE VOCABULARY OF FRACTIONS

The fractions which were used in the commerce of ancient times were distinguished from the sexagesimal fractions used in astronomy. The termi-nology used for this distinction has come down to us through the Latin

Types of fractions

fractiones vulgares (the vulgar fractions of the English) to give us the expression *common fractions*. The contrast, however, is no longer with sexagesimal fractions (those with powers of 60 as denominators), but with *decimal* fractions (those with powers of 10 as denominators). All common fractions may be grouped into two distinct classes: *proper fractions* (those whose numerator $<$ the denominator) and *improper fractions* (those whose numerator $\geq$ the denominator). Any improper fraction may be converted into a *mixed number* (a whole number and a proper fraction) or into a whole number.

The parts of a fraction are called *terms*. The upper number is called, after the Latin custom, the *numerator* (or numberer) while the lower number is called the *denominator* (namer). With the different interpretations of a fraction the numerator and denominator take on different meanings. When the fraction is thought of as a part of a whole, the denominator tells the number of equal parts into which the whole has been divided while the numerator tells how many parts have been taken. When the fraction is thought of as a fractional part of a group, the denominator tells the number of equal parts into which the group has been divided while the numerator tells the number in the group. When the fraction is thought of as a quotient, the denominator is the divisor while the numerator is the dividend. In this sense the fraction is sometimes thought of as an "expressed division." When the fraction is thought of as a ratio, the denominator simply gives quantitative expression to one of the groups being compared while the numerator gives quantitative expression to the other.

When the numerator and denominator of a fraction are multiplied or divided by the same nonzero number, another fraction is obtained whose value is the same as that of the first fraction. For this reason they are called *Equivalent fractions* *equivalent fractions*. Fractions with a common (like) denominator are called *like fractions*, and those with unlike denominators are called *unlike fractions*. Two unlike fractions such as $\frac{5}{6}$ and $\frac{3}{8}$ may be changed to like fractions by using as a common denominator a number which has both given denominators as factors and converting the two fractions to equivalent fractions, both having this denominator. In this case, 24 is the smallest number which has both 6 and 8 as factors.

$$\frac{5}{6} = \frac{5 \times 4}{6 \times 4} = \frac{20}{24} \quad \text{and} \quad \frac{3}{8} = \frac{3 \times 3}{8 \times 3} = \frac{9}{24}$$

It is usually best, though not necessary, that this common denominator be *Least common denominator* the smallest possible common denominator, or *least common denominator*. The fact that both terms of a fraction can be multiplied or divided by the same nonzero number without changing the value of the fraction can be justified also by an appeal to the multiplicative identity property of the field of rational numbers. Since $\frac{4}{4} = \frac{3}{3} = \frac{1}{1}$ it follows from this property that

$$\frac{5}{6} = \frac{5}{6} \times \frac{4}{4} = \frac{5 \times 4}{6 \times 4} = \frac{20}{24} \quad \text{and} \quad \frac{3}{8} = \frac{3}{8} \times \frac{3}{3} = \frac{3 \times 3}{8 \times 3} = \frac{9}{24}$$

In the reduction of fractions to lower terms a technique known as cancellation is sometimes used. For example,

$$\frac{6}{8} = \frac{\cancel{2} \times 3}{\cancel{2} \times 4} = \frac{3}{4}$$

The validity of this technique is justified by two facts: (1) a fraction may be considered as the expressed quotient of the numerator divided by the denominator, and (2) multiplication and division are inverse processes, and so one process neutralizes or annuls the effect of the other. In the above example, one may consider the fact that $\frac{6}{8}$ can be obtained from $\frac{3}{4}$ by multiplying by 2 and dividing by 2. The order in which the two operations are performed is immaterial; the net effect is to produce *no change in the value of the fraction.* $\frac{3}{4}$ and $\frac{6}{8}$ are equivalent fractions. The word "cancel" means "to annul or destroy; to neutralize or counterbalance." The "canceling" of a factor common to both numerator and denominator of a fraction is the mere physical act of indicating the fact that one factor destroys the effect of the other. Since addition and subtraction are inverse processes, the same device can at times be used to advantage in simplifying computations involving these two operations. For example, $8 - \cancel{5} + \cancel{5} = 8$. Such a technique can be used *only* in computing the result of inversely related operations. In the simplification of the expression $\frac{5 + 4}{4}$, the 4s cannot be canceled, because addition and division are not inversely related operations and one operation does not annul or destroy the effect of the other.

At more advanced levels of mathematical discussion, the term *fraction* is used to represent *any* type of algebraic expression which has both a numerator and a denominator. The types of fraction we have been discussing—namely, those with integral numerators and denominators—are then *Rational fractions* called *rational fractions.*

In any computation, rather than use mixed numbers as such, it is usually much simpler to convert them to equivalent improper fractions before attempting to carry out the computation. This is particularly true when multiplication or division is involved.

Example   Find the sum and product of $5\frac{1}{2}$ and $7\frac{1}{3}$.

Sum

Mixed numbers

$$5\frac{1}{2} + 7\frac{1}{3} = \left(5 + \frac{1}{2}\right) + \left(7 + \frac{1}{3}\right)$$

$$= (5 + 7) + \left(\frac{1}{2} + \frac{1}{3}\right)$$

$$= 12 + \frac{1 \cdot 3 + 2 \cdot 1}{2 \cdot 3}$$

$$= 12 + \frac{5}{6} = 12\frac{5}{6}$$

Improper fractions

$$5\frac{1}{2} + 7\frac{1}{3} = \frac{11}{2} + \frac{22}{3}$$

$$= \frac{11 \cdot 3 + 2 \cdot 22}{2 \cdot 3}$$

$$= \frac{33 + 44}{6}$$

$$= \frac{77}{6} = 12\frac{5}{6}$$

Product

$$5\tfrac{1}{2} \times 7\tfrac{1}{3} = \left(5 + \frac{1}{2}\right) \times \left(7 + \frac{1}{3}\right)$$

$$= \left[\left(5 + \frac{1}{2}\right) \times 7\right] + \left[\left(5 + \frac{1}{2}\right) \times \frac{1}{3}\right]$$

$$= \left[(5 \times 7) + \left(\frac{1}{2} \times 7\right)\right] + \left[\left(5 \times \frac{1}{3}\right)\right.$$

$$\left. + \left(\frac{1}{2} \times \frac{1}{3}\right)\right]$$

$$= \left(35 + \frac{7}{2}\right) + \left(\frac{5}{3} + \frac{1}{6}\right)$$

$$= \left(35 + \frac{7}{2}\right) + \frac{30 + 3}{18}$$

$$= \left(35 + \frac{7}{2}\right) + \frac{11}{6}$$

$$= 35 + \left(\frac{7}{2} + \frac{11}{6}\right)$$

$$= 35 + \frac{42 + 22}{12}$$

$$= 35 + \frac{64}{12}$$

$$= 35 + 5\tfrac{1}{3} = 40\tfrac{1}{3}$$

$$5\tfrac{1}{2} \times 7\tfrac{1}{3} = \frac{11}{2} \times \frac{\overset{11}{\cancel{22}}}{3}$$

$$= \frac{121}{3}$$

$$= 40\tfrac{1}{3}$$

EXERCISES

1   What justification can you give for the use of the terms "proper fraction" and "improper fraction" to indicate a fraction less than 1 and greater than 1, respectively?

2   What is meant by the statement: A fraction may be thought of as representing a number?

3   Give various illustrations of each of the following concepts of a fraction: (*a*) part of a unit; (*b*) part of a group; (*c*) quotient; (*d*) ratio; (*e*) number.

4   Explain the meanings to be given the terms "numerator" and "denominator" in each of the uses of fraction given in Exercise 3.

5   Criticize the number $5\tfrac{2}{7}$ given as an answer to the problem: Find the quotient of $37 \div 7$ and express the remainder as a fraction.

6   What property of rational numbers may be used in changing a fraction either to higher terms or to lower terms?

7   Why do the concept of equality and the processes of addition, subtraction, multiplication, and division have to be defined for fractions?

8   The introduction of rational numbers is necessary to give a set of numbers which will have the property of closure with respect to what operation?

9 The set of positive rational numbers has closure with respect to what operations?

10 Arrange these fractions in order from largest to smallest with the largest fraction on the left: $\frac{2}{3}, \frac{5}{5}, \frac{1}{2}, \frac{3}{4}, \frac{5}{8}, \frac{3}{7}$.

11 Explain why finding the sum of two rational numbers by applying Definition 6-3 is equivalent to finding the sum of two common fractions by first changing the fractions to fractions with common denominators and then adding their numerators.

12 Show that the associative and commutative laws hold for these sums: (a) $\frac{3}{4} + \frac{5}{6} + \frac{1}{2}$; (b) $\frac{7}{5} + \frac{2}{3} + \frac{8}{15}$. Why does this *not* prove that these two laws hold for the addition of fractions?

13 Show that the commutative law does not hold for finding the difference of two fractions. Use the fractions $\frac{3}{4}$ and $\frac{7}{8}$. Why does this prove that the commutative law does not hold for the subtraction of fractions?

14 Explain why one does not have to change fractions to a common denominator before they can be multiplied or divided.

15 Would it be necessarily incorrect to change two fractions to fractions with common denominators before finding their product or the quotient of one divided by the other?

16 Show that the commutative law holds for these products: (a) $\frac{3}{4} \times \frac{8}{27}$; (b) $\frac{5}{6} \times \frac{7}{9}$. Why does this *not* prove that multiplication of fractions is commutative?

17 Why would the practice of finding common denominators in multiplication and division of fractions be, in general, an undesirable procedure?

18 Use unit fractions to rationalize multiplication by the inverse of the divisor when it is a fraction in a division which has a fraction as the divisor.

19 Use the fractions $\frac{5}{8}, \frac{2}{3}$, and $\frac{3}{4}$ to construct examples which show that rational numbers are neither associative nor commutative under division. Does this statement hold true also for integers? For natural numbers?

20 Use each set of numbers to illustrate each of the three fundamental laws of arithmetic: associative, commutative, and distributive.

(a) $\{5,7,12\}$ (b) $\{3,4,6,9\}$ (c) $\left\{\frac{1}{2}, \frac{3}{4}, \frac{2}{3}\right\}$

(d) $\left\{\frac{5}{6}, \frac{4}{9}, \frac{7}{25}\right\}$ (e) $\{3\frac{1}{2}, 2\frac{1}{4}, 5\frac{1}{3}\}$ (f) $\left\{-\frac{3}{5}, \frac{1}{2}, -\frac{4}{7}\right\}$

21 Explain why you can use cancellation in the reduction of each of these fractions to lower terms, and then reduce each fraction to its lowest terms.

(a) $\frac{15}{20}$ (b) $\frac{7}{21}$ (c) $\frac{52}{28}$ (d) $\frac{16}{24}$

22 Explain why you cannot use cancellation in each of these fractions:

(a) $\dfrac{8 + 3}{8}$ (b) $\dfrac{5 - 2}{2}$ (c) $\dfrac{8}{3 + 2}$ (d) $\dfrac{10}{5 - 4}$

23 Explain how cancellation may be used to reduce each of these fractions to lower terms, and then reduce each fraction to its lowest terms.

(a) $\dfrac{6 - 4}{8}$ (b) $\dfrac{9 + 12}{3}$ (c) $\dfrac{5}{22 - 7}$ (d) $\dfrac{8}{20 + 4}$

24 Find these sums. Use the associative property to provide a check.

(a) $\dfrac{7}{8} + \dfrac{5}{24} + \dfrac{3}{16}$ (b) $\left(-\dfrac{5}{9}\right) + \left(-\dfrac{3}{4}\right) + \left(-\dfrac{2}{3}\right)$

(c) $5\frac{1}{2} + 3\frac{1}{4} + 8\frac{1}{5}$ (d) $11\frac{2}{3} + 16\frac{5}{9} + 8\frac{1}{3}$

(e) $(-8\frac{1}{2}) + (-2\frac{1}{4}) + (-3\frac{1}{8})$ (f) $9\frac{1}{5} + (-15\frac{1}{4}) + (-3\frac{1}{2})$

25 Find these differences. Use addition to provide a check.

(a) $\dfrac{8}{9} - \dfrac{5}{6}$ (b) $11\frac{7}{8} - 4\frac{3}{4}$

(c) $0 - \left(-\dfrac{2}{3}\right)$ (d) $7\frac{1}{2} - 8\frac{1}{3}$

(e) $(-15\frac{2}{3}) - 4\frac{5}{6}$ (f) $\left(-\dfrac{7}{8}\right) - \left(-\dfrac{3}{4}\right)$

26 Find these products. Use the associative property to provide a check.

(a) $\dfrac{2}{3} \times \dfrac{5}{7} \times \dfrac{1}{2}$ (b) $7\frac{1}{2} \times 3\frac{1}{3} \times 7\frac{1}{6}$

(c) $\left(-\dfrac{3}{4}\right) \times \left(-\dfrac{2}{3}\right) \times \left(-\dfrac{5}{6}\right)$ (d) $(-8\frac{1}{4}) \times (-2\frac{2}{3}) \times \dfrac{3}{4}$

(e) $8\frac{1}{3} \times 12\frac{4}{5} \times 4\frac{7}{8}$ (f) $9\frac{6}{11} \times 10\frac{2}{3} \times 6\frac{7}{8}$

27 Find these quotients. Use multiplication to provide a check.

(a) $\dfrac{8}{9} \div \dfrac{2}{3}$ (b) $\dfrac{7}{11} \div \dfrac{3}{5}$

(c) $7\frac{1}{2} \div 3\frac{1}{3}$ (d) $12\frac{2}{3} \div 10\frac{6}{7}$

(e) $(-8\frac{1}{2}) \div 3\frac{1}{5}$ (f) $(-6\frac{2}{3}) \div (-3\frac{7}{15})$

28 If $r = \dfrac{\frac{1}{2} + \frac{1}{3}}{2}$, compute r and show that $\frac{1}{3} < r < \frac{1}{2}$.

29 Use the same technique for finding r such that $\frac{7}{16} < r < \frac{5}{8}$.

6-8 DECIMAL FRACTIONS

The concept of place value, so important in the structure of our numeral system, places great significance on the decimal fraction. The analysis of any numeral representing a natural number will reveal that the place value of any position can be obtained by successive multiplication by 10 or by successive division by 10. Multiplication is used if the reference position of known value is to the right of that of unknown value, and division is used if it is to the left. Thus, if the ones place is the position of reference, successive values to the left will be 10×1, 10×10, 10×100, and so on. Similarly, successive values to the right of the ones place would call for division by powers of 10. Because of this fact it is evident that the efficiency of our numeral system was seriously restricted until the introduction of the concept of decimal fractions by Simon Stevin in 1585. With the introduction of this concept, positional values such as tenth $\left(0.1 = \dfrac{1}{10}\right)$, hundredth $\left(0.01 = \dfrac{1}{100} = \dfrac{1}{10^2}\right)$, and thousandth $\left(0.001 = \dfrac{1}{1,000} = \dfrac{1}{10^3}\right)$ have just as much significance as do ten (10), hundred (10^2), and thousand (10^3). With such flexibility of positional value and the 10 digital symbols (0,1,2,3,4,5,6,7,8,9) we have the means to write any numeral no matter how large or how small. (Review Sec. 1-8.)

6-9 READING AND WRITING DECIMALS

Too frequently, the decimal point receives undue emphasis. It is merely a device used to indicate the ones place and is not usually used if the numbers are whole numbers, because in such cases, it is easy to recognize the ones place at the extreme right-hand position. The ones place is the center (so to speak) of our notational scheme. There are still some differences in the device to be used for what we call the decimal point. The decimal which in the United States we write as 3.14 is written as 3 · 14 in England, while in other countries it may appear as 3,14 or 3_{14}. Even in the United States, many people in writing a check for three dollars and fourteen cents would write it 3^{14}. In each of the above numbers, however, the positional value of the 1 in the number is $\frac{1}{10}$ the value of the ones place, and that of the 4 is $\frac{1}{100}$ the value of the ones place.

Types of decimals   There are decimals which have only a fraction part and there are others which have a whole-number part as well as a fraction part. Decimal fractions which have only a fraction part are called *pure decimals*. There may be two kinds of pure decimals, as shown in Fig. 6-2. They may either end in a common fraction (*complex decimal*) or not (*simple decimal*). Pure decimal fractions correspond to proper common fractions; their numerators are smaller than their denominators.

	Pure	Mixed
Simple	0.25	2.5
Complex	$0.16\frac{2}{3}$	$1.6\frac{2}{3}$

FIGURE 6-2 Kinds of decimals

In decimal fractions there is no exact counterpart of the improper common fraction. The *mixed decimal* 3.14, mentioned previously, can be written as the equivalent improper fraction $\frac{314}{100}$ or as the equivalent mixed number $3\frac{14}{100}$. In reading any mixed decimal the word "and" should be used to bridge the decimal point. Thus, 3.14 is read as three *and* fourteen hundredths. This is exactly the same way that the symbol $3\frac{14}{100}$ is read. The word "and" should not be used in reading numerals in any way other than that indicated here. Frequently in business and industry a short method for reading decimals is used. In such cases 3.14 is read "three-point-one-four."

As is indicated in Fig. 6-2, mixed decimals can also be either simple or complex. The mixed complex decimal of the figure is read "one and six-and-two-thirds tenths." In any such case, whether decimals are used or not, the common fraction indicates a fractional part of the value of the position to which it is attached ($0.00\frac{1}{4}$ is read as one fourth of one hundredth; its value may be written as the common fraction $\frac{1}{400}$).

Finite decimals

All the decimal fractions we have used up to this point are called *finite*, or *terminating*, decimals. These two words mean the same thing about the decimal. They tell us that when we start writing the fraction we come to a stopping place. In Fig. 6-2, for example, the decimal fractions are all finite decimals; the writing stops after two digits have been written. There is a slight exception in the case of the common fraction of the complex decimals, but, even here, the writing stops with the writing of the common fraction.

Infinite decimals

There are decimal fractions for which the above is not true. Such decimals are called *infinite*, or *nonterminating*. They do not stop. If we should wish to write a finite decimal fraction which would be equal to $\sqrt{2}$ (the positive square root of 2), we could not do so (see Fig. 7-2, page 249). In the expression $\sqrt{2} = 1.41421 \cdots$ the three dots mean that there are digits omitted from the decimal. If we multiply 1.41421 by itself we get a number which is very close to 2 in value, but we do not get 2. It is not possible to fill in enough decimal places to give a number which when multiplied by itself will give the value 2. We might say that $\sqrt{2}$ is approximately equal to 1.41421. Note that here the three dots after the decimal are missing. In this case 1.41421 is a finite decimal which is an approximation to the positive square root of 2. Since 1.41421 is a rational number, it may be called a *rational approximation* to the positive square root of 2.

Repeating decimals

There are two types of nonterminating, or infinite, decimals. There are some like $0.333 \cdots$, which just keep on repeating the same digits in the same pattern. They are called *repeating decimals*. The decimals $0.142857142857 \cdots$ and $0.1666666 \cdots$ are two other repeating decimals. The standard notation used to indicate repeating decimals is a bar over the *repetend* (the part which repeats):

$$0.142857142857 \cdots = 0.\overline{142857}$$
$$0.333 \cdots \qquad\qquad = 0.\overline{3}$$
$$0.16666 \cdots \qquad\quad = 0.1\overline{6}$$

Nonrepeating decimals
Infinite decimals such as 1.41421 ··· representing $\sqrt{2}$ can be proved never to follow a pattern of repetition. They are called *nonrepeating decimals.*

6-10 USING DECIMALS

The concept of decimal fraction does not necessitate another extension of our number system. In fact, when we restrict our considerations to decimal fractions we are, indeed, simplifying our operations with fractions. So far we have been dealing with fractions for which nonzero integers might serve as denominators. Now, we are dealing with fractions whose denominators can be only powers of 10. No new definitions are needed to tell us how to operate with decimals. The definitions for equality, less than, greater than, and the fundamental operations that were given previously for fractions are sufficient here.

Two finite decimals are equal if they are composed of identically the same digits, each with identically the same place value. This means that, if these decimals were written as common fractions, they would be like fractions with the same numerators. In comparing two decimals, remember that one unit in any position is equal to 10 units of the next position to its right.

Ordering decimals
For this reason, the decimal 2.64 is larger than the decimal 2.639; in fact, 2.64 would still be the larger decimal regardless of how many 9s might be annexed to 2.63. Such statements can be checked by noting, for example, that 2.64 and 2.639 can be written as the following like common fractions: $\frac{2,640}{1,000}$ and $\frac{2,639}{1,000}$. There is an essential point that must be observed in this connection. The fraction $\frac{2,640}{1,000}$, written as a decimal, becomes 2.640, which can be obtained from 2.64 by annexing a zero to the right. Frequently, the statement is made that "annexing a zero to the right of the decimal point does not affect the decimal fraction." This is an incorrect statement. The value of the decimal fraction is not changed, but as we shall see in a later chapter (see Sec. 9-8), the significance of the fraction is changed considerably.

In the use of decimal fractions it is frequently desirable to *round* the numerals to some specified positional value. This is particularly true when dealing with infinite decimals or very large numbers. For example, the repeating decimal $0.\overline{3} = 0.333$ ··· rounded to the nearest hundredth becomes 0.33, while the repeating decimal 0.6666 ··· rounded to the nearest hundredth becomes 0.67. The distance from the earth to the sun, correct to the nearest tenth of a mile, is 92,956,524.4 miles. When rounded to the nearest million miles we have 93,000,000 miles, the figure frequently used in referring to this distance. These have been illustrations of applications of the rule to be followed in rounding numbers.

Rule for rounding numbers If the extreme left digit of those being dropped is 5 or more, increase by 1 the extreme right digit of those being retained. If the extreme left digit is less than 5, make no change in the digits retained.

Attention should be called to the fact that when whole numbers are rounded, zeros are required as placeholders for the digits dropped in order to maintain the relative place value for the digits which are not dropped. This is neither necessary nor desirable when fractions are rounded. For example, if 28,672 were rounded to retain only two nonzero digits, it would become 29,000, while 0.28672 would become 0.29.

Other rules are sometimes given. The results from all rules are essentially the same, and the rule given here submits itself to concise and simple statement. Whatever rule is given, the basic idea is to balance that which is dropped in a given computation with that which is retained. In electronic computation, or any pattern of computation which involves a sequence of operations with approximations, this problem becomes a very serious one in which more careful refinements are necessary than are possible to investigate here. Such approximations become not only desirable but also necessary when dealing with many forms of measurement. (See Chap. 9.)

In the addition and subtraction of finite decimals a very convenient practice to follow as a guide for getting digits of like positional value aligned is to align the decimal points. This is by no means a necessary practice; it is merely an accepted crutch that is very convenient to use. The actual process of adding is then carried out as if the members were all integers.

There is nothing new in the processes of multiplication and division with decimals except the proper placing of the decimal point in the results. The details of the operations are exactly the same as those for whole numbers. The rationalization of the placing of the decimal point in multiplication can be accomplished by different methods, two of which are illustrated in the following example.

Example Find the product 3.63 × 27.9.

1 Find the product of the two integers 363 × 279. Place the decimal point in the product so that there will be as many decimal places as there are in both the multiplier and the multiplicand. The rationalization of this scheme can be accomplished by writing both multiplier and multiplicand as common fractions and carrying through the detail of finding such a product.

$$3.63 \times 27.9 = \frac{363}{100} \times \frac{279}{10} = \frac{363 \times 279}{100 \times 10} = \frac{101,277}{1,000} = 101.277$$

2 Estimate the product:

$$3.63 \text{ is approximately equal to } 4 \qquad 3.63 \approx 4$$
$$27.9 \approx 28$$
$$3.63 \times 27.9 \approx 4 \times 28 = 112$$

$$
\begin{array}{r}
37.9 \approx \quad 38 \\
\underline{2.63} \approx \quad \underline{3} \\
1\ 137 \qquad 114 \\
22\ 74 \\
\underline{75\ 8} \\
99.677
\end{array}
$$

FIGURE 6-3

The estimate not only provides a rough check on the product but also tells us that there will be three places *to the left* of the decimal point. Now multiply as if the two numbers were integers and then place the decimal point three places from the left.

The example was chosen deliberately so that attention may be called to a seeming difficulty with the second suggested technique for placing the decimal point.

Compare the second method of the example with the same technique applied to finding the product 2.63×37.9 (Fig. 6-3). This time the product is approximately $3 \times 38 = 114$. The actual product, however, has only two places, rather than three, to the left of the decimal.

This difficulty is easily taken care of by remembering that the value at which a number changes from a two-figure number to a three-figure number is 100. If the final product has the digit 1 to the extreme left, there will be *three* digits to the left of the decimal; otherwise there will be *only two*.

This same precaution will have to be kept in mind when the products are near any transition point (10; 100; 1,000; 10,000; etc.).

Probably the simplest pattern for division of decimals is to multiply both divisor and dividend by the power of 10 necessary to convert the divisor into a whole number. After this has been done, the only new problem that remains is that of placement of the decimal point in the quotient. The actual mechanics of division are the same as if one were dividing one whole number by another. Here again custom has evolved, as in the case of addition and subtraction of decimals, a "rule without authority." The rule states: When dividing a decimal by a whole number, place the decimal point in the quotient directly over the decimal point of the dividend. This is in no sense a bona fide rule or law that must of necessity be followed. It is no more than a crutch of convenience to assist in placing the digits of the quotient in their respective places of value. A meaningful procedure for placing the decimal point in the quotient is illustrated in the example that follows:

Example Find the quotient of $0.2826 \div 6.28$.

Rewrite the exercise as $28.26 \div 628$ (see Fig. 6-4).

By inspection observe that $0.1 \times 628 = 62.8$, which is larger than the dividend 28.26. On the other hand, $0.01 \times 628 = 6.28$, which is smaller than the dividend 28.26. It therefore follows that the quotient is smaller than

$$6.28 \overline{)0.2826}$$

becomes

$$
\begin{array}{r}
.045 \\
628 \overline{)28.260} \\
25\ 12 \\
\hline
3\ 140 \\
3\ 140 \\
\hline
\end{array}
$$

FIGURE 6-4 Division by decimal

0.1 but larger than 0.01, so the first figure of the quotient must be placed in the hundredths position.

Now the placing of the decimal point in the quotient directly over the decimal point in the dividend becomes a convenient device for making sure that the figures of the quotient will be in their proper relative positions. This makes it possible to place the first quotient figure in the hundredths place by merely writing it directly above the figure in the hundredths place in the dividend. In this particular division it is necessary to use a placeholder in the tenths place. Why? Since we are seeking only the numerical value of the quotient, zeros may be annexed to the dividend if it is desirable in carrying out the division. Later (Sec. 9-11) we shall see that, under certain conditions, there are precautions that must be observed in such practice.

*6-11 SCIENTIFIC NOTATION

The use of exponents provides a very effective means for simplifying the writing of very large or very small numbers. *Numerals representing such numbers are expressed as being greater than or equal to one but less than ten, multiplied by an appropriate power of ten.* When numerals are in this

Scientific notation

form, they are said to be written in *scientific notation.* The appropriate power of ten is that power necessary to move the decimal point to its proper place in the positional notation form of the numeral. A positive exponent of ten indicates that the decimal point should be shifted that many places to the right. A negative exponent indicates that the decimal point should be moved that many places to the left. A few examples will illustrate the principle.

Positional notation	*Scientific notation*
1,000,000	$1. \times 10^6$
0.0000001	$1. \times 10^{-7}$
27,600,000	2.76×10^7
602,300,000,000,000,000,000,000	6.023×10^{23}
0.0000316	3.16×10^{-5}

Scientific notation is particularly helpful when computing with very large or very small numbers. To find such products as 27,600,000 × 0.0000316, it would be much simpler to use scientific notation. The problem then becomes that of finding the product $(2.76 \times 10^7) \times (3.16 \times 10^{-5})$ which we may write in the form

$$(2.76 \times 3.16) \times (10^7 \times 10^{-5}) = 8.7216 \times 10^2 = 872.16$$

The answer is given in two acceptable forms. In scientific notation it is 8.7216×10^2; in positional notation it is 872.16.

Similarly, 0.005781 ÷ 2,460,000 may be expressed in the simpler form

$$5.781 \times 10^{-3} \div 2.46 \times 10^6 = \frac{5.781 \times 10^{-3}}{2.46 \times 10^6}$$

$$= \frac{5.781}{2.46} \times 10^{(-3)-6}$$

$$= 2.35 \times 10^{-9}$$

$$= 0.00000000235$$

Example

1 The smallest electric charge contains 0.000000000000000000160 coulomb. One coulomb contains 3,000,000,000 electrostatic units. How many electrostatic units are there in the smallest electric charge?

$$0.000000000000000000160 = 1.60 \times 10^{-19}$$
$$3,000,000,000 = 3 \times 10^9$$
$$(1.60 \times 10^{-19}) \times (3 \times 10^9) = (1.60 \times 3) \times (10^{-19} \times 10^9)$$
$$= 4.80 \times 10^{-10}$$

or 0.000000000480 electrostatic units.

2 A motion picture image remains on the screen about 0.063 second. How many such images are there in a motion picture which requires 1 hour 45 minutes for showing?

$$1 \text{ hour } 45 \text{ minutes} = 6300 \text{ seconds}$$
$$= 6.300 \times 10^3 \text{ seconds}$$
$$0.063 = 6.3 \times 10^{-2}$$
$$(6.30 \times 10^3) \div (6.3 \times 10^{-2}) = \frac{6.30 \times 10^3}{6.3 \times 10^{-2}}$$
$$= \frac{6.30}{6.3} \times 10^{3-(-2)}$$
$$= 1.0 \times 10^5 \text{ images}$$
$$= 100,000 \text{ images}$$

EXERCISES

1 Give examples of each of the following types of decimals: (*a*) pure simple; (*b*) mixed complex; (*c*) mixed simple; (*d*) pure complex; (*e*) finite; (*f*) infinite.

2 Distinguish between terminating, repeating, and nonrepeating decimals.

3 What is the basic reason for aligning decimal points in arranging numerals for column addition or subtraction?

4 Find these sums. Use the commutative property to provide a check.

(*a*) 25.612	(*b*) 217.03	(*c*) 0.00236
3.078	901.68	0.28730
4.965	75.32	0.98612
32.872	123.79	0.97503
10.009	305.06	0.96123

5 Find these differences. Use addition to check.

(*a*) 8160.321	(*b*) 900.0764	(*c*) 600.002
709.812	813.9618	598.108

6 First estimate each product, then find the product, and then use your estimate to place the decimal point.

(*a*) 8.162 × 2.8   (*b*) 5.73 × 46.34
(*c*) 69.21 × 18.3   (*d*) 76.145 × 40.2
(*e*) 47.62 × 19.8   (*f*) 58.4 × 16.7
(*g*) 341.82 × 3.58   (*h*) 1.3478 × 0.0643
(*i*) 2.2765 × 0.449   (*j*) 0.0342 × 0.823

7 Use the commutative property to provide a check for each example of Exercise 6.

8 Check each example of Exercise 6 by dividing the product by the multiplier. The result should be the multiplicand.

9 Find these quotients. As an approximate check on the division first estimate the quotient.

(*a*) 23.4⟌40.95   (*b*) 2.345⟌716.8665

(*c*) 42.16⟌3.03552   (*d*) 0.0658⟌4.771158

10 Check each example of Exercise 9 by multiplying the quotient by the divisor to get the dividend.

11 Arrange these decimals in order from left to right with the smallest number on the left. 0.021; −0.5; 0.401; 0.514; 0.410; 0.$\overline{6}$; −0.49; 0.400; 0.012; −0.05; 0.395; 0.666; −0.514; 0.499; −0.504; 0.5; −0.012; 0.667

*12   Write each of these numerals in positional notation.

   (a) 6.0×10^{12} (b) 7.9×10^{-9} (c) 5.88×10^{6}
   (d) 1.58×10^{-13} (e) 2.93×10^{7} (f) 8.7×10^{-6}

*13   Write each of these numerals in scientific notation.

   (a) 483,900,000   (b) 0.453   (c) 0.000026
   (d) 253,000   (e) 0.00000125   (f) 29 billion

In Exercises 14 to 25 use scientific notation as an aid in the computation.

*14   93,000,000 × 30   *15   4,710 × 0.03
*16   0.000000182 × 0.00645   *17   0.0000456 × 786,000
*18   113,520,000 ÷ 26,400,000   *19   7,725,000 ÷ 10,300,000,000
*20   0.00002665 ÷ 0.00065   *21   0.000000000480 ÷ 3,000,000,000

*22   Correct to the nearest 10,000 square miles, the total surface area of
   the earth is 196,950,000 square miles. The land surface area is
   57,470,000 square miles. What percent of the earth's surface is land?
*23   In its annual report a car manufacturer stated that during the first six
   months of the year, a total of 2,572,000 cars had been manufactured.
   Of this total 0.8445 were passenger cars. Correct to the nearest
   thousand, how many passenger cars were manufactured during this
   period?
*24   In its semiannual report a large corporation reported that, correct to
   the nearest $100,000, its cash on hand amounted to $347,700,000,
   which was 0.0442 of the total current assets. Correct to the nearest
   $100,000, what were the total current assets of the corporation?
*25   The volume of the earth is 260,000,000 cubic miles. With the exception
   of Pluto, for which no factor has yet been determined, a multiplicative
   factor is given for each planet.† Use this factor to find the volume of
   each planet.

   Mercury, 0.06   Venus, 0.86
   Mars, 0.15   Jupiter, 1,318
   Saturn, 769   Uranus, 50
   Neptune, 59

6-12   CONVERTING COMMON FRACTIONS TO DECIMALS

Any common fraction will convert into a simple decimal which is either finite
or repeating. The division algorithm affords a simple explanation of why
this is so. The fraction a/b may be considered as synonymous with the
expressed division in which a is the dividend and b is the divisor. By the

---

† Factors are taken from Kenneth F. Weaver, Voyage to the Planets, *National Geographic*,
138:152–153 (August, 1970).

$$
\begin{array}{r}
0.75 \\
4\overline{\smash{\big)}\,3.00} \\
\underline{2\ 8} \\
20 \\
\underline{20} \\
\end{array}
$$

FIGURE 6-5　Decimal conversion of $\frac{3}{4}$

Common fractions equivalent to finite decimals
division algorithm, $a = bq + r$ where $0 \leq r < b$. Thus, in the conversion of the fraction a/b, the only possible remainders in the division are the numbers 0, 1, 2, 3, . . . , $b - 1$. At any point in the process when 0 is obtained as a remainder, the division terminates and the fraction has been converted into a finite decimal. When this does not occur, there are only $b - 1$ distinct remainders. Therefore, if it does not occur sooner, repetition will have to take place after $b - 1$ divisions.

The fraction $\frac{3}{4}$ will convert into the finite decimal 0.75 (see Fig. 6-5). To see why this is true, let us examine what takes place.

$$
\frac{3}{4} = \frac{3 \times 10 \times 10}{4 \times 10 \times 10} = \frac{3 \times \overset{5}{\cancel{10}} \times \overset{5}{\cancel{10}}}{\cancel{2} \times \cancel{2} \times 10 \times 10} = \frac{3 \times 5 \times 5}{100} = \frac{75}{100} = 0.75
$$

The fraction $\dfrac{3 \times 10 \times 10}{4 \times 10 \times 10}$ can be written in the form $\dfrac{3.00}{4}$. Each zero in the numerator represents a multiplication by 10. Each decimal place in the numerator represents a division by 10. The quotient is not changed if both dividend (numerator) and divisor (denominator) are multiplied by the same number. But why use two 10s, no more and no less? Notice that the denominator $4 = 2 \times 2$. These two 2s are to be removed from the divisor if the division is to terminate. $10 = 2 \times 5$, so that one 10 will take care of one 2. Why will it take three decimal places before $\frac{3}{8}$ will terminate when converted to a decimal? How many decimal places will be required to convert $\frac{5}{16}$ into a decimal fraction? Why will it take only one decimal place for $\frac{2}{4}$ to terminate and two places for $\frac{2}{25}$ to terminate?

Each 10 used as a multiplier in the numerator of a fraction will remove one and only one 2 and one and only one 5 from the denominator. When a fraction *which is in its lowest terms* is converted to a decimal, the fraction will produce a terminating decimal if and only if each prime factor of the denominator is a 2 or a 5. It will require as many decimal places as there are either 2s or 5s, whichever occurs the larger number of times. For example,

$$
\frac{3}{200} = \frac{3}{2 \times 2 \times 2 \times 5 \times 5}
$$

There are three 2s and only two 5s in the denominator, and so this fraction will convert into a three-place terminating decimal.

```
        0.33 . . .
    3 | 1.00
         9
        ──
        10
         9
        ──
         1
```

FIGURE 6-6 Decimal conversion of $\frac{1}{3}$

What about $\frac{1}{3}$? Since 10 does not contain 3 as a factor, there is no way the division of the numerator by the denominator can terminate. Also, since 3 is the divisor, and the division cannot come out even, there are only two possible remainders that can be obtained: 1 and 2. An examination of the division in Fig. 6-6 shows that actually only the 1 is obtained. Thus this division gives a repeating quotient: $\frac{1}{3} = 0.\overline{3}$. When a common fraction a/b in its lowest terms converts into a repeating decimal, we can say that the quotient will have no more than $b - 1$ digits in the repetend. For example, $\frac{5}{7}$ can have no more than six digits in the part that repeats; $\frac{7}{13}$ can have no more than twelve; etc. We cannot say exactly how many digits there will be without dividing. Actually there are techniques for discovering more precise statements which can be made in some such reductions. However, for our purposes it will suffice to restrict our conclusions to the statement that *no more than* b − 1 *digits can occur in the repetend.*

Now consider a fraction such as $\dfrac{5}{44} = \dfrac{5}{4 \times 11}$. In the denominator there are two 2s, which can be removed by two 10s, and one 11, which cannot be removed. This fraction, when converted to a decimal, will go two places before it starts the repeating pattern and there will be no more than 10 places in the repetend. Why? Actually there are only two (see Fig. 6-7). The fraction $\frac{9}{56}$ will go three decimal places before starting the repeating pattern. Why will $\frac{7}{75}$ go two places before starting the repeating pattern, and $\frac{14}{36}$ only one?

```
        0.1136 . . .
    44 | 5.0000
         4 4
         ──
          60
          44
         ──
         160
         132
         ──
         280
         264
         ──
          16
```

FIGURE 6-7 Decimal conversion of $\frac{5}{44}$

When investigating any common fraction to determine into what type of decimal it will convert, it is first desirable to reduce the fraction to its lowest terms.

6-13 CONVERTING DECIMALS TO COMMON FRACTIONS

It is a very simple matter to convert any terminating simple decimal to a common fraction in its lowest terms. All that is necessary is to use the proper power of 10 as the denominator and to use the numerator of the decimal as the numerator, then reduce.

Finite decimals equivalent to common fractions

$$0.075 = \frac{75}{1,000} = \frac{3}{40}$$

The technique for converting repeating decimals, while intuitively acceptable, really calls for proof. It will not be given here. There are two types of repeating decimals to consider, namely, those like $0.\overline{6}$, which have nothing but a repeating pattern, and those like $0.1\overline{6}$ or $3.2\overline{25}$, which have both a nonrepeating part and a repeating part.

Case I   Repeating decimals which have only a repeating part:

1   Multiply by the power of 10 sufficient to move one *repetend* (one group of digits which repeat) to the left of the decimal point.
2   Subtract the repeating decimal from the newly obtained mixed repeating decimal. The desired result is to have one repetend expressed as an integer.
3   Solve the resulting equation for the value of the repeating decimal.
4   Reduce the common fraction to its lowest terms.

Example   Convert $0.\overline{6}$ to a common fraction in its lowest terms (Fig. 6-8).

1   There is one digit in the repetend, and so the fraction is multiplied by 10 to move one repetend to the left of the decimal point.
2   The repeating decimal is subtracted from the newly obtained mixed decimal to obtain the integer 6.
3   The resulting equation is solved for the value of the repeating decimal.
4   The common fraction is reduced to its lowest terms.

A simple extension of the procedure used in Case I provides the technique for Case II.

$$
\begin{aligned}
10(0.\overline{6}) &= 6.\overline{6} \quad &(1)\\
1(0.\overline{6}) &= 0.\overline{6}\\
\hline
9(0.\overline{6}) &= 6 \quad &(2)\\
0.\overline{6} &= \tfrac{6}{9} \quad &(3)\\
0.\overline{6} &= \tfrac{6}{9} = \tfrac{2}{3} \quad &(4)
\end{aligned}
$$

FIGURE 6-8   Fractional conversion of $0.\overline{6}$

$$100(0.01\overline{6}) = 1.\overline{6}$$
$$= 1\tfrac{2}{3} = \tfrac{5}{3}$$
$$0.01\overline{6} = \tfrac{5}{300} = \tfrac{1}{60}$$

FIGURE 6-9 Fractional conversion of $0.01\overline{6}$

Case II Repeating decimals which have both a repeating part and a nonrepeating part:

1 Multiply by the power of 10 sufficient to move the nonrepeating part to the left of the decimal. The fraction part of the obtained mixed decimal will thus have only a repeating part.
2 Use Case I to convert the mixed decimal into an improper common fraction in its lowest terms.
3 Solve the resulting equation for the value of the repeating decimal.

Example Convert $0.01\overline{6}$ to a common fraction in its lowest terms.

1 There are two digits in the nonrepeating part, so that the fraction is multiplied by 100 to obtain $1.\overline{6}$, a mixed decimal whose fraction part contains only the repeating part of the given fraction (see Fig. 6-9).
2 By Case I, $1.\overline{6} = 1\tfrac{2}{3} = \tfrac{5}{3}$.
3 Solve the equation to get $0.01\overline{6} = \tfrac{5}{300} = \tfrac{1}{60}$.

There is a very simple rule which can be followed for the ready conversion of any repeating decimal of the type of Case I. While this rule can be developed in a rigorous manner, we shall merely state it for use here.

Repeating decimals equivalent to common fractions **Conversion rule for repeating decimals** A repeating decimal, containing only the repeating part, can be converted to a common fraction by writing one complete repetend as the numerator and then writing as the denominator a number composed of as many 9s as there are digits in the numerator.

$$0.\overline{6} = \frac{6}{9} = \frac{2}{3} \qquad 0.\overline{063} = \frac{63}{999} = \frac{7}{111}$$

In summary, the discussion of this last section has illustrated in a great deal of detail the truth of a very important theorem in mathematics. The proof of this theorem is beyond the scope of this book, but since it is of such fundamental significance, it will be stated here as a postulate.

Rational numbers expressed in decimal form **Postulate 6-3** Any rational number $\frac{a}{b}$ $(b \neq 0)$ may be expressed either as a finite decimal or as an infinite repeating decimal. Conversely, any finite or infinite repeating decimal may be expressed as a rational number.

In many mathematical discussions it is desirable to express a finite decimal in the form of an infinite repeating decimal. For example, 0.5 might be expressed either as $0.5\bar{0}$ or as $0.4\bar{9}$ (see Exercise 8, page 235). The latter form is the one which is more generally used when this pattern of representation is used.

6-14 PERCENTAGE

In discussing percentage no really new topic is being considered. The first concern of this discussion was with common fractions, i.e., fractions for which any integer might serve as a numerator and any nonzero integer might serve as a denominator. It was necessary to define the processes for comparing and operating with these new numbers. Second, decimal fractions were discussed, i.e., fractions for which only powers of 10 can serve as denominators. Now the discussion is directed to the use of percents, decimal fractions for which only the second power of 10, or 100, can serve as a denominator. No new definitions are needed to make it possible for us to use percents in making computations. All the rules for the use of fractions apply.

Percents as decimal fractions

The word *percent* is derived from the Latin *per centum*. It means *by the hundred*. Thus a percent, or *rate percent*, is always the ratio of some number to 100. Herein lies the great importance of percentage as an individual topic of elementary mathematics. The concept of the ratio of two integers is very basic to the practices of business, industry, and everyday living. Since this is so generally true it is very desirable to be able to express all such ratios in terms of some convenient denominator. The fact that our number system is decimal in nature makes some power of 10 a very natural choice for this denominator. The first power of 10, or 10, is too small to allow for desirable flexibility in making comparisons. The third power of 10, or 1,000, while desirable for certain types of comparisons—baseball standings, for example—is larger than is necessary for general practice. Thus the second power of 10, or 100, has been selected as the basis for making such comparisons.

Percents as ratios

6-15 USING PERCENTS

Since percents are fractions with a common denominator, the comparisons for equality, greater than, and less than reduce to the simple comparisons of integers. Thus 35% is larger than 20% because $35 > 20$. Here, as in dealing with all fractions, certain precautions are necessary. Although the number $\frac{3}{4}$ is larger than $\frac{1}{2}$, it still is quite possible that $\frac{3}{4}$ of one apple might be much smaller than $\frac{1}{2}$ of another. Similarly, $\frac{3}{4}$ of 20 is much smaller than $\frac{1}{2}$ of 60. In like manner, while 75% is larger than 50% in an abstract sense, it is absolutely essential that the objects or groups to which the percents are applied be taken into consideration before intelligent comparisons can be made (75% of 40 is 30, while 50% of 90 is 45, and 30 < 45).

Percents are always "like fractions" since they always have a common denominator. Thus, they technically meet the conditions for addition or subtraction. Here again it is essential that the base be taken into consideration. There are similar essential considerations of the base in problems involving percents and the processes of multiplication and division. The following example illustrates a basic error that is frequently made, namely, that of averaging percents.

Example   Mr. Thomas was able to invest four sums of money for 1 year as follows: $4,000 at 3.5%; $1,400 at 5%; $3,600 at 2.5%; and $1,000 at 3%. What average percent did he realize on his investment?

The total yield for the year on the total investment of $10,000 was $330 (see Fig. 6-10). Thus the average rate percent is 3.3%. The average of the percents given is 3.5%. If the interest on each amount is computed at 3.3%, the total yield is found to be $330, as would be expected. If the interest on each amount is computed at 3.5%, the total yield found is $350, giving an error of $20. Of course, if Mr. Thomas had invested exactly the same amounts at each of the four different rates of interest, then the average of the percents, 3.5%, would have been correct.

In dealing with percents it is frequently desirable either to convert from the percent form to the fraction form or vice versa. Conversion in either direction is a simple process if one only remembers that any percent represents a decimal fraction with two decimal places, or a common fraction with 100 as the denominator.

3.5% of $4,000 = $140		3.5%
5.0% of $1,400 = 70		5.0%
2.5% of $3,600 = 90		2.5%
3.0% of $1,000 = 30		3.0%
$10,000 $330		4 | 14.0%
		3.5%

Average percent

3.5% of $4,000 = $140
3.5% of $1,400 = 49
3.5% of $3,600 = 126
3.5% of $1,000 = 35
$350

Average yield

3.3% of $4,000 = $132.00
3.3% of $1,400 = 46.20
3.3% of $3,600 = 118.80
3.3% of $1,000 = 33.00
$330.00

FIGURE 6-10   Average percent on investment

Example
$$25\% = 0.25 = \frac{25}{100} = \frac{1}{4}$$

$$37\tfrac{1}{2}\% = 0.37\tfrac{1}{2} = \frac{37\tfrac{1}{2}}{100} = \frac{75}{200} = \frac{3}{8}$$

$37\tfrac{1}{2}\%$ may also be written as 37.5%, which is to be interpreted as $\frac{37.5}{100} = \frac{375}{1,000} = 0.375$.

6-16 THE PERCENTAGE FORMULA

One of the basic relationships that pervades all mathematics, and particularly arithmetic, is the product relationship, in which one number is the product of two other numbers. In this relationship one number serves as the number upon which the relationship is based. We call it the multiplicand or, better, the *base*. A second number serves as the *multiplier* which operates on the base to give the *product*. In formula form

$$\text{Product} = \text{multiplier} \times \text{base} \qquad \text{or} \qquad p = m \times b = mb$$

In the language of percentage, the multiplier is the rate percent r and the product is the percentage p. In such product relationships there are three numbers involved. If any two of these numbers are known, the third one can be determined.

Thus, $p = rb$ states that there is a certain unknown number which is a given percent of a given number; $b = p/r$ states that there is a certain unknown number of which a given number is a given percent; and $r = p/b$ states that one given number is a certain unknown percent of another given number.

Since rate percent always implies a ratio, all percents can be stated in the form of proportions. The proportion form combines all three types of problems into one, a single proportion in which one number is unknown. The basic proportion is

$$\frac{p}{b} = \frac{r}{100}$$

The above problems expressed in this proportion form are as follows:

1 What number is 25% of 600? $\dfrac{p}{600} = \dfrac{25}{100}$

2 150 is 25% of what number? $\dfrac{150}{b} = \dfrac{25}{100}$

3 150 is what percent of 600? $\dfrac{150}{600} = \dfrac{r}{100}$

It is thus evident that the whole topic of percentage is merely a consideration of a very special type of fraction, namely, a decimal fraction consisting of two decimal places or a common fraction with 100 as the denominator.

Furthermore, it should be evident that the frequently-referred-to three problems of percentage are by no means three distinct and unrelated problems. They are merely three related aspects of a basic arithmetical relationship; there are three numbers so related that one is the product of the other two, and the problem is to find one of the numbers when the other two are known.

EXERCISES

In Exercises 1 to 4, assume that each fraction is in its lowest terms and describe the type of decimal fraction into which it will convert.

1 $\dfrac{p}{2^m}$: (a) $p < 2^m$, (b) $p > 2^m$

2 $\dfrac{p}{5^n}$: (a) $p < 5^n$; (b) $p > 5^n$

3 $\dfrac{p}{2^m 5^n}$: (a) $p < 2^m 5^n$; (b) $p > 2^m 5^n$

4 Assume that $q \neq 0$ has neither 2 nor 5 as a factor: (a) $\dfrac{p}{q}$; (b) $\dfrac{p}{2^m 5^n q}$.

5 Convert each of these fractions to a terminating decimal. First state how many places will be necessary before each terminates.

(a) $\dfrac{5}{8}$ (b) $\dfrac{7}{125}$ (c) $\dfrac{9}{2,500}$ (d) $\dfrac{6}{32}$

(e) $\dfrac{15}{75}$ (f) $\dfrac{9}{72}$ (g) $\dfrac{60}{250}$ (h) $\dfrac{49}{560}$

6 Why will these fractions not convert into terminating decimals? Convert each fraction to a decimal.

(a) $\dfrac{3}{7}$ (b) $\dfrac{5}{6}$ (c) $\dfrac{9}{54}$

(d) $\dfrac{11}{56}$ (e) $\dfrac{6}{36}$ (f) $\dfrac{12}{450}$

7 Convert each of these decimals to common fractions in their lowest terms.

(a) 0.045   (b) $0.04\overline{5}$ (c) $0.0\overline{45}$ (d) $0.\overline{045}$

8 Show that $0.5\overline{0} = \frac{1}{2} = 0.5$; also that $0.4\overline{9} = \frac{1}{2} = 0.5$.

9 Arrange these numbers in order from left to right with the smallest number on the left: $0.\overline{6}$, 31%, 0.032, $\frac{2}{5}$, 0.403, 0.3%, $\frac{3}{4}$, 0.3, $\frac{2}{7}$, 0.430, $\frac{1}{2}$%, 0.302, $\frac{3}{5}$, 0.285, $\frac{3}{7}$.

10 (*a*) What is the numerical error in using 0.667 to represent $\frac{2}{3}$?

(*b*) What is the approximation to $\frac{2}{3}$ that has an error which is $\frac{1}{100}$ of the error in 0.667?

11 Write each of these percentage problems as an exercise in proportion:

(*a*) 13 is 15% of what number?

(*b*) 13 is what percent of 15?

(*c*) 13% of 15 is what number?

12 The figure presents hypothetical data relative to two situations of daily attendance in school. Note that the percents are identical in both situations. The average of these percents is 70%.

(*a*) Show that this is correct for the average percent of attendance in situation B but incorrect in situation A. Why is this so?

(*b*) If 70% is used as the average percent of attendance in situation A, what is the error in number of pupils?

(*c*) Compute the average percent of attendance in situation A correct to the nearest tenth percent.

(*d*) Prove that your answer gives the correct representation.

A. FIVE GRADES FOR ONE DAY				B. ONE GRADE FOR FIVE DAYS			
GRADE	ENROLL-MENT	ATTEND-ANCE	PER-CENT	DAYS	ENROLL-MENT	ATTEND-ANCE	PER-CENT
4	80	72	90	Mon.	80	72	90
5	70	28	40	Tues.	80	32	40
6	100	70	70	Wed.	80	56	70
7	90	90	100	Thurs.	80	80	100
8	40	20	50	Fri.	80	40	50
Total	380	280		Total	400	280	

EXERCISE 12 Records of daily attendance

13 What number is 150% of 124?

14 What number is 150% more than 306?

15 Mr. Smith's salary was increased from $12,000 to $12,600. What was the percent of increase?

16 A suit of clothes priced at $85.75 was sold at a discount of 20%. At what price did the suit sell?

17 A department store sells a certain type of merchandise at a price such that the cost is $66\frac{2}{3}$% of the selling price. What should be the selling price of an article which cost $49.50?

18 A certain article sold for $38.85 including a 5% sales tax. What was the selling price of the article?

19 An article priced to sell for $208 was later sold for $176.80. What was the discount rate percent?

20 A camera priced to sell for $59.94 was discounted $16\frac{2}{3}$%. What was the marked down price?

∗21 Use the fact that $\frac{1}{7} = 0.\overline{142857}$ to prove that 1 may be written as the periodic decimal $0.\overline{9}$.

∗22 If $0 < a < 15$ there are three distinct decimal fraction patterns into which the proper fraction $\frac{a}{15}$ can be converted. What are these patterns?

6-17 REVIEW EXERCISES

1–28 Write a carefully constructed answer to each of the guideline questions for this chapter.

29 What relation ($=$, $<$, or $>$) does the first number in each pair of numbers have to the second?

(a) $\left(\frac{2}{3}, \frac{30}{45}\right)$ (b) $\left(\frac{5}{8}, \frac{14}{16}\right)$ (c) $\left(\frac{-4}{5}, \frac{-7}{9}\right)$

(d) $\left(0, \frac{-1}{3}\right)$ (e) $\left(\frac{1}{1}, \frac{25}{25}\right)$ (f) $\left(\frac{4}{3}, \frac{4}{4}\right)$

(g) $\left(\frac{75}{100}, \frac{3}{4}\right)$ (h) $\left(\frac{1}{2}, 0\right)$ (i) $\left(\frac{6}{7}, \frac{8}{9}\right)$

30 Write ten elements in the equivalence class represented by each symbol.

(a) $\frac{3}{7}$ (b) $\frac{5}{2}$ (c) 0   (d) 1   (e) $\frac{-8}{9}$ (f) $\frac{15}{5}$

31 Use each pair of rational numbers to illustrate the fact that rational numbers are commutative under addition and multiplication.

(a) $\left(\frac{7}{8}, \frac{5}{3}\right)$ (b) $\left(\frac{-2}{3}, \frac{4}{7}\right)$ (c) $\left(\frac{-3}{5}, \frac{-6}{9}\right)$

32 Use each pair of rational numbers of Exercise 31 to prove that rational numbers are not commutative under either subtraction or division.

33 Use each set of three rational numbers to illustrate that rational numbers are associative under addition and multiplication.

(a) $\left\{\frac{1}{2}, \frac{1}{3}, \frac{1}{4}\right\}$ (b) $\left\{\frac{-3}{5}, \frac{2}{7}, \frac{1}{4}\right\}$

(c) $\left\{\frac{4}{3}, \frac{-5}{2}, \frac{-8}{7}\right\}$ (d) $\left\{-\frac{1}{7}, -\frac{1}{2}, -\frac{1}{5}\right\}$

34 Use the sets of rational numbers of Exercise 33 to prove that rational numbers are not associative under either subtraction or division.

35 Why is it correct to use the word *prove* in Exercises 32 and 34 while It would be incorrect in Exercises 31 and 33?

36 Use the sets of rational numbers of Exercise 33 to illustrate that multiplication is distributive over either addition or subtraction.

37　(a) Show that $(\frac{2}{3} + \frac{5}{6}) \div \frac{3}{4} = (\frac{2}{3} \div \frac{3}{4}) + (\frac{5}{6} \div \frac{3}{4})$.

(b) Show that $\frac{3}{4} \div (\frac{2}{3} + \frac{5}{6}) \neq (\frac{3}{4} \div \frac{2}{3}) + (\frac{3}{4} \div \frac{5}{6})$.

38　In each case give the value of the rational number r.

(a) $\frac{2}{3} + r = 0$ 　　　　　　(b) $\frac{-3}{5} + r = 0$

(c) $\frac{5}{9} \cdot r = 1$ 　　　　　　(d) $\frac{-3}{4} \cdot r = 1$

(e) $r + \frac{0}{2} = \frac{7}{8}$ 　　　　　(f) $r - \frac{0}{3} = \frac{3}{4}$

(g) $r + \frac{-15}{4} = 0$ 　　　　(h) $r - \left(\frac{-7}{2}\right) = 0$

(i) $\frac{7}{3} \cdot \frac{10}{10} = r$ 　　　　(j) $\frac{-2}{5} \cdot \frac{8}{8} = r$

(k) $r \cdot \frac{2}{3} = 0$ 　　　　　(l) $\frac{-5}{6} \cdot \frac{-1}{1} = r$

(m) $\frac{-4}{4} \cdot \frac{7}{8} = r$ 　　　　(n) $\frac{-7}{2} \cdot r = 0$

39　Is the set of all nonzero rational numbers with addition and multiplication a field?

40　Find these sums. Use the associative property to provide a check.

(a) $5\frac{2}{3} + 7\frac{1}{2} + \frac{1}{4} + \frac{5}{6}$ 　　　　　　(b) $8\frac{3}{8} + (-6\frac{3}{4}) + (-2\frac{1}{2})$

(c) $16\frac{3}{5} + 15\frac{1}{2} + 4\frac{7}{10} + (-3\frac{1}{5})$ 　　(d) $(-7\frac{3}{4}) + (-5\frac{1}{2}) + (-4\frac{5}{6})$

41　Find these products. Use the associative property to provide a check.

(a) $4\frac{1}{2} \times 3\frac{1}{3} \times 6\frac{1}{5}$ 　　　　　(b) $\left(-\frac{2}{3}\right) \times 7\frac{1}{2} \times 5\frac{1}{7}$

(c) $(-6\frac{3}{4}) \times (-3\frac{1}{9}) \times \frac{4}{5}$ 　　(d) $(-8\frac{1}{3}) \times (-4\frac{1}{5}) \times (-6\frac{1}{2})$

42　Find these quotients. Use multiplication to provide a check.

(a) $8\frac{5}{8} \div 4\frac{3}{5}$ 　　　　　(b) $-7\frac{5}{16} \div 3\frac{1}{4}$

(c) $16\frac{7}{8} \div (-1\frac{14}{16})$ 　　(d) $(-6\frac{8}{14}) \div (-9\frac{6}{7})$

43　Find six rational numbers between each pair of numbers.

(a) 0 and 1 　　　(b) $\frac{2}{3}$ and $\frac{3}{4}$ 　　　(c) $\frac{-7}{8}$ and $\frac{-5}{6}$

(Hint: See Exercise 28, Sec. 6-7.)

44　Solve and check each example. In each case use the inverse operation for the check.

(a) 235.63 + 740.09 + 18.71 + 61.78

(b) 0.02348 + 0.00310 + 0.00209 + 0.10064

(c) 8906.02 − 7859.81

(d) 70.002 − 41.876

(e) 0.0567 − 0.0289

(f) 27.16 × 7.03

(g) 0.157 × 0.083

(h) 0.00019152 ÷ .042

(i) 297.585 ÷ 3.89

Use scientific notation in the remaining examples.

∗(j) 7,685,000 × 34,200

∗(k) 62,800,000 × 0.00345

∗(l) 0.00000674 × 0.00000236

∗(m) 14,688,000 ÷ 24,000,000

∗(n) 28,512,000,000 ÷ 891,000,000,000

∗(o) 0.0000006846 ÷ 0.00042

45   Arrange these numbers in order from left to right with the smallest on the left. $\frac{2}{5}$; 0.8572; $\frac{1}{3}$; 0.401; $\frac{3}{7}$; $\frac{9}{11}$; 0.429; 33%; $\frac{6}{7}$; 0.5%; 0.428; 0.8571; $\frac{4}{5}$; 5%; 0.399

In Exercises 46 to 51 first use the percentage formula to set up the problem and then solve it.

46   What number is 15% of 96?

47   54 is what percent of 81?

48   71.3 is $5\frac{3}{4}$% of what number?

49   What will be the interest on $1,500 for 4 months at 7% annually?

50   A dress originally priced to sell for $45.99 was discounted to sell for $30.66. The discount was what percent of the original price?

51   A color television set was advertised to sell for $598.50, a 25% reduction from the original sales price. What was the original sales price?

In Exercises 52 to 55 use scientific notation.

∗52   During a recent year it was estimated that the revenue for the United States budget would come from these sources: individual income tax, 42%; employment tax, 17%; excise tax, 8%; miscellaneous revenue, 13%; and the balance from corporation income tax.

(a) What percent of the budget came from the corporation tax?

(b) If the total expected budget for the year was $181,100,000,000, how much money would be needed from each source of income to meet this budget?

(c) What is the final check on your answers to all the questions?

∗53   The total surface area of the earth is 196,951,000 square miles, and the total water surface area is 139,481,000 square miles. Each measure is given correct to the nearest thousand miles. To the nearest tenth what percent of the total surface area is water surface?

∗54   Of the water surface area, the Pacific Ocean has 63,985,000 square miles; the Atlantic Ocean, 31,529,000 square miles; the Indian Ocean,

28,357,000 square miles; and the Arctic Ocean, 5,542,000 square miles. What percent of the total water surface is in each ocean? Give your answers correct to the nearest tenth percent.

*55   What percent of the total water surface is not contained in the oceans?

56   On a vacation trip Mr. Byrd stopped to get gas. He noticed that his odometer registered 12,462.7 miles. At the next stop for gas it registered 12,802.1 miles, and 19.3 gallons were required to fill his tank. For this portion of his trip, correct to the nearest tenth of a mile how many miles per gallon did Mr. Byrd average? Correct to the nearest cent what was the total amount for this purchase of gas at 39.9 cents per gallon?

57   The gas tickets for the entire trip showed these amounts: 10.0; 10.3; 19.6; 14.5; 16.5; 14.9; 12.7; 15.3; 18.8; 18.0; 19.3; 13.0; 9.8; 16.2; 22.8; 19.6. How many gallons of gas did Mr. Byrd use on his trip?

58   If Mr. Byrd's total gas bill for his trip was $97.76, what average price per gallon did he pay for the gas used? Compute your answer correct to the nearest tenth of a cent.

59   A certain kind of brass is composed of 61.5% copper, 3.0% lead, 0.4% tin, and the balance is zinc. How many pounds of each component must be mixed with 2 pounds of tin to produce this type of brass?

*INVITATIONS TO EXTENDED STUDY

1   If a set S is an abelian group with respect to addition and the nonzero elements of S form an abelian group with respect to multiplication, is the system a field? Why or why not?

In Exercises 2 to 9 argue the truth of each statement.

2   If $r,s \in Q$ and $r \times s = 0$, then $r = 0$ or $s = 0$.

3   The order relation for rational numbers is transitive.

4   Addition and multiplication of rational numbers are unique.

5   The additive identity for rational numbers is unique.

6   The multiplicative identity for rational numbers is unique.

7   The additive inverse for rational numbers is unique.

8   The multiplicative inverse for rational numbers is unique.

9   If a,b are nonzero integers, then $\dfrac{1}{a} + \dfrac{1}{b} \neq \dfrac{1}{a + b}$.

10   Present a rigorous argument to prove the *conversion rule for repeating decimals* as stated on page 231.

11   Prove the theorem stated as Postulate 6-3.

12   Extend the study of decimal fractions to the investigation of *radix* fractions (fractions in numeral systems with different bases) and the conversions of fractions in one base to equivalent fractions with a different base.

THE NUMBER FIELD OF
ELEMENTARY MATHEMATICS

GUIDELINES FOR CAREFUL STUDY

The field of rational numbers provides the basic number structure sufficient for the intelligent servicing of a variety of the demands for quantitative inquiry and computational performance. This number field is characterized by a great amount of operational freedom due to the fact that the set of rational numbers is closed under each of the four operations addition, subtraction, multiplication, and division by nonzero divisors. In spite of this fact there still remain a few restrictions which prevent the exercise of complete procedural freedom. These restrictions will be considered in this chapter.

The following questions will serve as helpful guidelines for the careful study of Chap. 7:

1   What is the definition of a field?
2   In what decimal forms may rational numbers be expressed?
3   What is meant by the word "factor"?
4   What is meant by raising to a power?
5   What does a positive integer used as an exponent indicate?
6   What is meant by the extraction of a root?
7   What is the essential difference between a direct proof and an indirect proof?
8   What is an irrational number?
9   How may an irrational number be expressed as a decimal fraction?
10   What is the process for rounding approximate numbers?
11   What is a real number?
12   What is a number line and how may one be constructed?
13   What is meant by the absolute value of a real number?
14   What is a general condition that may be used for placing any two real numbers in their proper order with respect to each other?
15   What is the imaginary unit?
16   What is a complex number?
17   Why may the set of real numbers be considered as a proper subset of all complex numbers?
18   What are the definitions of equality, addition, subtraction, multiplication, and division for complex numbers?
19   What are the additive and multiplicative identities for complex numbers?
20   What are the essential differences between these number fields: rational numbers, real numbers, complex numbers?
21   Can you, without reference to the book, construct the diagram of the number systems of elementary mathematics?

INTRODUCTION

In Chap. 4, when subtraction was introduced and defined as the inverse operation of addition, it became necessary to extend the natural number system to embrace the integers. This extension gave us the domain of integers, which was closed not only under addition and multiplication but also under the newly defined operation of subtraction. A further extension to the field of rational numbers (Chap. 6) became necessary to provide closure under the operation of division with nonzero divisors, defined as the inverse of multiplication.

Involution

The closure property for multiplication has still other consequences which need to be investigated. A special form of multiplication, called *raising to powers (involution)*, occurs when all the factors in a given product are the same. As previously noted (Sec. 1-6), such multiplication may be expressed in a condensed form using exponents. For example, $2^3 = 2 \cdot 2 \cdot 2 = 8$ and $\left(\frac{3}{4}\right)^2 = \frac{3}{4} \cdot \frac{3}{4} = \frac{9}{16}$. The closure property states that whether the equal factors are natural numbers, integers, or rational numbers, the product and the repeating factor are elements of the same set. Since $N \subset Z \subset Q$, this last statement may be summarized in these words: If we are given the fact that $a^n = b$ where a is a rational number and n is a positive integer, then we know that b is necessarily a rational number. Does it also follow that if b is a rational number and n a positive integer, then a is necessarily a rational number? Although the example exhibits a value for a in each of the two cases $a^3 = 8$ and $a^2 = \frac{9}{16}$, it does not follow that either of these values is necessarily the only value of a for which the respective equality holds. For example, if $a = -\frac{3}{4}$, then $a^2 = \frac{9}{16}$ since $\left(-\frac{3}{4}\right) \cdot \left(-\frac{3}{4}\right) = \frac{9}{16}$. That there are values other than 2 such that $a^3 = 8$ will be demonstrated later. (See Exercise 8d and i, Sec. 7-6.) Furthermore, there are rational numbers b for which there exist no rational numbers a such that $a^n = b$ for n an integer greater than 1. The truth of this statement will be established in Theorem 7-1 where it is proved there is no rational number a such that $a^2 = 2$.

Evolution

Another form in which to make statements about a, equivalent to $a^3 = 8$ and $a^2 = \frac{9}{16}$, is to use radicals to indicate *extraction of roots (evolution)*. In this form $\sqrt[3]{8} = 2$ is read "a cube root of 8 is 2" or "the principal cube root of 8 is 2," and $\sqrt{\frac{9}{16}} = \frac{3}{4}$ is read "the positive square root of $\frac{9}{16}$ is $\frac{3}{4}$" or "the principal square root of $\frac{9}{16}$ is $\frac{3}{4}$." The difference in the wording of the two statements enclosed in quotation marks is important but not of too great concern here. The symbol $\sqrt[3]{8}$ may be interpreted also in the form of the question: What is a number which used three times as a factor will give 0 as the product? Similarly the symbol $\sqrt{\frac{9}{16}}$ may be interpreted as the question: What positive number used two times as a factor will give $\frac{9}{16}$ as the product?

Unary operations

The operations of involution and evolution are *unary* operations since they affect only one number. They are inversely related just as are addition and subtraction, multiplication and division. The field of rational numbers has closure under involution, since it is a form of multiplication. For the number system to have closure under evolution also, further extensions are necessary.

7-1 THE CONCEPT OF IRRATIONAL NUMBER

One illustration of the necessity for an additional extension of the number system is established by Theorem 7-1. In the proof of this theorem the pattern of indirect proof is used. In such an argument the desired conclusion is assumed to be false. This assumption is then used as a basis for the argument. If it can be shown that it leads to a contradiction of a statement that is already accepted as true, the assumption is then known to be false. From this it follows that the conclusion, as stated in the theorem, must be true.

In this case the theorem states that there exists no rational number whose square is 2. As the contradiction of this desired conclusion we shall assume that there does exist a rational number whose square is 2. The argument will then show that this assumption implies a conclusion which contradicts a statement that must be true if the hypothesis of the theorem is true. Since this is an impossible situation, the assumption must be false. It then follows that the stated conclusion of the theorem is true.

Theorem 7-1   There exists no rational number whose square is 2.

Hypothesis   There exists the set of all rational numbers.

Conclusion   The square of no rational number is 2.

Proof

Statements	*Reasons*
1 There exists the rational number a/b where a and b are relatively prime integers and $b \neq 0$.	Hypothesis and the reduction of fractions
2 Assume $a^2/b^2 = 2$.	The assumed contradiction of the conclusion
3 $a^2 = 2b^2$	E-5 (multiply both sides by b^2)
4 Since 2 is a factor of the square of a, it must be a factor of a.	Since $a^2 = a \cdot a$ by Postulate 3 3. any prime factor of a^2 must also be a factor of a.

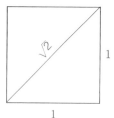

FIGURE 7-1 A pythagorean discovery

5 $a = 2m$, where m is an integer and $a^2 = 4m^2$.	Definition of factor
6 $4m^2 = 2b^2$	Substitution in step 3
7 $2m^2 = b^2$	E-5 (multiply both sides by $\frac{1}{2}$)
8 $b = 2k$, where k is an integer.	Same reasoning as in steps 4 and 5
9 a and b have a common factor 2, which contradicts step 1.	Steps 5 and 8
10 Therefore 2 cannot be expressed as the square of a rational number.	

While there exists no rational number whose square is 2, the pythagoreans exhibited the fact that there does exist such a number. It expresses the value of the ratio of the diagonal of a square to the length of a side (see *Irrational numbers* Fig. 7-1). Such numbers, which are not rational numbers, are called *irrational numbers.*† Many other types of irrational numbers exist. In addition to a few square roots and cube roots, the only such numbers significant here are π, the ratio of the circumference of a circle to its diameter, and, later in Chap. 9, the values of a few ratios related to triangles.‡

7-2 DECIMAL REPRESENTATION OF IRRATIONAL NUMBERS

We have seen how every rational number may be expressed uniquely either as a finite decimal or as an infinite repeating decimal (Postulate 6-3). Similarly, we shall postulate a provable theorem concerning irrational numbers:

Irrational numbers as infinite decimals

Postulate 7-1 Every irrational number may be expressed in one and only one way as an infinite nonrepeating decimal.

---

† The "ir" of the word "irrational" is a derivative from the Latin prefix "in," which means *not*, and so the word "irrational" means simply "not rational."

‡ To prove that π is an irrational number is a much more involved process than to prove that $\sqrt{2}$ is irrational. That π is irrational was first shown by J. H. Lambert in 1767, and later supported by F. Lindeman in 1882.

$$1^2 = 1 < 2 < 4 = 2^2$$
$$(1.4)^2 = 1.96 < 2 < 2.25 = (1.5)^2$$
$$(1.41)^2 = 1.9881 < 2 < 2.0164 = (1.42)^2$$
$$(1.414)^2 = 1.999396 < 2 < 2.002225 = (1.415)^2$$
$$(1.4142)^2 = 1.99996164 < 2 < 2.00024449 = (1.4143)^2$$
$$(1.41421)^2 = 1.9999899241 < 2 < 2.0000182084 = (1.41422)^2$$

FIGURE 7-2 Approximation to the positive square root of 2

Although techniques exist for determining the successive digits in the decimal representation of irrational numbers, they do not submit to simple explicit formalization as in the case of rational numbers. In Fig. 7-2 two step-by-step approximations to the positive square root of 2 are given.

None of the rational numbers exhibited in Fig. 7-2 gives 2 when squared. Each is a rational approximation to the square root of 2. From the last line we know that 1.41421 is a rational approximation to $\sqrt{2}$, correct to five decimal places. As can be seen, its square differs from 2 by a very small amount, and by a smaller amount does the square of 1.41422 differ from 2. The rule for rounding numbers (Sec. 6-10) may be used to obtain from this value other approximations. For example, 1.414 is the rational approximation to $\sqrt{2}$ correct to three decimal places. The approximations of Fig. 7-2 were obtained from a table of square roots. If such a table is not available, division may be used to compute any reasonable approximation.

Example What is $\sqrt{3}$ correct to four decimal places?

1   Select a number n between the two integers whose squares are closest to 3: $1^2 = 1$ and $2^2 = 4$, and so n is between 1 and 2 and closer to 2. Select $n = 1.8$.

2   Divide 3 by the selected number: $3 \div 1.8 = 1.66667$. Five places are used to protect the fourth decimal place. As long as the digit in the fifth decimal place is 5 or more it can affect the fourth decimal place.

3   Average the divisor and quotient of step 2 to obtain a new divisor: $\frac{1}{2}(1.8 + 1.66667) = 1.73334$.

4   Divide 3 by this new divisor: $3 \div 1.73334 = 1.73076$. Continue to repeat steps 3 and 4 until the desired approximation has been obtained.

5   $\frac{1}{2}(1.73334 + 1.73076) = 1.73205$

6   $3 \div 1.73205 = 1.73205$. This last division establishes that, correct to four decimal places, $\sqrt{3} = 1.7321$.

The chronology of π

The chronology[†] of π is an interesting and eventful story dating from the

[†] Howard Eves, "An Introduction to the History of Mathematics," pp. 88–95, Holt, Rinehart and Winston, Inc., New York, 1969.

H. C. Schipler, The Chronology of Pi, *Mathematics Magazine*, January–February, 1950, pp. 165–170; March–April, 1950, pp. 216–228; May–June, 1950, pp. 279–283.

first scientific efforts of Archimedes (ca. 240 B.C.) to the modern age of the electronic computer. It has been given distinctive character by Biblical edict,[†] efforts at state legislation,[‡] the peculiar peregrinations of circle squarers,[§] and inquiring research of competent mathematicians using the techniques of infinite series and, in recent years, the facilities of electronic computers. A most recent such computation gives the value of π correct to 500,000 decimal places.[¶]

As with the square root of 2, rational approximation of the value of π can be given correct to any reasonably desired number of decimal places. Probably the most familiar approximations used are $\frac{22}{7}$, $3\frac{1}{7}$, 3.14, and 3.1416. Many mnemonics have been given from time to time to aid in quoting a value of π correct to a specified number of decimal places.

EXERCISES

1 Give three illustrations each of the operations of involution and evolution. Explain how these two operations are inversely related.

2 Distinguish between rational and irrational numbers.

3 Are 0 and 1 rational or irrational numbers?

4 Prove that each of these numbers is irrational: (a) $\sqrt{3}$; (b) $\sqrt{8}$; (c) $\sqrt[3]{4}$.

5 Use division to find each of these square roots correct to four decimal places: (a) $\sqrt{5}$; (b) $\sqrt{7}$; (c) $\sqrt{8}$; (d) $\sqrt{15}$; (e) $\sqrt{150}$.

6 The irrational numbers π and e are very important in mathematics and science. Rational approximations to their values are: $\pi \approx 3.14159$ and $e \approx 2.71828$, each correct to five decimal places. Show that $r = \dfrac{3.14159 + 2.71828}{2}$ is between these approximations to π and e.

Is this a rational number or an irrational number? Why?

7 Is the number of Exercise 6 a rational number between π and e? (*Hint:* Correct to six decimal places $\pi \approx 3.141593$ and $e \approx 2.718282$.)

* 8 It can be argued that $\dfrac{\pi + e}{2}$ is an irrational number. Prove that $\pi > \dfrac{\pi + e}{2} > e$.

9 Let a, b, c represent three rational numbers such that $1 < a < b < c < 2$. Find three other such numbers between 1 and 2: (a) if c is the first one determined; (b) if a is the first one determined.

[†] "Then he made the molten sea; it was round, ten cubits from brim to brim, and five cubits high, and a line of thirty cubits measured its circumference." (I Kings 7:23 and II Chronicles 4:2.)
[‡] House Bill No. 246 of the Indiana State Legislature (1897): "Be it enacted by the General Assembly of the State of Indiana: It has been found that a circular area is to the square on a line equal to the quadrant of the circumference, as the area of an equilateral rectangle is to the square on one side . . . ." (Eves, *ibid.*)
[§] Among the more ambitious of such efforts are the voluminous writings of one author in support of the claim that the value of π is given exactly by the rational number $3\frac{13}{81}$.
[¶] Eves, *ibid.*

10 Use Exercises $4b$ and $5c$ as aids in showing that $\sqrt{8}$ is an irrational number between π and e.

11 Find five rational numbers between $\sqrt{2}$ and $\sqrt{5}$, if $\sqrt{2} \approx 1.414$ and $\sqrt{5} \approx 2.236$, each correct to three decimal places.

12 Find five rational numbers between $\sqrt{3}$ and $\sqrt{7}$.

13 Find five rational numbers between $\sqrt[3]{4}$ and $\sqrt{4}$.

14 Find five rational numbers between $\sqrt[3]{5}$ and $\sqrt[3]{10}$, if $\sqrt[3]{5} \approx 1.7100$ and $\sqrt[3]{10} \approx 2.1544$, each correct to four decimal places.

15 The value of π, correct to five decimal places, is 3.14159. Which of the numbers 3.142, $\frac{22}{7}$, or $3\frac{13}{81}$ is the closest approximation to this value?

16 The formula for the circumference C of a circle in terms of the diameter d is $C = \pi d$. The equatorial diameter of the earth is 7,926.4 miles. Correct to the nearest mile what is the equatorial circumference of the earth?

17 The earth makes one complete revolution in 24 hours. Correct to the nearest tenth of a mile, at what speed does a point on the equator travel?

7-3 THE CONCEPT OF REAL NUMBER

In each of the previous extensions—from natural numbers to integers, from integers to rational numbers—it has been fairly simple to extend the definitions of the basic operations to accommodate the new numbers. Historically this has not been the case with the extension to real numbers. In fact, a good many centuries marked the time between the disturbing discovery of the existence of irrational numbers by the pythagoreans (ca. 540 B.C.) and the refinement of their definition late in the nineteenth century by Georg Cantor (1845–1918), Richard Dedekind (1831–1916), and Karl Weierstrass (1815–1897). These and subsequent researches have led to further refinements, all of which have given a sound philosophical basis for the generalization from the field of rational numbers of the operations of addition, subtraction, multiplication, and division as well as the relations of "equal to," "less than," and "greater than." All the properties which characterize these operations and relations in the field of rational numbers are retained under this new extension. Unlike the rational numbers, however, the irrational numbers are not closed under these operations. One counterexample in each case will suffice to establish the truth of this statement:

1 *Addition* $\sqrt{2} + (-\sqrt{2}) = 0$; definition of additive inverse.

2 *Subtraction* $\sqrt{2} - \sqrt{2} = \sqrt{2} + (-\sqrt{2}) = 0$; definition of subtraction.

3 *Multiplication* $\sqrt{2} \times \sqrt{2} = 2$; definition of square root.

4 *Division* $\sqrt{2} \div \sqrt{2} = 1$, since $\sqrt{2} \times 1 = \sqrt{2}$; definition of division.

The numbers 0, 1, and 2 are not irrational numbers.

Rational numbers as a proper subset of real numbers

The set of all rational numbers and the set of all irrational numbers are disjoint sets and their union is the set of all *real numbers.* Since every irrational number can be expressed uniquely as an infinite nonrepeating decimal (Postulate 7-1), we may state the following postulate concerning real numbers:

Postulate 7-2 Every real number may be expressed in one and only one way either as a finite decimal, an infinite repeating decimal, or an infinite nonrepeating decimal.

On page 232 the statement was made that, at times, mathematicians prefer to think of finite decimals as infinite repeating decimals. In Exercise 8, page 235, you were asked to show that the rational number $\frac{1}{2}$ can be expressed as an infinite repeating decimal in either of the two forms $0.5\overline{0}$ or $0.4\overline{9}$. Likewise, 0.43 can be expressed as $0.43\overline{0}$ or $0.42\overline{9}$. (As an exercise you might prove this.) A similar statement can be proved about any rational number. For certain good reasons the mathematician usually prefers to use the succession of 9s rather than the succession of 0s. If it is agreed that any finite decimal will be written as an infinite repeating decimal in a form such as $0.5 = 0.4\overline{9}$ or $0.43 = 0.42\overline{9}$, Postulate 7-2 may be written in the equivalent, but simpler, language:

Real numbers as infinite decimals

Postulate 7-2′ Every real number may be expressed uniquely as an infinite decimal.

Postulate 7-2, or its simplified form, provides the background for the proof of an important property of real numbers which will be postulated here:

The real number line

Postulate 7-3 There exists a one-to-one correspondence between the set of all real numbers and the set of points on a line; that is, to each real number there corresponds one and only one point, and to each point there corresponds one and only one real number.

This correspondence on the *number line* may be set up in a manner described in these six steps:

1 On a line of infinite extent, mark a reference point, *O* in Fig. 7-3, and place it in correspondence with the real number 0. This point will be referred to as the *origin* or *point of reference.*
2 Choose a positive direction on the line, to the right in Fig. 7-3. The opposite direction (to the left) then will be the negative direction on the line.
3 Mark a second point, *A* in Fig. 7-3, and place it in correspondence with the real number 1. *OA* then becomes a unit to be applied to the line.

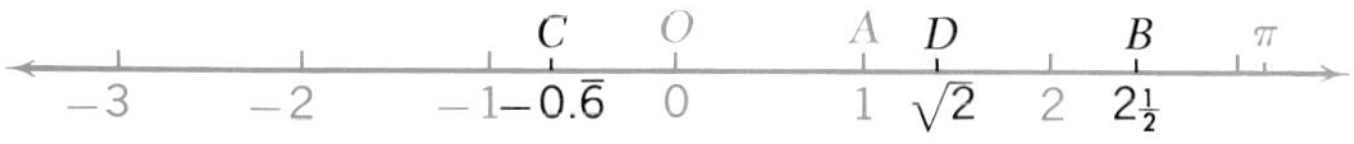

FIGURE 7-3 The real number line

4 Assume that this unit can be applied indefinitely along the line, as it were, without loss of identity. By such application to the right from O, points are selected which are in one-to-one correspondence with the positive integers. Application to the left from O will select the points in correspondence with the negative integers.

5 Points can now be placed in correspondence with the rational fractions, by constructing the appropriate fractional parts of the unit length OA and applying them in the proper intervals between integers. The rational number $2\frac{1}{2}$ can be placed in correspondence with the point B, which is one-half unit distance more than 2 units to the right of O. The rational number $-0.\overline{6}$ can be placed in correspondence with the point C, which is two-thirds unit distance to the left of O.

By the technique outlined in these five steps a unique point can be found on the number line to correspond to any rational number. Such points are *Rational points* referred to as *rational points*. An interesting and important observation to make here is that for any two distinct rational points, no matter how close together they may be located, there always exists at least one rational point which lies between them. For example, if a and b are distinct rational numbers, then their average $\left(\dfrac{a + b}{2}\right)$ is a rational number c which is between a and b. (See Exercise 9 of the previous section.) The rational point c (the point on the number line which corresponds to c) will be located between the rational points a and b. This fact, which seems intuitively correct, will be accepted here without formal argument. Similarly, the rational point $\dfrac{a + c}{2}$ will be located between the rational points a and c, hence between the rational points a and b. This can be continued indefinitely to exhibit an infinite number of rational numbers between any two given rational points a and b. Because of this fact and the fact that the set of *Density of rational numbers* rational numbers is a subset of the real numbers, the rational numbers are said to be *dense* in the reals. Synonymously, the rational points are said to be dense in the number line.

Example In Fig. 7-4 the rational points 1 and 3 are shown. The rational point $2 = \dfrac{1 + 3}{2}$ is located between them. The point $\dfrac{3}{2} = \dfrac{1 + 2}{2}$ is between 1 and 2 and, therefore, between 1 and 3. Similarly, $\dfrac{5}{4} = \dfrac{1 + \frac{3}{2}}{2}$, $\dfrac{9}{8} = \dfrac{1 + \frac{5}{4}}{2}$, $\dfrac{17}{16} = \dfrac{1 + \frac{9}{8}}{2}$ are all between 1 and 3. The process can be

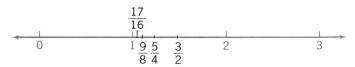

FIGURE 7-4   Rational points on the real number line

continued indefinitely to exhibit an infinite number of rational points between 1 and 3.

6   Points on the line can also be found to correspond to irrational numbers. We shall see later (Sec. 8-7) how the pythagorean theorem applied to right triangles such as triangle OAD of Fig. 7-5 establishes the fact that $\overline{OD}$ is $\sqrt{2}$ units in length if $\overline{OA}$ and $\overline{AD}$ are each 1 unit in length. The length of $\overline{OD}$ has been measured on the number line (Fig. 7-3), and point D is in correspondence with the real number $\sqrt{2}$. This is but one illustration. The techniques for finding other irrational points would not all be so simple as this illustration.

Real numbers
are ordered

With this correspondence established the real numbers may be identified with distances measured from the point of reference on the number line. A real number a is then said to be *less than* a real number b (that is, $a < b$) if and only if the point which corresponds to a on the number line is to the left of the point which corresponds to b. If the point corresponding to a is to the right of the one corresponding to b, then a is said to be *greater than b* ($a > b$). If the two points coincide, then $a = b$. Thus a positive real number a is one whose corresponding point on the number line is to the right of the reference point O and can be identified by the symbol $a > 0$. By like token a negative real number b is one whose corresponding point is on the number line to the left of the reference point O. It may be identified by $b < 0$, in which case $^-b > 0$.

7-4   THE FIELD OF REAL NUMBERS

With the definitions of the fundamental operations extended to apply to irrational numbers as well as rational numbers, it can be shown that real numbers satisfy the field postulates F-1 to F-8 (Sec. 6-3). As was pointed out in the last paragraph of the preceding section, the real numbers can

FIGURE 7-5   The pythagorean property

be ordered in increasing order of magnitude in terms of their relative position on the number line. This is equivalent to the previous techniques for ordering the integers and the rational numbers. In fact, for real numbers which are integers or, more generally, for rational numbers, either the previously developed tests or that presently mentioned may be used.

Example

1   $8 > 5$ since the point which corresponds to 8 is on the number line to the right of the point which corresponds to 5. This is equivalent to saying that $8 - 5 = 3$, a positive integer.

2   $-\frac{15}{4} < -\frac{7}{3}$ since the point which corresponds to $-\frac{15}{4}$ is on the number line to the left of the point corresponding to $-\frac{7}{3}$. This is equivalent to saying that $(-15)(3) < 4(-7)$, the test for rational numbers.

3   $\sqrt{2} > \frac{1}{2}$ since the point which corresponds to $\sqrt{2}$ is on the number line to the right of the point which corresponds to $\frac{1}{2}$. Another method for comparing two such numbers is by means of their decimal expansions.

The field of real numbers is an ordered field since the positive real numbers share with the positive integers and positive rational numbers the following three important postulated properties:

Properties of positive real numbers

1   *Addition*   The sum of two positive real numbers is a positive real number.

2   *Multiplication*   The product of two positive real numbers is a positive real number.

3   *Trichotomy*   For any given real number a, one and only one of the following three relations holds: a is positive $(a > 0)$, $a = 0$, or ^-a is positive $(^-a > 0)$.

In many situations one may be concerned merely with the numerical value of a number, and not whether it is positive or negative. If one person loses $5 and another gains $5, the amount of money is the same. True, for any given individual, it might make considerable difference whether he gains or loses the $5, but that does not alter the fact that the numerical value of the money involved is $5. If we symbolize the loss of $5 by -5, the gain of $5 by $+5$, and the numerical value involved by $|-5|$ or $|+5|$, as the case may be, we can make this evidently true statement: $|-5| = |+5| = 5$. This is an illustration of a very important concept in dealing with real numbers.

Absolute value   **Definition 7-1**   The *numerical*, or *absolute value* of the real number a is such that

$$|a| = a \qquad \text{if } a \geq 0$$
$$|a| = {}^-a \qquad \text{if } a < 0$$

Note that in the definition, *numerical value* and *absolute value* are used synonymously. Mathematicians, for good reasons which are not pertinent here, give preference in most instances to the phrase "absolute value" over "numerical value." In the context of this discussion, they may be used interchangeably.

Example

1   The condition $a \geq 0$ simply means that the real number is either positive or equal to zero. If a is $+9$, then $|a| = |+9| = 9$. If a is 0, then $|a| = |0| = 0$.

2   The condition $a < 0$ means that a is negative; that is, it is located on the number line to the left of the reference point O. If $a = -6$, then $|a| = |-6| = {}^-(-6)$, by definition. By Sec. 4-2 ${}^-(-6) = 6$. Whence $|-6| = 6$.

Interval on the number line

In Fig. 7-6 the portion of the number line from -2 to $+2$ has been indicated. Any portion of the number line thus specifically designated may be referred to as an *interval*. Several points in this interval have been labeled by the real numbers to which they correspond. The two points labeled in color to correspond to -2 and 2, respectively, are each two units' distance from the origin. Each of the other points of the interval is less than two units' distance from the origin. This means that each of the points labeled in black is to the right of the point labeled -2; that is, -2 is less than each of the corresponding numbers. (For example, $-2 < -\sqrt{2}$; $-2 < -1$; $-2 < -\frac{1}{2}$; $-2 < 0$; $-2 < \frac{1}{2}$; and $-2 < \frac{5}{3}$ are a few specific illustrations.) Similarly, each of these numbers is less than 2. (Thus, $-\sqrt{2} < 2$; $-1 < 2$; $-\frac{1}{2} < 2$; $0 < 2$; $\frac{1}{2} < 2$; $\frac{5}{3} < 2$.) We may combine these two statements about the numbers in the interval into this one statement: Each number of the interval is greater than -2 but less than 2. (In symbols, $-2 < -\sqrt{2} < 2$; $-2 < -1 < 2$; $-2 < -\frac{1}{2} < 2$; $-2 < 0 < 2$; $-2 < \frac{1}{2} < 2$; $-2 < \frac{5}{3} < 2$.) If we use some symbol, say, x, to represent any number in the interval, then we may state this fact by the symbol $-2 < x < 2$, which may be read either of two ways: x is greater than -2 but less than 2; or -2 is less than x, which is less than 2. The symbol $|x| < 2$ also means the same thing. [For example, $|-\sqrt{2}| = {}^-(-\sqrt{2}) = \sqrt{2} < 2$; $|-1| = {}^-(-1) = 1 < 2$; $|-\frac{1}{2}| = {}^-(-\frac{1}{2}) = \frac{1}{2} < 2$; $|0| = 0 < 2$; $|\frac{1}{2}| = \frac{1}{2} < 2$; $|\frac{5}{3}| = \frac{5}{3} < 2$.] If we wish to indicate that both -2 and $+2$ are possible values of x, then the two symbols should be written $-2 \leq x \leq 2$ (-2 is less than or equal to x, which is less than or equal to 2); $|x| \leq 2$ (the absolute, or numerical, value of x is less than or equal to 2).

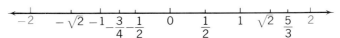

FIGURE 7-6   The interval $-2 \leq x \leq 2$

Example Express each of these inequalities in the equivalent form $a \leq x \leq b$: (1) $|x| \leq 5$; (2) $|x + 3| \leq 2$.

1 $|x| \leq 5$ means that the numerical value of x is a number less than 5 or at most equal to 5. This means that $-5 \leq x \leq 5$.

2 As in item 1, $|x + 3| \leq 2$ means $-2 \leq x + 3 \leq 2$. Since Theorem 3-3 (page 104) holds for real numbers as well as natural numbers, we may add -3 to each member of these inequalities to get

$$
\begin{array}{r}
-2 \leq x + 3 \leq 2 \\
-3 \qquad -3 \quad -3 \\
\hline
-5 \leq x \qquad \leq -1
\end{array}
$$

Over the field of real numbers it is always possible to solve linear equations of the form $ax + b = c$, where a, b, and c are real numbers with $a \neq 0$. While there are also equations of higher degree with real coefficients which can be solved over the field of real numbers, there are other such equations for which solutions do not exist until a further extension of the number system is made. As a simple illustration of this fact, consider the equation $x^2 = -4$. To solve this equation is to find a real number which multiplied by itself will give -4. Because the proposition of Theorem 4-8 (page 142) can be extended to real numbers, there exists no negative real number whose square is a negative real number. Furthermore, positive real numbers are closed under multiplication. Therefore, there exists no real number whose square is a negative real number.

EXERCISES

1 Prove that $4 + \sqrt{3}$ is an irrational number. (*Hint:* Assume that $4 + \sqrt{3}$ is rational.)

2 Prove that $\dfrac{\sqrt{2} + \sqrt{3}}{2}$ is an irrational number. [*Hint:* $(\sqrt{2} + \sqrt{3})^2 =$ $5 + 2\sqrt{6}$.]

* 3 Prove that if q is a rational number and k is an irrational number, then $q + k$ is irrational.

* 4 Give a counterexample to show that if k and l are two irrational numbers, it is not necessarily true that $\dfrac{k + l}{2}$ is an irrational number.

5 (*a*) To four decimal places $\sqrt{2} = 1.4142$. Find the decimal equivalent of $\dfrac{1}{\sqrt{2}}$, correct to four decimal places.

  (*b*) The product of these two decimal values should closely approximate what value? There is an expected error due to rounding.

  (*c*) What error is there in your product? It should be less than 0.00002.

6 Is the set of all irrational numbers with the usual definitions of addition and multiplication a field?

7   Express each of these real numbers as an infinite decimal: (a) $\frac{5}{7}$; (b) $\frac{7}{8}$; (c) $\frac{2}{3}$; (d) $7\frac{1}{2}$.

8   Draw a number line and mark a point on it to correspond to each of these rational numbers: 5, $-\frac{1}{2}$, $3\frac{1}{2}$, -6, 0.5, $0.\overline{3}$, 0.125, $-1.\overline{6}$.

9   Find four rational numbers between $3\frac{1}{4}$ and $3\frac{1}{2}$.

*10   Prove Theorem 7-2: If a and b are real numbers such that $a < b$, then
$$a < \frac{a+b}{2} < b. \quad \left(Hint: \frac{a+b}{2} = \frac{a}{2} + \frac{b}{2}. \right)$$

11   On the number line drawn for Exercise 8, mark a point to correspond to each of these irrational numbers: $\sqrt{2}$, $\sqrt{3}$, $\sqrt[3]{4} \approx 1.587$, $-\sqrt[3]{2} \approx -1.260$, $\pi \approx 3.1416$. ($Hint$: Round each number to the nearest tenth. Use this value as an aid in marking the approximate position for each number in the line.)

12   Arrange these numbers in ascending order from left to right: $\sqrt[3]{3}$, -2, -1.25, 0.857, $-1.2\overline{5}$, $-\sqrt[3]{2}$, 1.4, 1.414, $0.85\overline{7}$, $\frac{6}{7}$.

13   Express each example in its equivalent form using inequalities of the form $a < x < b$ or $a \leq x \leq b$.

(a) $|x| < 3$ $\qquad\qquad$ (b) $|x| \leq \dfrac{1}{2}$

(c) $|x + 2| \leq 4$ $\qquad$ (d) $|x - 1| \leq 3$

14   Express each inequality in its equivalent form using absolute value:

(a) $-1 \leq x \leq 1$ $\qquad$ (b) $3 \leq x \leq 5$
(c) $-5 < x < 5$ $\qquad$ (d) $-\sqrt{3} \leq x \leq \sqrt{3}$

15   Argue the truth of this statement: If x is a real number, then $\sqrt{x^2} = |x|$.

*16   Prove that the multiplicative inverse of an irrational number is also an irrational number.

7-5   THE FIELD OF COMPLEX NUMBERS

The needed extension of the number field to make possible the solution of such equations as $x^2 = -4$ calls for the introduction of a new unit. In the real field the unit is 1. We now define a new unit $i = \sqrt{-1}$, called the *Imaginary unit*, **imaginary unit**, which is such that $i^2 = -1$. With the introduction of this new unit we also postulate the existence of the set of complex numbers.

Complex numbers   **Postulate 7-4**   There exist complex numbers of the form $a + bi$ where a and b are real numbers and $i = \sqrt{-1}$ is such that $i^2 = -1$.

For complex numbers we then state these definitions:

Equality   **Definition 7-2**   Two complex numbers $a + bi$ and $c + di$ are equal if and only if $a = c$ and $b = d$.

Addition   **Definition 7-3**   The sum of the two complex numbers $a + bi$ and $c + di$ is the complex number $(a + c) + (b + d)i$.

Multiplication   **Definition 7-4**   The product of the two complex numbers $a + bi$ and $c + di$ is the complex number $[ac + {}^-(bd)] + (ad + bc)i$.

Just as in the case of real numbers, subtraction of complex numbers is the inverse of addition, and division is the inverse of multiplication.

Under these definitions of addition and multiplication the closure, associative, commutative, and distributive properties can be shown to hold for complex numbers. The additive identity exists and is the complex number $0 + 0i$ since $(a + bi) + (0 + 0i) = (a + 0) + (b + 0)i = a + bi$ by the field properties of real numbers. Similarly, the multiplicative identity is $1 + 0i$ since

$$(a + bi) \cdot (1 + 0i) = [a \cdot 1 + {}^-(b \cdot 0)] + (a \cdot 0 + b \cdot 1)i = a + bi$$

by the field properties of real numbers. Finally, additive and multiplicative inverses can be shown to exist. Hence the set of all complex numbers with
Field of complex numbers   addition and multiplication, as defined above, satisfy the field properties F-1 to F-8 (pages 207 and 208).
Additive inverse   The *additive inverse* of $a + bi$ is the complex number $({}^-a) + ({}^-b)i$. This statement can be verified as follows:

$$(a + bi) + [({}^-a) + ({}^-b)i] = [a + ({}^-a)] + [b + ({}^-b)]i$$
$$= 0 + 0 \cdot i$$

Multiplicative inverse   If a and b are not both zero, then the *multiplicative inverse* of $a + bi$ is the complex number $\dfrac{a}{a^2 + b^2} + \dfrac{{}^-b}{a^2 + b^2} i$. This may be verified as follows:

$$(a + bi)\left(\frac{a}{a^2 + b^2} + \frac{{}^-b}{a^2 + b^2}\right)i = \left(\frac{a^2}{a^2 + b^2} + \frac{b^2}{a^2 + b^2}\right)$$
$$+ \left(\frac{a \cdot ({}^-b)}{a^2 + b^2} + \frac{ba}{a^2 + b^2}\right)i$$
$$= \frac{a^2 + b^2}{a^2 + b^2} + \frac{{}^-(ab) + ab}{a^2 + b^2} i$$
$$= 1 + 0 \cdot i$$

Note that the condition that a and b are not both zero is equivalent to stating that $a + bi$ is not the additive identity. The condition for the existence of multiplicative inverses for complex numbers thus corresponds to that for the existence of multiplicative inverses for real numbers.

Example What are the additive and multiplicative inverses of $-3 + 2i$?

Additive inverse

Since $^-(-3) = 3$ and $^-2 = -2$, we have

$$^-(-3 + 2i) = 3 + (-2)i = 3 - 2i$$

This statement is verified by

$$(-3 + 2i) + [3 + (-2)i] = (-3 + 3) + [2 + (-2)]i$$
$$= 0 + 0 \cdot i$$

Multiplicative inverse

$$\frac{a}{a^2 + b^2} = \frac{-3}{(-3)^2 + 2^2} = \frac{-3}{13}$$

$$\frac{^-b}{a^2 + b^2} = \frac{^-(2)}{(-3)^2 + 2^2} = \frac{-2}{13}$$

$$(-3 + 2i) \cdot \left(\frac{-3}{13} + \frac{-2}{13}i\right) = \left[(-3)\left(\frac{-3}{13}\right) + {}^-\left((2)\left(\frac{-2}{13}\right)\right)\right]$$
$$+ \left[(-3)\left(\frac{-2}{13}\right) + 2\left(\frac{-3}{13}\right)i\right]$$
$$= \left(\frac{9}{13} + \frac{4}{13}\right) + \left(\frac{6}{13} + \frac{-6}{13}\right)i$$
$$= 1 + 0 \cdot i$$

The inverse relation between multiplication and division in the field of real numbers continues to hold for complex numbers. It thus follows that the quotient $(a + bi) \div (c + di)$ can be obtained by multiplying $a + bi$ by the multiplicative inverse of $c + di$.

$$(a + bi) \div (c + di) = (a + bi) \cdot \left(\frac{c}{c^2 + d^2} + \frac{^-d}{c^2 + d^2}i\right)$$

Example Find the quotient $\dfrac{2 + 3i}{1 + 2i}$.

The multiplicative inverse of $1 + 2i$ is the complex number $\dfrac{1}{1 + 4} -$
$\dfrac{2}{1 + 4}i = \dfrac{1}{5} - \dfrac{2}{5}i$. Therefore

$$(2 + 3i) \div (1 + 2i) = (2 + 3i)\left(\frac{1}{5} - \frac{2}{5}i\right)$$

$$= \left(\frac{2}{5} + \frac{6}{5}\right) + \left(\frac{-4}{5} + \frac{3}{5}\right)i$$

$$= \frac{8}{5} - \frac{1}{5}i$$

Check: $\quad (1 + 2i) \cdot \left(\frac{8}{5} - \frac{1}{5}i\right) = \left(\frac{8}{5} + \frac{2}{5}\right) + \left(\frac{-1}{5} + \frac{16}{5}\right)i$

$$= 2 + 3i$$

In this field there exists a number x such that $x^2 = -4$. There are indeed two values of x which meet the conditions of this equation. If $x = 2i$, then $x^2 = (2i)(2i) = (0 + 2i)(0 + 2i)$. From the definition of multiplication we have

$$(0 + 2i)(0 + 2i) = (0 \cdot 0 - 2 \cdot 2) + (0 \cdot 2 + 2 \cdot 0)i$$
$$= -4 + 0i = -4$$

whence $x^2 = -4$. This multiplication can be accomplished in a much simpler way if we recall that the definition of multiplication for complex numbers is such that it is both associative and commutative. From this fact we have $x^2 = (2i)(2i) = (2 \cdot 2)(i \cdot i) = 4i^2 = 4(-1) = -4$. Similarly, $x = -2i$ is a solution, since

$$x^2 = (-2i)(-2i) = (-2)(-2)(i \cdot i) = 4i^2 = 4(-1) = -4$$

Pure imaginary numbers Numbers such as $2i$ or $-2i$, which are of the form bi, are called *pure imaginary numbers*.

Not only are simple quadratic equations of this type solvable in the field of complex numbers, but no further extensions of our number system are necessary to solve any algebraic equations of the form

$$ax^n + bx^{n-1} + cx^{n-2} + \cdots + d = 0$$

where n is a positive integer, and a, b, c, d, . . . are complex numbers.

7-6 THE NUMBER SYSTEMS OF ELEMENTARY MATHEMATICS

In each of the number systems, two requisites that have persisted are that the previously established set of numbers must be a subset of the newly defined set and that the newly defined operational processes must give *Real numbers as a proper subset of complex numbers* results consistent with previously established definitions. These conditions are met in the extension from the field of real numbers to the field of complex numbers. The real number a may be represented as the complex number $a + 0i$, and thus the set of real numbers is a proper subset of the set of

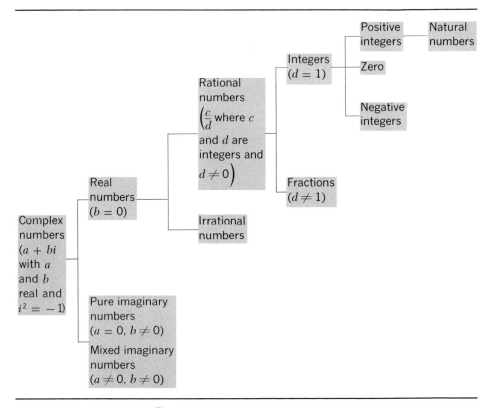

FIGURE 7-7 The number systems of elementary mathematics

complex numbers. Furthermore, from the definition of the sum of two complex numbers, we have

$$(a + 0i) + (b + 0i) = (a + b) + (0 + 0)i = (a + b) + 0i$$

which is the representation of the real number $a + b$. This is the sum of the two real numbers a and b in accordance with the definition of addition of real numbers. Similarly, the product

$$(a + 0i)(b + 0i) = [ab + {}^-(0 \cdot 0)] + (a \cdot 0 + 0 \cdot b)i = ab + 0i$$

which is the representation of the real number ab.

The outline of the entire discussion of Chaps. 3 to 7 is presented briefly, but significantly, in the diagram of Fig. 7-7.

The number experiences of the elementary school are confined to the field of real numbers. In fact, with the exception of the irrational number π and a few square roots, these experiences are restricted to the field of rational numbers. That these should be tremendously significant mathematical experiences is emphasized by the diagram of Fig. 7-7, since it places the field of real numbers and its subfield of rational numbers in their proper orientation in the structure of the complex number system, which is the self-sufficient number system of elementary mathematics. The diagram also lends emphasis to the fact that the surety of structure and the firmness of understanding in any extension of the number system not only are dependent upon the soundness of all substructure but also are essential to the length and elegance of all superstructure.

EXERCISES

Use the diagram of Fig. 7-7 as an aid in answering the questions of Exercises 1 to 4.

1 How may an integer be considered as being a rational number? A real number? A complex number?
2 How may a natural number be considered as being a complex number? A real number?
3 How may an irrational number be considered a complex number?
4 Can an integer be considered an irrational number?
5 Consider the sets of numbers: N, Z, Q, $I =$ the set of all irrational numbers, $R =$ the set of all real numbers, and $K =$ the set of all complex numbers. Of which sets is each given number an element: 6; $-\dfrac{7}{8}$; $\sqrt{2}$; -1.5; -4; $\sqrt[3]{4}$; π; $\dfrac{1}{\pi}$; $2i$; $3 - 4i$?

6 Write the additive inverse and multiplicative inverse of each of these complex numbers. Use field properties F-6 and F-8 to provide a check for your results.

   (a) i (b) $1 + i$ (c) $3 + 4i$ (d) $6i$
   (e) $7 + 0 \cdot i$ (f) $5 - 12i$ (g) $-i$ (h) $2 + 7i$
   (i) $\sqrt{2} + 4i$ (j) $-\sqrt{3}\,i$ (k) $9 - \sqrt{5}\,i$ (l) $\sqrt{2} + \sqrt{3}\,i$

7 Find the sum and product of each pair of complex numbers.

   (a) $2 + i, 3 + 2i$ (b) $1 + i, 1 - i$
   (c) $i, 2 + 4i$ (d) $3 - i, 4 - i$
   (e) $-4i, 3i$ (f) $4, 2 + 3i$

   (g) $2 + \sqrt{3}\,i, \sqrt{2} - \sqrt{3}\,i$ (h) $\dfrac{1}{2} + \dfrac{\sqrt{3}}{2}i, \dfrac{1}{2} + \dfrac{\sqrt{3}}{2}i$

   (i) $\dfrac{1}{2} + \dfrac{\sqrt{3}}{2}i, \dfrac{1}{2} - \dfrac{\sqrt{3}}{2}i$ (j) $\dfrac{\sqrt{3}}{2} + \dfrac{1}{2}i, 2 - i$

8 Raise each of these complex numbers to the indicated power.

   (a) $(1 + i)^2$ (b) $(3i)^2$ (c) i^4

   (d) $(-1 + \sqrt{3}\,i)^3$ (e) i^3 (f) $\left(-\dfrac{1}{2} + \dfrac{\sqrt{3}}{2}\right)^3$

$(g)\ (-5i)^2$ $(h)\ \left(-\dfrac{1}{2} - \dfrac{\sqrt{3}}{2}i\right)^3$ $(i)\ (-1 - \sqrt{3}\,i)^3$

9 Find these differences.

$(a)\ (7 + 2i) - (6 + i)$ $(b)\ (5 - 3i) - (-2 + 6i)$
$(c)\ (\sqrt{2} + i) - (4 + \sqrt{3}\,i)$ $(d)\ (-8 - i) - (4 - 2i)$

10 Find these quotients.

$(a)\ \dfrac{3 + i}{1 - i}$ $(b)\ \dfrac{2 - 3i}{3 + 4i}$ $(c)\ \dfrac{6 + 5i}{2 - 3i}$

$(d)\ \dfrac{7 + i}{i}$ $(e)\ \dfrac{5 + i}{1 - \sqrt{5}\,i}$ $(f)\ \dfrac{i}{1 + i}$

*11 Construct the multiplication table for the set $\{1, -1, i, -i\}$ where i is the imaginary unit.

*12 Prove that the set $\{1, -1, i, -i\}$ is an abelian multiplicative group.

7-7 REVIEW EXERCISES

1–21 Write a carefully constructed answer to each of the guideline questions of this chapter.

22 Prove that $\sqrt{11}$ is an irrational number.

23 Use division to find each of these positive square roots, correct to four decimal places: $(a)\ \sqrt{11}$; $(b)\ \sqrt{39}$; $(c)\ \sqrt{624}$.

24 These radical expressions represent what rational numbers: $\sqrt{16}$; $\sqrt{25}$; $\sqrt{\frac{4}{81}}$; $\sqrt{\frac{64}{9}}$?

25 (a) Find the multiplicative inverse of $\sqrt{11}$, correct to four decimal places.

(b) What product can you use to check your result?

(c) What error is there in your product? It should be less than 6.0×10^{-5}.

26 Find five rational numbers r in each of these intervals.

$(a)\ 2.5 < r < 3$ $(b)\ -\dfrac{1}{3} < r < 0$

$(c)\ |r| < 2$ $(d)\ 0 < r < 1$

27 Name the integers x, if any, in each of these intervals.

$(a)\ |x| < \pi$ $(b)\ |x| \le 1$ $(c)\ |x| \le \dfrac{\pi}{4}$

$(d)\ |x| < 1$ $(e)\ 0 < x < \dfrac{1}{2}$ $(f)\ -\dfrac{1}{3} \le x < 0$

$(g)\ 0 \le x < \dfrac{2}{3}$ $(h)\ -\dfrac{3}{4} \le x \le 0$ $(i)\ |x| < \dfrac{5}{2}$

Use scientific notation in Exercises 28 to 31. Give results both in scientific notation and in positional notation.

*28 The earth is approximately 92,567,000 miles from the sun. Light travels about 186,000 miles per second. Approximately how many seconds does it take light from the sun to reach the earth? How many minutes?

*29 The Grand Coulee Dam contains about 11,200,000 cubic yards of concrete, which weighs approximately 49,000 pounds per cubic yard. How many tons of concrete are there in this dam?

*30 The star Arcturus is 223,700,000,000,000 miles from the earth. How many light years is this? One light-year is 5,870,000,000,000 miles, which is the distance light will travel in one year's time. Give your answer correct to the nearest tenth light-year.

*31 The Grayson rulings are graduated markings used in microscopic work. If their average separation is 0.0000083 inch, approximately how many such rulings are there to one inch?

32 Using the rational approximation $\sqrt{17} \approx 4.123$ and $\sqrt{18} \approx 4.243$, find six rational numbers which lie between these two irrational numbers.

*33 Prove that if r is rational and k is irrational, then $r \cdot k$ is irrational.

34 Arrange these real numbers in descending order with the *largest* number on the left: $-\frac{\pi}{4}$; 0; 0.7143; $87\frac{1}{2}\%$; -1; $\frac{5}{7}$; $\frac{8}{9}$; 160%; $\frac{\pi}{2}$; $\sqrt{2}$.

35 In each of these problems associate each given number with the appropriate symbol in the percentage formula, then solve the problem.
  (a) 126 is 15% of what number?
  (b) 150 is what percent of 225?
  (c) What is $12\frac{1}{2}\%$ of 648?

36 A finance charge of $1\frac{1}{2}\%$ per month is equivalent to what yearly rate?

37 A person sold two lots for $6,000 each. On one he made a 25% profit while on the other he took a 20% loss. How much money did he gain or lose? Correct to the nearest tenth percent, what percent was this of his original investment?

38 Express each inequality in its equivalent form using either $a < x < b$ or $a \leq x \leq b$.

  (a) $|x| < 7$ (b) $|x| \leq \frac{1}{2}$ (c) $|x + 1| < 1$

  (d) $|x - 3| \leq 4$ (e) $|x + 5| < 2$ (f) $|x - 1| \leq 4$

39 Express each inequality in its equivalent form using absolute value.

  (a) $-2 < x < 2$ (b) $0 \leq x \leq 2$

  (c) $-\frac{1}{2} \leq x \leq \frac{3}{4}$ (d) $-\frac{5}{2} < x < 0$

40 For what values of x and y are these statements true?

  (a) $x + yi = 3 + 2i$ (b) $x + yi = -5 + 6i$

(c) $x + yi = 7i$ (d) $x + yi = \dfrac{\sqrt{3}}{2} + \dfrac{1}{2}i$

(e) $x + yi = 1$ (f) $x + yi = -4 - 3i$

(g) $x + yi = (4 + 2i) + (3 - 5i)$

(h) $x + yi = (2 - i) + (6 + 8i)$

(i) $x + yi = (9 + 2i) - (5 + 4i)$

(j) $x + yi = (7 - 6i) - (-3 - 2i)$

41   Find these quotients.

(a) $\dfrac{2 + i}{3 + i}$ (b) $\dfrac{1 - i}{1 + i}$ (c) $\dfrac{4}{i}$ (d) $\dfrac{7i}{3 - 4i}$

42   Write the additive and multiplicative inverse for each complex number. Check your results.

(a) $5 + 4i$ (b) $-i$ (c) $3 - 2i$ (d) $-1 + i$

*INVITATIONS TO EXTENDED STUDY

1   (a) Show that with addition and multiplication, numbers of the form $a + b\sqrt{3}$ where a and b are integers satisfy the properties of an integral domain.

(b) Do they satisfy the properties of a field?

(c) If $a,b \in Q$ do such numbers satisfy the properties of a field?

2   (a) Show that with addition and multiplication, complex numbers of the form $a + bi$ where a and b are integers satisfy the properties of an integral domain. Such numbers are called *gaussian integers*.

(b) Does this proper subset of the set of complex numbers form a field?

(c) Are there any proper subsets of the set of complex numbers which are fields?

3   The *conjugate* of the complex number $a + bi$ is $a + (-b)i$. Prove that the product of any complex number with its conjugate is a nonnegative real number. Under what conditions will such a product be a positive real number?

4   A rational approximation to the value of π, frequently given, is $\pi \approx 3.14159265$. If Q is the set of all rational numbers, let $P = \{q \,|\, q \in Q$ and $q < \pi\}$ and $R = \{q \,|\, q \in Q$ and $q > \pi\}$.

(a) Assume a real number line with a point corresponding to π located on it. Where would all points corresponding to elements of P be located on this line? Corresponding to elements of R?

(b) In the definitions of P and R could $<$ be replaced by $\leq$ or $>$ by $\geq$? Why or why not?

(c) Each of these frequently used approximations to π are elements of which set, P or R? 3.14; 3.142; 3.1416; $\frac{22}{7}$.

(d) Using the approximation stated in this exercise, select other appropriately related approximations to π which are elements of P. Elements of R.

(e) The value of π has been computed correct to 500,000 decimal places. Call this approximation a. The approximate value given in this exercise exhibits, along with other values, these rational numbers which are among the 500,000 that can be selected from a: 3.1; 3.14; 3.141; 3.1415; . . . . Are these 500,000 rational numbers elements of the replacement set for x where $3 < x < \pi$? Is there possibly a question about any of these numbers being a replacement for x?

(f) Assume that instead of merely listing values as in (e), rounding is used to obtain each value as a correct approximation to π. Give a general description of these selected values which would be elements of P. Elements of R.

(g) Is there a largest rational number in P? A smallest rational number in R?

(h) Are P and R disjoint sets? Is $P \cup R = Q$?

(i) How is each element of P related to all elements of R?

(j) Can you use these exercises as a basis for writing a definition of the concept of irrational number using only sets of rational numbers?

THE CONCEPTS OF POSITION, SHAPE, AND SIZE

GUIDELINES FOR CAREFUL STUDY

In our study of number systems we were at all times dealing with sets of numbers. Complex numbers were defined in terms of real numbers and a new unit which we called the imaginary unit; real numbers, in terms of rational numbers and a set of new numbers which we called irrational numbers; rational numbers, in terms of integers; and integers, in terms of natural numbers and a new number which we called zero, or the additive identity. No effort was made to define natural numbers; rather, we accepted them as the numbers used in counting—undefined concepts for which we could establish acceptable intuitive perceptions through the technique of one-to-one correspondence between sets of objects. Our familiarity with these undefined elements was of such a nature that we were able to postulate a set of basic properties which controlled their use within the demands of certain well-defined operations and relations. From these basic assumptions other properties were derived to give structure to the natural number system. Upon this system as a foundation we were then able to build the other systems.

In the study of geometry we shall follow a somewhat similar pattern, except that for this development the undefined elements will be "point," "line," and "plane" rather than "number." The universal set will be the set of all points, and it will be called "space." Subsets of this space in which we shall be interested are such familiar geometric configurations as lines, planes, angles, polygons, circles, prisms, pyramids, cylinders, cones, and spheres. Fundamental properties and relations which characterize such sets of points will constitute the subject matter of this chapter.

The following questions will serve as helpful guidelines to careful study of Chap. 8:

1 When is one set said to be a subset of another set? A proper subset?
2 What is meant by the union of two sets? The intersection of two sets?
3 What is the empty (null) set?
4 What name has been given to this set: $\{c \mid c$ is the cardinal number of a nonempty set$\}$?
5 What is the cardinal number of the empty set?
6 What are the definitions of addition and subtraction?
7 What is the real number line?
8 What is meant by the absolute value of a real number?
9 What is meant by the term *space?*
10 What is the basic distinction between a straight line and a curved line?
11 What are the distinctions between line, line segment, and ray?
12 What are the distinctions between parallel lines, skew lines, intersecting lines, and perpendicular lines?
13 When is a set of points said to be collinear?
14 When is the point C said to be between the points A and B?

15 What is a half line? A half plane? A half space?

16 What is the distance between two points on a number line?

17 When are two line segments said to be congruent? Two angles?

18 What are conditions which determine a plane?

19 What is the intersection of two planes?

20 When are two planes parallel?

21 When are points or lines said to be coplanar?

22 What is a plane angle? A dihedral angle?

23 What is meant by the interior and exterior of a plane angle? Of a dihedral angle?

24 What is meant by vertical angles? Adjacent angles? Complementary angles? Supplementary angles?

25 What is meant by a cartesian frame of reference?

26 What are the names of the different axes in a cartesian frame of reference in space? In a plane?

27 What are circles of longitude and latitude?

28 What is the prime meridian? The international date line?

29 What is a simple closed curve? A circle? A polygon? A regular polygon?

30 How may triangles be classified according to angles? According to sides?

31 How may quadrilaterals be classified?

32 What is a sphere? A prism? A pyramid? A cone?

33 What are the regular polyhedrons?

34 When are two geometric figures said to be similar?

35 When are two geometric figures said to be congruent?

36 What are the conditions for congruence of two triangles?

37 When is a plane region said to be symmetric with respect to a point? An axis?

38 What is meant by a euclidean construction?

INTRODUCTION

During our earlier study of number systems no effort was made to give a logically structured definition of number. Instead, we accepted the intuitive concept we have of number and its basic characteristics as sufficiently convincing and firm to serve as a foundation for the development of the systems which provide for its efficient use. In the study of geometry a similar situation exists concerning *points*. We shall make no effort to define the concept of "point." We shall assume, however, an intuitive familiarity with the concept sufficient to serve as an acceptable basis for the definition and discussion of different sets of points. For these sets of points we shall then proceed to formulate clear-cut ideas. Just as was the case in the study of number, some properties and relations will be postulated or merely

"Point" is
undefined

described, while others will be examined in somewhat more detail. One major difference between the two developments will be that we shall restrict our interests much more severely here than in the study of number. We shall be concerned primarily with those basic concepts and principles of geometry which are acceptable on a purely intuitive or experimental basis. This subject matter, though quite restricted in nature, constitutes a body of geometrical content which is of great significance not only to elementary mathematics but to mathematics in general.

8-1 THE CONCEPT OF POINT

A point has position

Universal set

The cardinal number of any set of objects which can be placed in one-to-one correspondence with the set of fingers on any normal hand is "five." If we wish to record this number we do so by means of the symbol 5. This symbol is not the number. Similarly, when we wish to identify a point we usually do so by a dot (.), maybe with a letter attached. This dot is not a point; it is merely a convenient symbol acceptable as the representation of a point. A point has no size. It has only *position*, and a dot, or any equivalent symbol, is only a recognized symbol for indicating such position. For example, this dot (.) indicates a different position from that of the dot previously used. They thus represent two different points within the context of this discussion. The set of all points is what we shall call *space*. We shall see later (Sec. 8-5) that there is a very simple, yet effective, technique for locating in space any particular point, or points, pertinent to a given discussion. The configurations of geometry are subsets of the points of space. Thus each different configuration is a particular set of points identified by a specific nonambiguous characterization.

8-2 THE CONCEPT OF LINE

Line is a set of points

One of the most familiar and the simplest of the nonempty sets of points is the configuration we call a *line*. There are essentially two different types of lines: (1) the *straight line* (or *line*), which may be drawn continuously without changing directions, as illustrated in Fig. 8-1a; and (2) the *curved line* (or *curve*), no part of which is straight, so that when drawn, it continuously changes direction, as illustrated in Fig. 8-1b. If a person is driving in a straight line along a highway, he is maintaining the same direction at all points along his line of travel. On the other hand, if he is driving along a curve, he is changing direction at all points along his line of travel.†

---

†The specifications of a rigorous definition would include both the straight line and the curved line as curves. The distinction between their respective characterizing properties would be couched in more precise language than the appeal to intuition which seems desirable here. See Sec. 8-7 for a discussion of these concepts which, although still intuitive in nature, is slightly closer to the more rigorous concept.

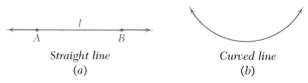

Straight line *(a)*   Curved line *(b)*

FIGURE 8-1 Two kinds of line

As we proceed in our discussion we shall use the terms "straight line" and "line," "curved line" and "curve," synonymously. A ruler or straightedge may be used to draw the representation of a line. In Fig. 8-1*a* the two arrowheads indicate that the line represented extends indefinitely in either direction. We shall designate such a line by a small letter, such as the letter l, or by the symbol $\overleftrightarrow{AB}$, using two points of the line. These two concepts find their justification in the following basic assumptions about points and lines.

Postulate 8-1 For any two distinct points in space there exists one and only one straight line which contains these two points.

Postulate 8-2 A straight line may be extended without break and indefinitely in either direction.

By Definition 2-7 the *intersection* of two sets is the set of elements common to both sets. As we have seen, such an intersection may be either nonempty or empty (see the example on page 43). When the sets are lines (sets of points), it is convenient to speak of *intersecting* or *nonintersecting lines*, depending on whether they do or do not have points in common.

Intersecting lines **Definition 8-1** Intersecting straight lines are lines which have at least one point in common.

As an immediate consequence of Postulate 8-1 we have this property of intersecting lines.

Theorem 8-1 If two distinct straight lines intersect, they have one and only one point in common.

Hypothesis Lines s and t are distinct lines which intersect (Fig. 8-2).

Conclusion The two lines have one and only one point in common.

FIGURE 8-2   Intersecting lines

From the hypothesis we have two distinct sets of points, the lines s and t which intersect. We can thus use the properties of lines and of intersection of sets. This means that the two lines have at least one point in common.

We are to prove that they can have only one point in common since they are distinct. The immediate question which arises is: Under what conditions will it be impossible for two lines to be distinct? This question brings Postulate 8-1 to mind.

Proof

Statements	*Reasons*
Lines s and t are nonempty sets of points.	Definition of line
$s \cap t$ is nonempty and hence contains at least one point.	Hypothesis; Definition 2-7
s and t are distinct lines.	Hypothesis
s and t cannot have more than one point in common.	Postulate 8-1

Any two points with fixed positions on a line, such as the points A and B in Fig. 8-1a, determine a measurable portion of the line, called a *line segment*, which is designated $\overline{AB}$. It is frequently advantageous to think of a line segment as a set of points in accordance with this definition:

Line segment   **Definition 8-2**   The line segment $\overline{AB}$ consists of two points A and B on a straight line and all points between them. The points A and B are called end points of the segment.

In this definition, and previously in the discussion of the number line (page 253), the undefined concept of "between" has been used. One thing that both usages have in common is the fact that the points with which the concept of "betweenness" was associated were all on the same straight line. It is very familiar usage and a clearly understood statement to say that three points A, B, and C are situated so that C is *between* A and B if the three points are on a straight line in one of the two patterns of Fig. 8-3. Also the points A, B, and C of the figure are said to be collinear in accordance with this definition:

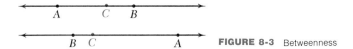

FIGURE 8-3 Betweenness

Collinear points

Definition 8-3 A set of points is said to be *collinear* if and only if all points of the set are contained in the same straight line. Otherwise they are said to be *noncollinear*.

Half line

Ray

On line *l* of Fig. 8-4*a* a point *P* has been marked. This point divides the line *l* into two *half lines*. The point *P* is not a point of either half line but is the *boundary* of each. Since the two half lines have no point in common, they do not intersect. On line *r* of Fig. 8-4*b* a segment $\overline{OM}$ has been marked. The point *P* has been placed on the line so that *M* is between *O* and *P*. These three points may be used as an aid in defining the *ray* $\overrightarrow{OP}$ as the union of the points of the line segment $\overline{OM}$ and of the set of all points *P* such that *M* is between *O* and *P*. The arrow of the figure indicates that the ray $\overrightarrow{OP}$ consists of the point *O* and all points of the line to the right of *O*. The point *O* is called the end point of the ray. While the ray $\overrightarrow{OP}$ may have any direction, the arrow of the symbol is always drawn from left to right. Note that, on each of the lines of Fig. 8-3, the set of points consisting of *C* and either of the two half lines defines a ray. Consequently, in each case the point *C* may be considered as the end point of either of the two rays $\overrightarrow{CA}$ or $\overrightarrow{CB}$. On each line the two rays have opposite directions from *C*. In such a case the rays are said to be *oppositely directed*

Opposite rays

or, more simply, *opposite rays*.

The number line of Chap. 7 (page 253) used the concept of a measurable line segment in the designation of "a unit to be applied to the line." This unit, whether considered as some specified standard unit of measure (Sec. 9-6) or merely as an unspecified unit, may be used as in Sec. 7-3 to set up a one-to-one correspondence between the set of real numbers and the set of points on a line. For example, in Fig. 8-5, whatever the unit may be, the point *A* is 1 unit removed from the reference point *O* and is located to the right of *O*; the point *D* is $3\frac{3}{4}$ units to the right of *O*; and the point *E* is 5 units to the right of *O*. Similarly, the point *F* is 2 units removed from *O*, the minus sign (−) simply indicates that the 2 units are measured to

Half line
(a)

Ray
(b)

FIGURE 8-4

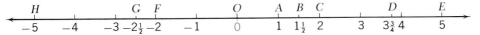

FIGURE 8-5   Distance on the number line

the left of the reference point O; the point G is $2\frac{1}{2}$ units to the left of O; and H is 5 units to the left of O.

There is at least one other familiar and important concept implied by the discussion of the preceding paragraph. Not only is the number 5 associated with the position of the point E on the line of Fig. 8-5, but it is also associated with the line segment $\overline{OE}$ in that it indicates the length of the segment, or the *distance* between the two points O and E. Whatever the unit of measure may be, the length of the segment $\overline{OE}$ is 5 units; or, synonymously, the distance from the point O to the point E is 5 units. Supporting argument should not be necessary for the statement that, since the distance from O to E is 5 units, it follows immediately that the distance from E to O is also 5 units. Likewise the distance from O to H, or from H to O, is 5 units. The minus sign used with each of the numbers to the left of the reference point O has nothing to do with the distance each point is from O. The sign merely indicates that the point is located on the line to the left of O at a distance designated by each respective real number. Similarly, the distance CE,† or the length of the segment $\overline{CE}$, is given by the number $3 = 5 - 2$. The number 5 corresponds to the point E, and the number 2 corresponds to the point C; and the distance corresponds to the number 3, which is the difference between the numbers 5 and 2. By analogy, the distance EC would correspond to the difference $2 - 5 = 2 + (-5) = -3$. This seems to contradict what we expected to find, namely, that the length of segment $\overline{EC}$ = the length of segment $\overline{CE}$. Such possibilities of conflict and contradiction in the concepts of "distance between two points" and "length of a line segment" are removed by these two definitions:

Definition 8-4   The *distance* between two points on a number line is the absolute value of the difference between their corresponding numbers.

Length of a line segment   **Definition 8-5**   The *length* of a line segment is the distance between its two end points.

The assumption in step 4 of the discussion of the number line (page 253) is equivalent to the following distance postulate.

---

† At this point it is well to call attention to the need for careful distinction of notation: $\overleftrightarrow{AB}$ represents a line which extends indefinitely in either direction, as indicated by the double arrow; $\overrightarrow{AB}$ is the ray whose end point is A and extends indefinitely in one direction from A; $\overline{AB}$ is the line segment whose end points are A and B; AB represents the length of the line segment $\overline{AB}$, the distance from A to B along the line joining them.

Postulate 8-3 To every pair of distinct points on a number line there corresponds a unique positive real number, called the *measure* of the length of the line segment joining the two points.

Example If the points P and Q correspond to the real numbers p and q, respectively, then the length of the line segment $\overline{PQ}$ is given by the formula

$$PQ = |p - q| = |q - p|$$

In Fig. 8-5

$$CE = |5 - 2| = 3 = |2 - 5| = EC$$
$$OE = |5 - 0| = 5 = |0 - 5| = EO$$
$$OH = |(-5) - 0| = 5 = |0 - (-5)| = HO$$
$$HE = |5 - (-5)| = 10 = |(-5) - 5| = EH$$

To review the meaning of $|a|$, where a is a real number, see Definition 7-1 (page 255).

Now that the concepts of line segment and length of a line segment have been introduced, it is possible to present a more precise concept of *betweenness*, illustrated previously in the discussion of Fig. 8-3.

Betweenness **Definition 8-6** The point C is *between* points A and B if and only if A, B, and C are distinct collinear points such that $AC + CB = AB$.

From this definition it follows that if C is between A and B, the three points are situated as in Fig. 8-3. On the other hand, if the three points are distinct and yet so related that $AC + CB = AB$, then it follows that C is between A and B.

In the previous example note that $OE = OH = 5$. These two line segments are said to be *congruent*. Since the point O divides the line segment $\overline{HE}$ into two congruent segments, it is said to be the *midpoint* of $\overline{HE}$.

Congruent line segments **Definition 8-7** Two line segments of the same length are said to be *congruent*.

If $\overline{AB}$ and $\overline{CD}$ are congruent line segments, we may indicate this fact in either of two equivalent forms:

$$AB = CD \quad \text{(length of } \overline{AB} = \text{length of } \overline{CD}\text{)}$$
$$AB \cong CD \quad \text{(segment } AB \text{ } is \text{ } congruent \text{ } to \text{ segment } CD\text{)}$$

Midpoint of a line segment **Definition 8-8** If the point R is located between the points P and Q so that $PR = RQ$, then R is the *midpoint* of the line segment $\overline{PQ}$. The point R is said to *bisect* the line segment $\overline{PQ}$.

Attention is called to the fact that Definition 8-4 allows for the concept of zero distance. If, in the example, the numbers p and q are equal, then the points P and Q are the same point and $PQ = |p - p| = 0$.

8-3 THE CONCEPT OF PLANE

Plane is a set of points

Another very important set of points is that of a *flat surface*, or *plane*, of which a tabletop, a windowpane, and a room floor are familiar illustrations. An accepted test to apply to a surface to determine whether or not it is a plane is to select arbitrary pairs of points anywhere on the surface and fit a straightedge or ruler to these pairs of points. If, in every case, the straight line represented by the edge of the instrument lies wholly within the surface then it may be said to be a flat surface, or plane (see Fig. 8-6). This basic relation between points of a line and points of a plane is formalized in Postulate 8-4.

Postulate 8-4 If a straight line contains two points of a plane, then the line lies entirely within the plane.

Any drawing used to represent a plane, as in Fig. 8-6, must be recognized as just that. A plane extends indefinitely as a flat surface. Just as a line segment is used frequently to represent a line, so a three- or four-sided figure usually is drawn to represent a plane. The test previously described may be used to identify a surface as a plane; it does not distinguish one plane from another. This fact may be illustrated by using the cover of a book to represent a plane. It can be brought to rest in any one of many different positions, as indicated in Fig. 8-7a. The edge of the cover, lettered AB in the figure, represents the line $\overleftrightarrow{AB}$ which lies wholly within any one

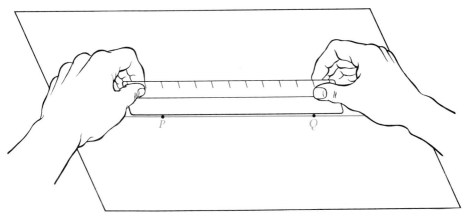

FIGURE 8-6 Using a straightedge to test a plane surface

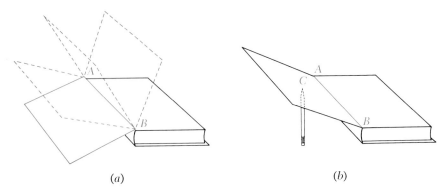

FIGURE 8-7 Examples of planes

of the planes pictured. Figure 8-7*b* illustrates that the line $\overleftrightarrow{AB}$ and a point not on the line are sufficient to identify a particular plane.

Plane uniquely determined **Postulate 8-5** For any given line and a point not on the line there is one and only one plane that contains both the point and the line.

As an immediate consequence of this postulate we have this theorem:

Theorem 8-2 Any three noncollinear points lie in one and only one plane.

Hypothesis Given three points A, B, and C not all lying on the same line.

Conclusion There exists one and only one plane containing the three points A, B, and C.

What conditions are sufficient to determine a plane? Is there any way the given three noncollinear points can be used to get the needed conditions? These questions bring to mind Postulates 8-5 and 8-1, respectively.

Proof

Statements	*Reasons*
The three points A, B, C do not lie on the same line.	Hypothesis
Any two of the points, say A and B, lie on one and only one line.	Postulate 8-1
There is one and only one plane which contains the line $\overleftrightarrow{AB}$ (hence points A and B) and the point C.	Postulate 8-5

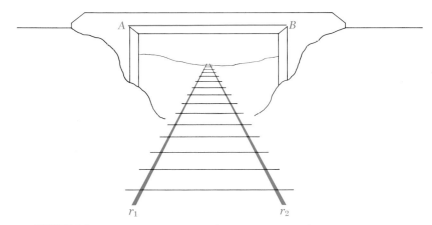

FIGURE 8-8 Parallel lines are coplanar. Skew lines are noncoplanar.

Corollary Two intersecting lines lie in one and only one plane.

Coplanar **Definition 8-9** Lines or points which lie in the same plane are said to be *coplanar*. Otherwise they are said to be *noncoplanar* or *nonplanar*.

If two lines have no points in common, that is, if their intersection is the empty set, they are either *parallel lines* or *skew lines*. In Fig. 8-8 the rails of the railroad track may be used to represent parallel lines r_1 and r_2. An edge of the bridge, such as the one emphasized and lettered AB in the figure, then can represent a line skew both to r_1 and to r_2.

Parallel lines **Definition 8-10** *Parallel lines* are coplanar lines which do not intersect.

Skew lines **Definition 8-11** *Skew lines* are lines which are not coplanar.

Parallel planes **Definition 8-12** *Parallel planes* are planes which do not intersect.

Postulate of parallelism **Postulate 8-6** Given a line l and a point P not contained in l. In the plane determined by l and P there exists one and only one line through P and parallel to l.

Intersection of two planes **Postulate 8-7** If two planes intersect, their intersection is a straight line.

Figure 8-9 shows planes M and N intersecting in the line $\overleftrightarrow{AB}$, and the line l, which is not parallel to either plane, intersecting the plane M in the point C and the plane N in the point D.

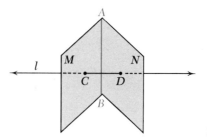

FIGURE 8-9 Two intersecting planes

Half plane

Just as a point on a line is the boundary of each of the two half lines into which the point divides the line, so a line l in a given plane will divide the plane into two *half planes*. The line l is in neither half plane, but is the *edge* of each. The two half planes do not intersect, that is, they have no points in common. In Fig. 8-10, the line l divides the plane into the two half planes α (alpha) and β (beta). Two points (A and D) are said to be in different half planes if there is a point of l between A and D. Such a point is the point C. Two points (A and B) are said to be in the same half plane if there is no point of l between A and B. Sometimes a point which is in a particular half plane is used to designate it. For example, the half plane β of Fig. 8-10 might be designated either as the A side or as the B side of l.

Half space

In a similar manner, a plane π (pi) may be considered as dividing space into two *half spaces* which do not intersect. The plane π is not in either half space, but is called the *face* of each.

It is to be noted that a line is an edge of an infinite number of half planes, but a plane is the face of only two half spaces. Why is this so?

A half plane is an illustration of another very important type of point set, namely, a *convex set*.

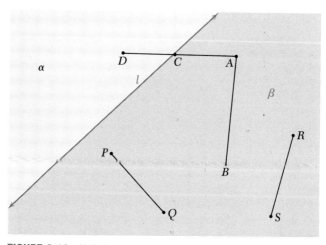

FIGURE 8-10 Half planes

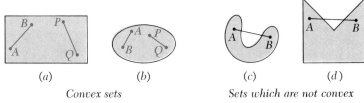

| (a) | (b) | (c) | (d) |

Convex sets   Sets which are not convex

FIGURE 8-11

Convex set **Definition 8-13** A set of points is said to be a *convex set* if it is such that for any two points A and B selected anywhere in the set it is true that the segment $\overline{AB}$ lies entirely within the set.

In Fig. 8-10 the points A and B of the half plane β are such that $\overline{AB}$ lies entirely within the half plane. It should be evident that this would be true for any two points selected anywhere in the half plane, such as P and Q or R and S. The same illustration can be extended to α or any other half plane. The condition of the definition is met, so each half plane is a convex set. Two other illustrations of convex sets are given in Fig. 8-11a and b, and two illustrations of sets of points which are not convex sets are given in Fig. 8-11c and d. In Fig. 8-11c and d, while two points may be chosen such that the line segment joining them lies entirely in the set, it is clear from the picture that this is not true for *any two* points selected anywhere in the set.

EXERCISES

1 Find different illustrations of point, line, and plane.
2 Which of these statements are true?
   (a) A line is a subset of space.
   (b) Space is the universal set of points.
   (c) A plane is a subset of a line.
   (d) A point is a subset of space.
   (e) A line is a subset of a plane.
   (f) A point is a subset of a line.
3 Use the picture of the box to give illustrations of each of the following:
   (a) Intersecting lines; (b) parallel lines; (c) skew lines; (d) parallel planes; (e) intersecting planes; (f) line intersecting a plane.

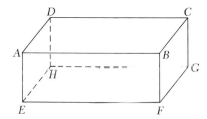

EXERCISE 3 E F

The dashed lines of the figure indicate lines which cannot be seen. They are behind planes.

4 Given that the three noncollinear points P, Q, and R are in plane M and also in plane N, what can you say about planes M and N?

5 Given that the three noncollinear points P, Q, and R are in plane M and the three noncollinear points P, Q, and S are in plane N, what can you say about planes M and N?

6 Draw a line and label four points A, B, C, and D on it in alphabetical order from left to right. Carry out these instructions:

    (a) Give as many illustrations as you can of the concept of "between-ness" by using subsets of the given four points.

    (b) Name two line segments whose union is a line segment.

    (c) Name two line segments whose intersection is a point.

    (d) Name two line segments whose intersection is empty.

    (e) Name two rays whose intersection is C.

    (f) Name two pairs of opposite rays.

    (g) Name two rays whose union is the line.

7 Use a stiff piece of cardboard and three sharp pencils to illustrate that one and only one plane contains three points which are not collinear.

8 Use the equipment of Exercise 7 to show that two points are not sufficient to determine a plane. What does this say about one line?

9 Prove the corollary to Theorem 8-2.

10 Given four noncoplanar points, how many distinct lines and planes do they determine? Identify each line and plane by naming the points which determine it.

11 Repeat Exercise 10 for five noncoplanar points, no three of which are collinear. There are two distinct cases.

12 Draw pictures to illustrate the four relative positions of three distinct coplanar lines.

13 Use sheets of paper to illustrate the relative positions of three distinct planes in space.

14 Describe the union of two collinear half lines; of two coplanar half planes; of two half spaces.

15 Give an illustration of a plane which separates space into two half spaces.

16 Use the figure as an aid in solving these exercises.

    (a) What is the length of each of the segments $\overline{OG}$, $\overline{OJ}$, $\overline{MA}$, $\overline{CF}$, $\overline{LC}$?

    (b) Show that H is the midpoint of the segment $\overline{MC}$.

    (c) What is the midpoint of each of the segments $\overline{OG}$, $\overline{LD}$, $\overline{LB}$, $\overline{JF}$?

    (d) What is the length of each of the segments $\overline{OE}$, $\overline{OK}$, $\overline{KE}$, $\overline{BE}$, $\overline{MK}$?

N	M	L	K	J	I	H	O	A	B	C	D	E	F	G
-3	$-2\frac{1}{2}$	-2	$-\sqrt{3}$	$-1\frac{1}{3}$	-1	$-\frac{1}{2}$	0	$\frac{2}{3}$	1	$1\frac{1}{2}$	2	$\sqrt{6}$	$2\frac{2}{3}$	3

17 Use a sheet of paper as the representation of a plane π. Draw a line l in this plane. Now place three points P, Q, and R in the plane π in such a manner that the point P will be on the Q side of l but not on the R side of l.

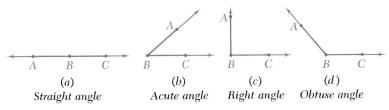

FIGURE 8-12

8-4 THE CONCEPT OF ANGLE

Angle

The union of two noncollinear rays with a common end point is called an *angle*. Such a union might take the form of any one of the four typical angles of Fig. 8-12. The two rays are the *sides* of the angle and the common end point is its *vertex*. The symbol $\angle$ is used to indicate an angle; the plural form is $\measuredangle$. Each of the angles of the figure would be indicated by the symbol $\angle ABC$, read "angle A, B, C," or $\angle CBA$, read "angle C, B, A." At times when there is no ambiguity as to what rays are its sides an angle may be designated by reading merely the letter at its vertex. For example, each angle of the figure might be designated as $\angle B$ and read "angle B." The positions of the points marked on each side, as an aid in labeling an angle, are immaterial. The angle of Fig. 8-13 is the same whether read as $\angle PQR$, $\angle PQV$, $\angle TQS$, or $\angle Q$. Name other ways of reading the same angle.

In Fig. 8-12a the sides of $\angle ABC$ are opposite rays. In such a case the angle is called a *straight angle*. In a sense this is an exceptional angle and will not enter into our discussions a great deal. There are a few cases, however, where the concept makes a very pertinent contribution.

For every line segment there exists a positive real number which is the measure of the length of the segment. Its meaning becomes significant once the unit is determined. Similarly, for every angle there exists a positive real number called *the measure of the angle*. While there are several different units of angular measure, the one most commonly used is the *degree*, with subunits of minutes and seconds such that 60 seconds = 1 minute $(60'' = 1')$, and 60 minutes = 1 degree $(60' = 1°)$. A familiar instrument for measuring angles is the protractor, which has a scale from 0 to 180 marked as in Fig. 8-14. Each number on the scale indicates the measure of the angle in degrees. In the figure the center of the protractor is placed at the vertex B of the angle ABC, and the diameter (the 0 to 180 line) is

Measure of an angle

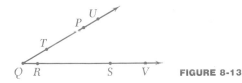

FIGURE 8-13

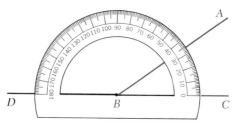

FIGURE 8-14 Protractor measures angle.

placed along $\overrightarrow{BC}$, arbitrarily selected as the *initial side* of the angle. The measure of the angle is then given by the number which corresponds to $\overrightarrow{BA}$, the *terminal side* of the angle. The symbol $m\angle ABC$ is used to represent "the measure of the angle ABC." In the figure, $m\angle ABC = 35$, or the angle ABC is a 35-degree angle, or the angle ABC is a 35° angle.

Postulate 8-8 The measure of a straight angle is 180. ($m\angle DBC = 180$.)

Conversely, if the ray $\overrightarrow{OP}$ is contained in the edge of a half plane M, then for every positive real number p between 0 and 180 there exists one and only one ray $\overrightarrow{OQ}$ in M such that the measure of $\angle POQ$ is p (Fig. 8-15).

Congruent angles **Definition 8-14** Angles with the same measure are called *congruent angles.*

If $\angle ABC$ and PQR are congruent angles, we may indicate this fact in either of two equivalent forms: $m\angle ABC = m\angle PQR$ (measure of $\angle ABC =$ measure of $\angle PQR$), or $\angle ABC \cong \angle PQR$ (angle ABC *is congruent to* angle PQR).

Interior and exterior of an angle With the exception of the straight angle, any angle separates a plane into three disjoint sets of points: the points of the angle, those *interior* to the angle, and those *exterior* to the angle. In Fig. 8-16, points A, B, and C are points of the angle, points P and Q are in the interior of $\angle ABC$ since they are each on the A side of ray $\overrightarrow{BC}$ *and also* on the C side of ray $\overrightarrow{BA}$.

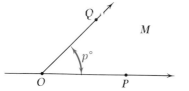

FIGURE 8-15

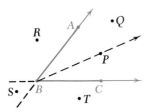

FIGURE 8-16 Interior and exterior of $\angle ABC$

The points R, S, and T are in the exterior of the angle for the respective reasons listed here:

R is on the A side of ray $\overrightarrow{BC}$ *but not also* on the C side of ray $\overrightarrow{BA}$.

T is on the C side of ray $\overrightarrow{BA}$ *but not also* on the A side of ray $\overrightarrow{BC}$.

S is neither on the A side of ray $\overrightarrow{BC}$ nor on the C side of ray $\overrightarrow{BA}$.

The ray $\overrightarrow{BP}$ of Fig. 8-16 lies within the interior of $\angle ABC$ since it joins the vertex B to the interior point P. Thus the two angles ABP and CBP have a common vertex and a common side, formed by the ray $\overrightarrow{BP}$ which is entirely within the interior of $\angle ABC$, and are such that $m \angle ABP + m \angle CBP = m \angle ABC$. Two such angles are called *adjacent angles*.

Adjacent angles

The point of intersection of two intersecting lines may be considered as the common end point of four rays which are oppositely directed in pairs. In Fig. 8-17 the rays $\overrightarrow{OA}$ and $\overrightarrow{OB}$ are opposite rays, as are the rays $\overrightarrow{OC}$ and $\overrightarrow{OD}$. Thus there are formed the two straight angles AOB and COD. Other pairs of significant angles are also formed. Angles AOD and BOC have a common vertex and their sides are respective opposite rays: $\overrightarrow{OD}$ of one angle is opposite to $\overrightarrow{OC}$ of the other, and $\overrightarrow{OA}$ of the first angle is opposite to $\overrightarrow{OB}$ of the second. Such angles are called *vertical angles*. Name another pair of vertical angles to be found in the figure. The angle AOC and the angle BOC have the ray $\overrightarrow{OC}$ as a common side, while the other two sides are the opposite rays $\overrightarrow{OA}$ and $\overrightarrow{OB}$. They are related in such a manner that $m \angle AOC + m \angle BOC = m \angle AOB$ and thus are adjacent angles. When the sum of the measures of two angles, whether they are adjacent or not, is equal to the measure of a straight angle, the angles are said to be *supplementary angles*. Each angle is said to be the *supplement* of the other. $\angle AOC$ and BOC are supplementary adjacent angles.

Vertical angles

Supplementary angles

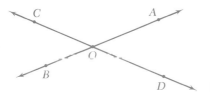

FIGURE 8-17 Vertical angles

Theorem 8-3 Angles which are supplements of the same angle are congruent angles.

Hypothesis $\measuredangle AOD$ and BOC each a supplement of $\angle AOC$ (Fig. 8-17).

Conclusion $\angle AOD \cong \angle BOC$.

From the hypothesis we have that $\measuredangle AOD$ and AOC are supplementary and also that $\measuredangle BOC$ and AOC are supplementary. This means that the sum of the measures of each supplementary pair of angles is 180. Definition 4-1 implies that the measure of each angle of a supplementary pair of angles can be expressed in terms of 180 and the measure of the other angle.

The conclusion we wish to draw is that $\angle AOD$ is congruent to $\angle BOC$. From Definition 8-14 we know that this conclusion can be drawn if we can establish that $m\angle AOD = m\angle BOC$. How can we use the hypothesis and what we know about angle measure to establish this relation of equality between the measures of these two angles?

Proof

Statements	*Reasons*
$m\angle AOD + m\angle AOC = 180$	Supplementary $\measuredangle$
$m\angle AOD = 180 - m\angle AOC$	Definition 4-1
$m\angle BOC + m\angle AOC = 180$	Supplementary $\measuredangle$
$m\angle BOC = 180 - m\angle AOC$	Definition 4-1
$\angle AOD \cong \angle BOC$	Definition 8-14

Corollary 1 Angles which are the supplements of congruent angles are congruent.

Right angle **Corollary 2** Vertical angles are congruent.

If two supplementary angles are congruent, each angle is called a *right angle*. As a result of this definition and Postulate 8-8 we have Theorem 8-4 and its corollary, the proofs of which are left to the reader.

Theorem 8-4 The measure of a right angle is 90.

Corollary All right angles are congruent.

Acute and obtuse angles **Definition 8-15** An *acute angle* is an angle whose measure is less than 90. An *obtuse angle* is an angle whose measure is greater than 90 but less than 180. (See Fig. 8-12.)

The statement of Theorem 8-4 is sometimes used as the definition of a right angle. It is equivalent to the one used here. Theorem 8-5, which states an important property of two intersecting lines, follows as an immediate consequence of Theorem 8-4. Two angles the sum of whose measures is 90 are *complementary angles*. Each angle is the *complement* of the other.

Complementary angles

Theorem 8-5   If one of the angles formed by two intersecting lines is a right angle, then all the angles formed are right angles.

The proof of this theorem is also left to the reader.

Perpendicularity   **Definition 8-16**   Two lines which intersect at right angles are called *perpendicular lines*.

Definition 8-17   If a line *l* intersects a plane π in a point P in such a manner that it is perpendicular to every line lying in the plane π and passing through P, then and only then is the line *l* said to be perpendicular to the plane π.

Postulate 8-9   Given a line l and a point P not on l. In the plane determined by P and l one and only one line can be drawn through P perpendicular to l. Whether P is or is not a point of l, one and only one plane can be passed through P perpendicular to l. (See Figs. 8-18 and 8-19.)

Up to this point in the discussion of angles, the only ones in which we have been interested are those formed by the union of two rays with a common end point. Such angles are frequently referred to as *plane angles*, since two intersecting rays are always sufficient to fix or determine a plane. There are angles of interest which might by contrast be called *space angles*,

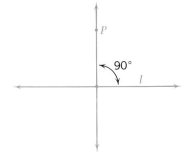

FIGURE 8-18   Mutually perpendicular lines

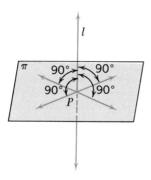

FIGURE 8-19   Line l perpendicular to plane π

since they are not formed by lines in planes but by planes in space. Of these the *dihedral angle* is important here.

Dihedral angle   **Definition 8-18**   A *dihedral angle* is the union of two noncoplanar half planes and their common edge. This common edge is called the *edge*, and its union with each half plane is called a *face*, or *side*, of the angle.

In Fig. 8-20 a letter has been placed in each face of the dihedral angle to facilitate the reading of the angle. The dihedral angle C-AB-D of the figure has the line $\overleftrightarrow{AB}$ as its edge and the planes CAB and DAB as its two faces. A point P in space is interior to $\angle C$-AB-D if it is on the D side of the half plane C and also on the C side of the half plane D. Under what conditions would a point be said to be exterior to $\angle C$-AB-D?

In Fig. 8-21 a point P has been marked on the edge $\overleftrightarrow{AB}$ of the dihedral angle. By Postulate 8-9 one and only one plane can be passed through P perpendicular to the line $\overleftrightarrow{AB}$. Furthermore, this plane will intersect the planes forming the dihedral angle in lines of which rays $\overrightarrow{PR}$ and $\overrightarrow{PQ}$ will be subsets. These rays with the common end point P form the plane angle RPQ. By Definition 8-17 the rays $\overrightarrow{PR}$ and $\overrightarrow{PQ}$ are each perpendicular to the edge $\overleftrightarrow{AB}$ of the dihedral angle. The angle RPQ is called a *plane angle* of the dihedral angle C-AB-D.

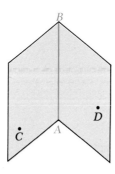

FIGURE 8-20   Dihedral angle C-AB-D

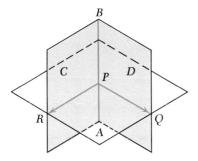

FIGURE 8-21 Plane $\angle RPQ$ of dihedral $\angle C\text{-}AB\text{-}D$

Properties of dihedral angles

Postulate 8-10 All plane angles of a given dihedral angle are congruent.

As a consequence of Postulate 8-10 the measure of a dihedral angle is the same as the measure of any one of its plane angles. There are two very important properties of dihedral angles which are derived immediately from this fact:

1 Two dihedral angles are said to be *congruent* if and only if their plane angles are congruent.

2 A dihedral angle is said to be a *right dihedral angle* if and only if its plane angle is a right angle.

Just as two lines intersect in a point to form pairs of vertical plane angles, so two planes intersect in a line to form pairs of *vertical dihedral angles*. An immediate consequence of the previously stated properties of dihedral angles is this theorem:

Theorem 8-6 Vertical dihedral angles are congruent.

The proof of this theorem is left as an exercise for the reader.

EXERCISES

1 Locate and label three noncollinear points as A, B, and C. Draw the lines which these points determine. Are $\angle ABC$ and $\angle CBA$ the same angle or different angles? Why? Are $\angle ABC$ and $\angle BCA$ the same angle or different angles? Why?

2 How many different angles can you identify in the figure determined by the three lines of Exercise 1?

3 Use the accompanying picture of two parallel lines l and m to describe (*a*) the intersection of the half plane on the P side of l with the half plane on the P side of m; (*b*) the intersection of the half plane on the

Q side of l with the half plane on the Q side of m; (c) the intersection of the half plane on the P side of m with the half plane on the Q side of l.

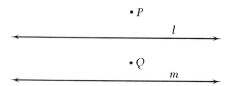

EXERCISE 3

4　Use a clean sheet of paper to represent a plane. In this plane draw two perpendicular lines and locate a point P as in the accompanying figure. Now shade the portion of the figure which represents the intersection of the half plane on the P side of $\overleftrightarrow{X'X}$ with the half plane on the P side of $\overleftrightarrow{Y'Y}$. Identify this portion of the figure by using the letters of the figure other than P.

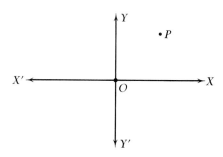

EXERCISE 4

5　Use the pages of a book or two sheets of paper as an aid in describing the interior and exterior of a dihedral angle.

6　For each three-dimensional configuration of the accompanying figure, carry out these instructions:

(a) Count the points, lines, and planes represented.

(b) Count and name the plane angles.

(c) Count and name the dihedral angles.

(d) If the line $\overleftrightarrow{AD}$ is perpendicular to the plane represented by the figure $HABE$, is there a significant relation between the plane angle BAH and the dihedral angle G-AD-C?

(e) If, in addition to the relation of (d), the lines $\overleftrightarrow{AH}$ and $\overleftrightarrow{AB}$ are also perpendicular to each other, what can be said about the dihedral angle G-AD-C?

(f) If the line $\overleftrightarrow{KM}$ is not perpendicular either to the line $\overleftrightarrow{ML}$ or to the line $\overleftrightarrow{MN}$, is the plane angle LMN a plane angle of the dihedral angle L-KM-N? Why?

(g) How many planes are represented as passing through each line of each figure?

(h) Select one line from each figure. Explain why there exist planes other than those represented which might be passed through the selected line.

(i) What is the minimum number of planes necessary to determine a line in space? Why?

(j) How many planes are represented as passing through each point of each figure?

(k) Select one point from each figure. Explain why there exist planes other than those represented which might be passed through the selected point.

(l) What is the minimum number of planes necessary to determine a point in space? Why?

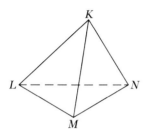
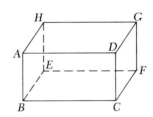

EXERCISE 6

* 7   Prove the two corollaries of Theorem 8-3.
* 8   Prove Theorem 8-4. Also prove its corollary.
* 9   Prove Theorem 8-5.
  10   What are the measures of the complement and supplement of each of these plane angles: (a) 10°; (b) 45°; (c) 60°; (d) $(90 - A)°$?
  11   The measure of one of the angles formed by two intersecting lines is 35. What is the measure of each of the remaining angles?
  12   In the figure the lines $\overleftrightarrow{AB}$ and $\overleftrightarrow{CD}$ intersect at P. From P the rays $\overrightarrow{PM}$ and $\overrightarrow{PN}$ are drawn in the plane determined by $\overleftrightarrow{AB}$ and $\overleftrightarrow{CD}$. $m \angle APM = 90$ and $m \angle DPN = 90$.

(a) Are there any other right angles in the figure?
(b) Name pairs of complementary angles.
(c) Name pairs of supplementary angles.
(d) Name pairs of congruent angles.

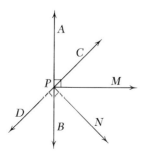

EXERCISE 12

13 Given four plane angles A, B, C, and D so related that A and B are supplementary, B and D are complementary, and $m \angle C - m \angle D = 45$. How do the measures of angles A and C compare?

*14 Prove Theorem 8-6.

*15 *Prove Theorem 8-7:* Plane angles which are the complements of the same angle, or of congruent angles, are congruent angles.

*16 *Prove Theorem 8-8:* If two plane angles are complementary, then each angle is an acute angle.

*17 *Prove Theorem 8-9:* If two unequal plane angles are supplementary, then one and only one is an acute angle.

*18 *Prove Theorem 8-10:* In a plane, one and only one line can be drawn perpendicular to a given line at a fixed point of the line.

8-5 THE CARTESIAN CONCEPT OF POSITION

Take a look at the corner of your room where the floor meets two of the walls. You see a picture such as that of Fig. 8-22 in which the ray $\overrightarrow{OZ}$ represents the intersection of the two walls, and each of the rays $\overrightarrow{OX}$ and $\overrightarrow{OY}$ represents the intersection of a wall and the floor. Properly constructed walls and floors are flat surfaces and thus represent planes. Let your imagination help you to visualize the extension of both walls and the floor in all directions, each retaining its characteristics of a flat surface. The picture visualized should then be of the form of Fig. 8-23, where the line $\overleftrightarrow{Z'Z}$ now represents the intersection of the planes of the two walls, and each of the lines $\overleftrightarrow{X'X}$ and $\overleftrightarrow{Y'Y}$ represents the intersection of the plane of the floor with that of a wall. In such a configuration the lines are said to be *mutually perpendicular* (each line is perpendicular to the other two), and also the planes are mutually perpendicular.

If now a convenient unit is chosen and used, as in Sec. 7-3, to construct a number line on each of the three lines $\overleftrightarrow{X'X}$, $\overleftrightarrow{Y'Y}$, and $\overleftrightarrow{Z'Z}$, we have three mutually perpendicular number lines with a common point as the zero point,

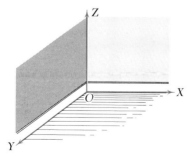

FIGURE 8-22 Room corner

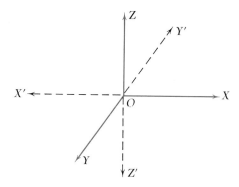

FIGURE 8-23 Cartesian frame of reference

Frame of reference   or *origin*. Such a configuration is called a *cartesian frame of reference.*†
In such a frame of reference the line $\overleftrightarrow{X'X}$ is called the *x axis;* the line $\overleftrightarrow{Y'Y}$,
the *y axis;* and the line $\overleftrightarrow{Z'Z}$, the *z axis.* In Fig. 8-23 the colored portion
of each line represents the positive real numbers, while the dashed portion
represents the negative real numbers. With such a frame of reference any
set of three real numbers will locate a unique point in space, our universal
set of points. Thus, if x is used to represent the number of units to be
measured along the x axis, y the number along the y axis, and z the number
Ordered triple   along the z axis, then the *ordered triple* of real numbers (x,y,z) locates
a specific point in space. This is called an ordered triple since the order
of listing the numbers is always x-axis number, y-axis number, z-axis
number. Each number serves as a coordinate on its respective axis. The
entire set of three axes constitutes a *cartesian coordinate system.* Since
the lines and planes are respectively mutually perpendicular, the system
Rectangular   is also called a *rectangular coordinate system.* In such a frame of reference
coordinate
system   the symbol $P(x,y,z)$ is to be read "the point P whose coordinates are
x, y, and z." In Fig. 8-24 the locations of two points $P(3,1,4)$ and
$Q(-5,-4,2)$ have been indicated. A frame of reference with axes identified
and lettered as in Figs. 8-23 and 8-24 is called a *left-handed system.* This
name is due to the fact that if you use the index finger of your left hand
to point in the direction of the positive x axis, the middle finger can then
point along the positive y axis, and the thumb along the positive z axis.
If the x and y axes were interchanged, you would then have a *right-handed*
system.

Our principal use of a cartesian frame of reference will be in the plane.
In this case there are two perpendicular number lines with a common zero
point. The two coordinate axes will be the x axis $(\overleftrightarrow{X'X})$ and the y axis
$(\overleftrightarrow{Y'Y})$, drawn as shown in Fig. 8-25. Each axis is frequently called by two
other names: the x axis is also called the *horizontal axis* or the axis of

†Named after the Frenchman, René Descartes, who first suggested such an idea in one of three
appendices to a book published in 1637.

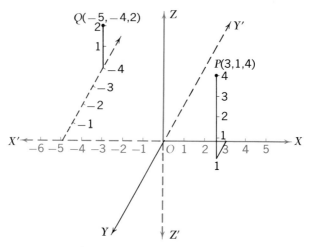

FIGURE 8-24 Cartesian coordinates in space

abscissas, and the y axis is at times called either the *vertical axis* or the axis of *ordinates*. In the plane a point is located by an *ordered pair* of real numbers in the form (x,y), with the x coordinate written first. In the figure the locations of the two points $P(4,3)$ and $Q(-3,-2)$ are indicated. Note that the positive x values are always to the right of the origin and the negative values to the left. On the y axis the positive values are always above the origin and the negative values below. In such a frame of reference any point in the plane is uniquely located. The two axes divide the plane

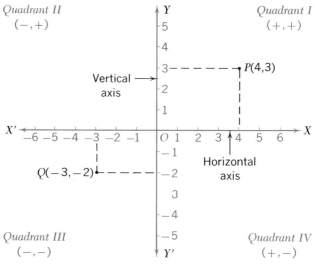

FIGURE 8-25 Cartesian coordinates in a plane

Quadrants of the plane

into four sections called *quadrants*. They are identified as the first, second, third, and fourth quadrants and are labeled with the respective Roman numerals I, II, III, IV. The method used to identify each quadrant is to label it with the signs of the coordinates which will locate points in that quadrant. As indicated in Fig. 8-25, the signs for each quadrant are I($+,+$), II($-,+$), III($-,-$), and IV($+,-$).

*8-6 POSITION ON THE EARTH'S SURFACE

Angular measure affords a very significant adaptation of the cartesian method to the location of points on the surface of the earth. For this purpose the earth is assumed to be a true sphere (Definition 8-22). Any circle on the surface of a sphere whose center coincides with the center of the sphere is called a *great circle*. All other circles on the sphere are called *small circles*. Geometry tells us that three points, not all on the same straight line, will fix the position of a specific circle and that no other distinct circle will pass through these same three points. Since this is true there is one and only one great circle which will pass through both the North Pole and the South Pole and through any other one point on the surface of the earth. The equator is the great circle which lies in the plane perpendicular at the center of the sphere to the diameter joining the North Pole to the South Pole (see Fig. 8-26).

Frame of reference on the earth's surface

The frame of reference for locating points on the earth's surface consists of the equator and the great circle which passes through Greenwich, England, and through the two poles. This great circle is called the *prime meridian*, or sometimes the *Greenwich meridian*. All other great circles on the earth's surface that pass through the two poles are called *meridians*. *Longitude* is measured by the number of degrees, minutes, and seconds in the central angles† of the equator, and is either east (E) or west (W),

† A central angle of a circle is an angle whose vertex is at the center of the circle (see Fig. 8-31).

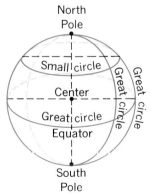

North
Pole

South
Pole

FIGURE 8-26 Great and small circles on earth's surface

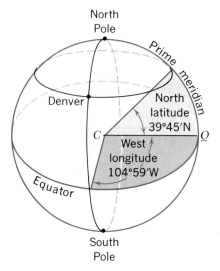

North
Pole

Denver

North
latitude
39°45′N

West
longitude
104°59′W

C

O

Equator

South
Pole

FIGURE 8-27 Longitude and latitude on earth's surface

depending on which direction it is measured from the prime meridian. The number of degrees, minutes, and seconds in the central angle of a meridian is called *latitude*. It is either north (N) or south (S), depending on whether it is measured from the equator in the direction of the North Pole or the South Pole. In Fig. 8-27 a west longitude and a north latitude are indicated.

The origin of this frame of reference is the point O, where the prime meridian and the equator intersect. Figure 8-28 lists the longitudes and latitudes, correct to the nearest minute, of several different localities. For example, Denver, Colorado, is located at 104°59′ west longitude and 39°45′ north latitude. This means that its location on the earth's surface is 104°59′ west of the prime meridian and 39°45′ north of the equator. Such information is extremely important in air and sea travel and world communications.

CITY	LONGITUDE	LATITUDE
Camden, New Jersey	75° 7′ W	39° 57′ N
Denver, Colorado	104° 59′ W	39° 45′ N
Dutch Harbor, Alaska	166° 33′ W	53° 33′ N
Fresno, California	119° 47′ W	36° 44′ N
Honolulu, Hawaii	157° 52′ W	21° 18′ N
Natal, Brazil	35° 13′ W	5° 47′ S
Rome, Italy	12° 30′ E	41° 54′ N
St. Louis, Missouri	90° 12′ W	38° 38′ N
Sydney, Australia	151° 12′ E	33° 52′ S
Yokohama, Japan	139° 40′ E	35° 27′ N

FIGURE 8-28 Longitude and latitude of selected cities

Natal and Sydney are in the *Southern Hemisphere* (half sphere), while all other cities of the table are in the *Northern Hemisphere*. Rome, Sydney, and Yokohama are in the *Eastern Hemisphere*, and the others are in the *Western Hemisphere*.

An added fact of great significance is that the key for uniformity of patterns of time throughout the world is the basic relationship that 15° of longitude is equivalent to 1 hour of time. This is based on the fact that there are 360° of longitude around the world and 24 hours in one day (360° ÷ 24 = 15°). Time zones are established in terms of this relationship; of course there are many adjustments in the basic pattern to adapt to intrinsic characteristics of local situations. Each zone is approximately 15° "wide," and its time is fixed in accordance with that meridian passing through it which is an exact multiple of 15°. There are seven time zones in the United States: *Eastern*, which uses the 75th west meridian; *Central*, the 90th west meridian; *Mountain*, the 105th west meridian; *Pacific*, the 120th west meridian; *Yukon*, the 135th west meridian; *Alaska-Hawaii*, the 150th west meridian; and *Bering*, the 165th west meridian. Camden (Eastern), St. Louis (Central), Denver (Mountain), Fresno (Pacific), Honolulu (Alaska-Hawaii), and Dutch Harbor (Bering) are listed in the table as cities of the United States, each situated very near the time-standard meridian of its respective zone. Notice that the difference in longitude between any two of these cities approximates very closely an exact multiple of 15°. To find the time difference between any two cities one can find the difference between the two exact multiples of 15° which most nearly correspond to the respective longitudes of the cities. The difference in time in hours will then be the same as the obtained multiple of 15°. Although the Yukon time zone crosses a portion of the United States, the section is a very small and thinly populated area of Alaska. For this reason it is not too significant as a time zone of the United States.

The *international date line* is halfway around the world from the prime meridian. Thus it is the 180° meridian, whether measured westward or eastward from the prime meridian. This is where the new day starts. For a person crossing the international date line from the Western Hemisphere to the Eastern Hemisphere, the day of the week advances one day: for example, 8 A.M. Sunday becomes 8 A.M. Monday. Analogously, Monday becomes Sunday if the direction is reversed.

Example When it is Sunday in Honolulu, Hawaii, what day is it in Sydney, Australia?

Since Honolulu is near the 157° west meridian and Sydney is near the 151° east meridian, the international date line is between them. So when it is Sunday in Honolulu it is Monday in Sydney.

Example What is the difference in time between (1) Fresno, California, and Camden, New Jersey; (2) Dutch Harbor, Alaska, and Denver, Colorado;

Time zones

(3) Honolulu, Hawaii, and Yokohama, Japan; (4) St. Louis, Missouri, and Rome, Italy?

1   Fresno is near the 120° west meridian and Camden is near the 75° west meridian. There is a difference of 45° of longitude in the time-standardizing meridians of these two cities. Therefore, there is a difference of 3 hours in time.

2   Dutch Harbor is near the 165° west meridian and Denver is near the 105° west meridian. The difference of 60° of longitude corresponds to a difference of 4 hours of time.

3   Honolulu is near the 150° *west* meridian and Yokohama is near the 135° *east* meridian. The difference in longitude, in a case such as this where one city is in a west longitude zone and the other in an east longitude zone, is found by *adding* the two longitude measures. The difference in longitude between Honolulu and Yokohama is 150° + 135° = 285°. This difference in longitude corresponds to a difference of 19 hours in time.

4   St. Louis is near the 90° west meridian and Rome is near the 15° east meridian. The difference in longitude is 105°, and so there is a difference of 7 hours of time.

EXERCISES

1   Use two pages of an open book and the top of a table or desk to represent a cartesian frame of reference in space with x, y, and z axes, as indicated in Fig. 8-23.

2   Use the frame of reference of Exercise 1 to describe the position of each of these points: (0,0,0); (2,0,0); (0,−3,0); (0,0,1); (0,0,−4); (−5,0,0); (0,6,0).

3   Use the frame of reference of Exercise 1 to describe the position of each of these points: (3,1,0); (4,−2,0); (−3,5,0); (−4,−6,0); (4,0,2); (3,0,−3); (−5,0,4); (−1,0,−1); (0,1,3); (0,−1,2); (0,4,−3); (0,−5,−6).

4   Use the frame of reference of Exercise 1 to describe the position of each of these points: (2,1,4); (5,−2,3); (−8,4,1); (−3,−7,−5); (1,−3,−4); (−2,−5,4); (3,2,−4); (−8,6,−4).

5   Exercise 4 has exhibited points located in each of eight distinct regions into which space is separated by three coordinate planes. These regions are called *octants*. In a left-handed cartesian frame of reference the coordinate signs which will locate a point in an octant identify it by number in the following manner: Octant I(+,+,+); II(+,−,+); III(−,−,+); IV(−,+,+); V(+,+,−); VI(+,−,−); VII(−,−,−); VIII(−,+,−). Select a set of coordinates that will locate a point in each of the eight octants described.

6   Write four ordered pairs of numbers which are such that used as coordinates of points in a plane they will locate points, one in each of the four quadrants. Use a sheet of graph paper to draw a cartesian

frame of reference and then plot each of the four points whose coordinates you selected.

7 Plot the three points (1,6), (−2,0), and (0,4). These are three collinear points. How can you test to see whether they seem to be collinear?

8 Plot the points (2,5), (−4,−3), and (6,−1). These points are not collinear. How many distinct lines can you draw containing distinct pairs of these points? Draw them.

9 How many distinct lines can you draw containing distinct pairs of the points you plotted in Exercise 6? Draw them.

10 Select ordered pairs to give you five distinct points in a plane, no three of which are collinear. Draw the lines which are determined by distinct pairs of these points.

11 Which city of the United States, included in Fig. 8-28, is situated most nearly on the same circle of latitude with Yokohama, Japan?

12 Are Sydney, Australia, and Honolulu, Hawaii, on approximately the same meridian (circle of longitude)? Why?

13 How many hours' difference in time exist between each pair of cities: St. Louis and Denver; Dutch Harbor and Camden; Honolulu and Fresno; Honolulu and Denver?

14 Which city is farther east: Natal, Brazil, or Camden, New Jersey? Why?

8-7 THE CONCEPT OF SHAPE

Simple closed curve

One of the basic concepts of the geometry of a plane is that of a simple closed curve which lies entirely in the plane. Rather than give a rigorous definition of such a curve it will suffice for this discussion to present a few typical illustrations. This is done in Fig. 8-29, in which each picture also illustrates the fact that a simple closed curve which lies entirely in a plane separates the plane into three disjoint sets of points: the set of points in the *interior* of the curve, the set of points of the curve, and the set points in the *exterior* of the curve. The points A and B of Fig. 8-30a are situated so that one of them (A) is in the interior and the other (B) is in the exterior, since every line segment, whether straight or not, joining A and B has at least one point in common with the curve. The points C and D, both of which are in the interior, are joined by a line segment which has no points in common with the curve. The same is true for points E and F, both of

FIGURE 8-29   Simple closed curves in a plane

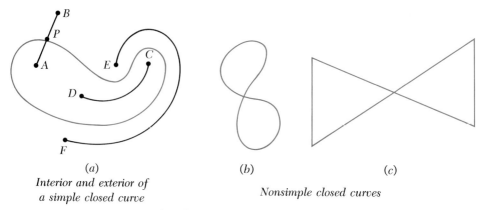

(a)
*Interior and exterior of
a simple closed curve*

(b)   (c)

Nonsimple closed curves

FIGURE 8-30 Closed curves

which are in the exterior. The figure also illustrates that a simple closed curve does not intersect itself. Figure 8-30*b* and *c* illustrates nonsimple closed curves.

As an aid to clearer comprehension, the intuitive descriptions of the previous paragraph are summarized in these statements:

Curve   A *curve* is any geometric figure which may be drawn without lifting the pencil. If the curve can be retraced repeatedly by continuing in the same direction, it is a *closed curve*. If a closed curve does not cross itself, it is called a *simple closed curve;* otherwise it is a *nonsimple closed curve.*

There are basically two types of simple closed curves in a plane with which we are to be concerned, namely, *circles* and *polygons*.

Circle   **Definition 8-19**   A *circle* is a simple closed curve in a plane all points of which are at a fixed distance from a fixed point. The fixed distance is called the *radius* and the fixed point the *center* of the circle. (See Fig. 8-31.)

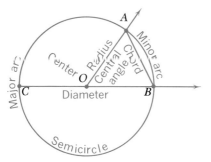

FIGURE 8-31   Elements of a circle

A *chord* of a circle is a line segment whose end points are points of the circle. A *diameter* of the circle is a chord which contains the center of the circle, and its length is always twice the length of the radius. A *central angle* is an angle whose vertex is the center of the circle. If the sides of a central angle, with measure less than 180, intersect the circle in points A and B, then these two points and all points of the circle interior to the angle form a *minor arc* $\overset{\frown}{AB}$, while the two points and all points of the circle exterior to the angle form a *major arc* $\overset{\frown}{ACB}$. Similarly, the end points of the chord $\overline{AB}$ can be said to separate the circle into a minor arc and a major arc. If the chord is a diameter of the circle, then the two arcs are congruent and each is called a *semicircle*. Two or more circles with the same center are said to be *concentric circles*.

Postulate 8-11   It is always possible to draw a circle with a given point as center and a fixed length as radius.

Polygon   **Definition 8-20**   A *simple polygon* is a simple closed curve in a plane which is the union of a finite number of line segments so situated that no two segments have a point of intersection other than a common end point and no two line segments with such a common end point are collinear. Each line segment is a *side* of the polygon, and each point common to two line segments is a *vertex*, and the angle formed is an *angle of the polygon*.

Some of the more familiar simple polygons are illustrated in Fig. 8-32. A *triangle* has three vertices and three sides; a *quadrilateral*, four vertices and four sides; a *pentagon*, five vertices and five sides; and a *hexagon*, six vertices and six sides. In general, an *n-gon* is a polygon with n vertices and n sides. The triangle and the quadrilateral are the most generally useful of the four polygons shown. Two other polygons which are used fairly frequently are the *octagon* (8-gon) and the *decagon* (10-gon). If the set of interior points of a polygon is a convex set, the polygon is a *convex polygon*. Each of the polygons of Fig. 8-32 is a convex polygon. The polygons of Figs. 8-11*d* and 8-30*c* are polygons that are not convex.

Classification of triangles   There are several basic classifications of triangles which are of significance. They may be classified according to characteristics of angles or according to characteristics of sides, as shown in Fig. 8-33. While it is not

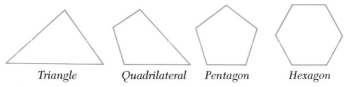

Triangle   Quadrilateral   Pentagon   Hexagon

FIGURE 8-32   Kinds of polygons

KIND OF TRIANGLE	CHARACTERISTICS		TYPICAL FIGURE
	ANGLES	SIDES	
Scalene	No two congruent.	No two congruent.	
Isosceles	Only two congruent.	Only two congruent.	
Equilateral or equiangular	All three congruent.	All three congruent.	
Acute angle	All acute angles. No two, only two, or all three may be congruent.	No two, only two, or all three may be congruent.	No 2 ≅ Only 2 ≅ All 3 ≅
Right	One angle is a right angle. The other two angles are acute and may or may not be congruent.	Two sides including the right angle may or may not be congruent. Hypotenuse is the longest side.	No sides ≅ 2 sides ≅ No angles ≅ 2 angles ≅
Obtuse angle	One angle is an obtuse angle. The other two angles may or may not be congruent.	Two sides including the obtuse angle may or may not be congruent.	No angles ≅ No sides ≅ 2 sides ≅ 2 angles ≅

FIGURE 8-33 Classification of triangles

too difficult to prove, we shall accept without proof the proposition that *the sum of the measures of the angles of any triangle is 180*. This, of course, implies that no triangle can have more than one of its angles either a right angle or an obtuse angle. Thus a triangle may have all its angles acute angles (*an acute-angle triangle*); or it may have one right angle (*a right triangle*); or it may have one obtuse angle (*an obtuse-angle triangle*). In an acute-angle triangle there are three possibilities: No two angles are congruent (*scalene triangle*); only two angles are congruent (*isosceles triangle*); or all three angles are congruent (*equiangular triangle*). In either a right triangle or an obtuse-angle triangle, there always are two acute angles, which may or may not be congruent.

The three sides of any triangle have three possible distinguishing relationships: No two sides are congruent (*scalene triangle*); only two sides are congruent (*isosceles triangle*); or all three sides are congruent (*equilateral triangle*). From the above discussion a scalene triangle is seen to be one that has no two angles and no two sides congruent; an isosceles triangle has only two angles and only two sides congruent; and an equilateral triangle, which is also equiangular, has all three sides and all three angles congruent. In an isosceles triangle the side which is not one of the two congruent sides is called the *base* of the triangle.

In a right triangle the side opposite the right angle is always the longest side and is called the *hypotenuse*. The other two sides are called *legs*. One of the most famous and significant theorems in mathematics relates the hypotenuse and legs of any right triangle. Although its proof is beyond the scope of this book, the theorem will be listed here because of its interest and general usefulness.

Pythagorean theorem

Theorem 8-11   The pythagorean theorem   The square of the length of the hypotenuse of a right triangle is equal to the sum of the squares of the lengths of the two legs.

In the right triangle ABC of Fig. 8-34 the angle at C is the right angle, c is the length of the hypotenuse, and the lengths of the two legs are a and b, respectively. The pythagorean theorem then states that

$$c^2 = a^2 + b^2$$

The theorem gets its name from the fact that the first general proof of it was by the outstanding Greek mathematician Pythagoras (ca. 540 B.C.). One book has been published† in which the author has compiled 370 proofs of this one theorem. It was through the application of this theorem to an isosceles right triangle that Pythagoras and his followers discovered the irrational number $\sqrt{2}$. For example, if the length of each leg is 1 unit, then $c^2 = (1)^2 + (1)^2 = 2$, and the length of the hypotenuse is $\sqrt{2}$ units.

Regardless of the type of triangle being discussed, an accepted, and frequently used, symbol for "triangle" is $\triangle$. The plural form of the symbol is $\triangle\!\!\!\triangle$.

† E. S. Loomis, "The Pythagorean Theorem," 2d ed., Edwards Brothers, Inc., Ann Arbor, Mich., 1940; later (1968) reissued by the National Council of Teachers of Mathematics as a title in its "Classics in Mathematics Education."

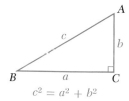

$$c^2 = a^2 + b^2$$

FIGURE 8-34   The pythagorean theorem

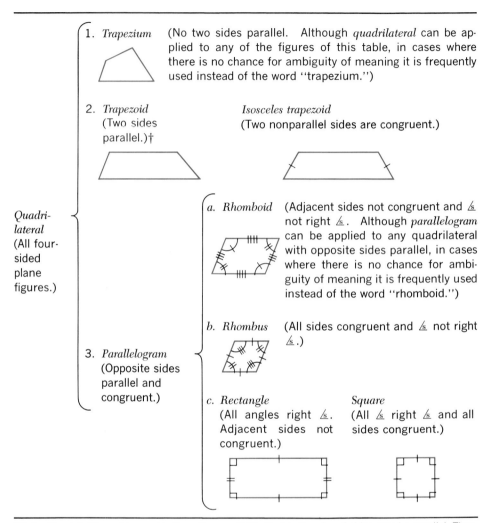

1. *Trapezium*   (No two sides parallel.   Although *quadrilateral* can be applied to any of the figures of this table, in cases where there is no chance for ambiguity of meaning it is frequently used instead of the word "trapezium.")

2. *Trapezoid*
   (Two sides
   parallel.)†

   Isosceles trapezoid
   (Two nonparallel sides are congruent.)

Quadri-lateral
(All four-sided plane figures.)

3. *Parallelogram*
   (Opposite sides parallel and congruent.)

   a. Rhomboid   (Adjacent sides not congruent and ∠ not right ∠.   Although *parallelogram* can be applied to any quadrilateral with opposite sides parallel, in cases where there is no chance for ambiguity of meaning it is frequently used instead of the word "rhomboid.")

   b. Rhombus   (All sides congruent and ∠ not right ∠.)

   c. Rectangle   *Square*
   (All angles right ∠.   (All ∠ right ∠ and all
   Adjacent sides not   sides congruent.)
   congruent.)

†Note: The definition of *trapezoid* allows the possibility that both pairs of sides may be parallel. Thus, a parallelogram is a trapezoid. The table departs from this classification deliberately to provide greater ease in reading and clearer presentation of general practice in interpretation.

FIGURE 8-35   Typical plane quadrilaterals

Classification of plane quadrilaterals

Just as all triangles fall into distinct classifications according to fundamental characteristics, so do all plane quadrilaterals. These classifications with their distinguishing characteristics are shown in Fig. 8-35.

The only distinction made in classifications of polygons of more than four sides which is of significance to this discussion has to do with whether or not they are regular polygons.

Regular polygon   **Definition 8-21**   A *regular polygon* is a polygon all of whose sides and all of whose angles are congruent.

Thus, there are pentagons and regular pentagons, hexagons and regular hexagons, octagons and regular octagons, decagons and regular decagons. Which of the three-sided and four-sided polygons are regular?

In space we are interested in solids and their surfaces. One convenient pattern to follow in the classification of solid figures is to put the *sphere* in a set by itself.

Sphere   **Definition 8-22**   A *sphere* is the set of all points in space at a given distance from a given point. The given distance is the *radius* of the sphere and the given point is the *center*. Two or more spheres with the same center are called *concentric spheres*. A *chord* of the sphere is a line segment whose end points are on the sphere. A *diameter* of the sphere is a chord which contains the center of the sphere and has length twice the length of the radius. (See Fig. 8-36.)

Theorem 8-12 is an immediate consequence of the definitions of circle and sphere. The proof of the theorem is left to the reader.

Theorem 8-12   The intersection of a sphere and a plane through its center is a circle.

The *interior of a sphere* is the set of points which contains the center and all those points of space whose distance from the center is less than the length of the radius of the sphere, and the *exterior of a sphere* is the set of all points of space whose distance from the center of the sphere is greater than the length of the radius. A sphere is a surface that is curved instead of being flat like a plane.

Classification of solid figures   With the sphere classified in one set, a second set of solids would then contain all various types of prisms and the *circular cylinder*. Four types of *prisms* are shown in Fig. 8-37. The polygons of the boundary surfaces

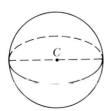

FIGURE 8-36 Sphere

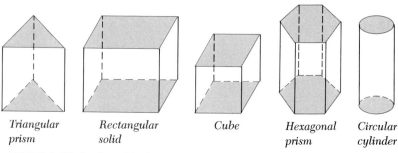

| *Triangular prism* | *Rectangular solid* | *Cube* | *Hexagonal prism* | *Circular cylinder* |

FIGURE 8-37 Familiar solids of two bases

of the prisms are called *faces*. The lines in which the faces intersect are called *edges*. Two of the faces of each prism are in parallel planes and are called the *bases* of the prism. In the figure the bases are in color. The faces between the parallel planes are called *lateral faces* and together make up the *lateral surface* of the prism. While the bases may be of any polygonal pattern, the lateral faces are parallelograms. In the *rectangular prism*, or *rectangular solid*, as it is frequently called, all faces are rectangles. The *cube* is a rectangular solid all of whose faces are squares or, equivalently, all of whose edges are equal. In common with prisms, the cylinder has two bases. For the cylinder, however, the bases are not polygons. They may be any type of curve, and the lateral surface is a corresponding curved surface. The only type of cylinder that will concern us here is the *circular cylinder*, i.e., one whose bases are circles.

A third set of the familiar solids would contain *pyramids* and *circular cones* (Fig. 8-38). This group has the common characteristic of having only one base, shown in color. The base of a circular cone is a circle, and its lateral surface is curved. While the base of a pyramid may be any type of polygon, its lateral faces are triangles. The common vertex of these triangles is called the *vertex* of the pyramid. If the base is a regular polygon, the pyramid is called a *regular pyramid*.

Because of their basic composition, prisms and pyramids belong to another important classification of solids known as *polyhedrons* (solids of

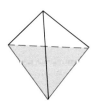  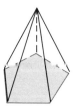 

Circular cone

Pyramids

FIGURE 8-38 Familiar solids of one base

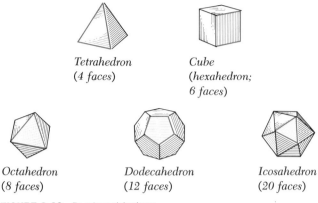

Tetrahedron
(4 faces)

Cube
(hexahedron;
6 faces)

Octahedron
(8 faces)

Dodecahedron
(12 faces)

Icosahedron
(20 faces)

FIGURE 8-39 Regular polyhedrons

Regular polyhedrons

many faces). Of great importance in this group of solids are the five regular polyhedrons. They are so called because their faces are all regular polygons (Fig. 8-39).

8-8 THE CONCEPT OF SIMILARITY

Fundamentally, two geometric figures are similar if they have the same shape. As illustrations of similar plane figures (Fig. 8-40) we may cite any two line segments, any two circles, and any two regular polygons of the same number of sides. In space, similar figures would be any two spheres and any two regular polyhedrons with the same number of faces (Fig. 8-41).

Similarity of polygons and polyhedrons is not necessarily restricted to those which are regular. The more general concept of similarity in the plane is in accordance with this definition:

Similar polygons

Definition 8-23 Two polygons for which a one-to-one correspondence between vertices has been established are said to be *similar* if and only if their corresponding angles are congruent and there exists a constant ratio between corresponding sides.

The symbol $\sim$ is used to indicate that two figures are similar to each other. In other words, if polygon P is similar to polygon Q we might indicate this fact by writing "polygon $P \sim$ polygon Q."

A B

C D

FIGURE 8-40 Similar plane figures

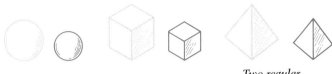

Two spheres of different size   *Two cubes of different size*   *Two regular tetrahedrons of different size*

FIGURE 8-41 Similar solid figures

It should be obvious that regular polygons of the same number of sides meet the criteria of the definition. It can be established that two triangles which satisfy either criterion will necessarily satisfy the other. In other words, two triangles whose corresponding angles are congruent are similar; also two triangles whose corresponding sides are proportional are similar. That this is not true for polygons in general is illustrated by an examination of the rectangle, square, and rhombus of Fig. 8-42. The angles of the rectangle and square are all congruent, but the ratios of corresponding sides are not constant ($\frac{12}{3} \neq \frac{4}{3}$). The ratios of corresponding sides of the square and rhombus are constant ($\frac{8}{3} = \frac{8}{3}$), but the corresponding angles are not congruent. (All angles of the square are right angles, while two angles of the rhombus are acute angles and two are obtuse angles.)

As a consequence of the special nature of triangles, there are two other significant conditions of similarity. One of these will be stated as a theorem whose proof is left to the reader. Although the second can be proved as a theorem, its proof is beyond the scope of this treatment and it will be stated as a postulate.

Similarity of triangles

Theorem 8-13 Two triangles are similar if two angles of one are congruent to two angles of the other.

Postulate 8-12 Two triangles are similar if two pairs of corresponding sides are proportional and the included angles are congruent.

Although there exist criteria for determining similarity of nonregular polyhedrons, they are not of concern to this treatment.

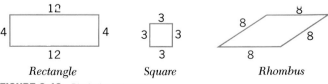

Rectangle   *Square*   *Rhombus*

FIGURE 8-42 Dissimilar polygons

Example

1 Given the triangles ABC and DEF such that $m \angle A = 24$, $m \angle B = 126$, and $m \angle C = 30$; $m \angle D = 30$, $m \angle F = 126$, and $m \angle E = 24$. The angles of the two triangles can be made to correspond in such a manner that the corresponding angles are congruent. This correspondence is $\angle A \leftrightarrow \angle E$, $\angle B \leftrightarrow \angle F$, and $\angle C \leftrightarrow \angle D$. Thus the similar triangles are ABC and EFD.

2 Given the triangles ABC and DEF such that $AB = 3$, $BC = 4$, and $CA = 5$; $DE = 12$, $EF = 16$, and $FD = 20$. If the vertices of the two triangles be made to correspond so that $A \leftrightarrow D$, $B \leftrightarrow E$, and $C \leftrightarrow F$, then the ratios of corresponding sides remain constant $(\frac{3}{12} = \frac{4}{16} = \frac{5}{20} = \frac{1}{4})$. Therefore ABC and DEF are similar triangles.

3 Given triangles ABC and DEF such that $m \angle A = 75$ and $m \angle B = 42$; $m \angle F = 75$ and $m \angle D = 42$. If $\angle A \leftrightarrow \angle F$ and $\angle B \leftrightarrow \angle D$, then $\triangle ABC$ and FDE are similar.

4 Given the triangles ABC and DEF such that $m \angle A = 52$, $m \angle D = 52$, and $\dfrac{AB}{DF} = \dfrac{AC}{DE} = 2$. Here $\angle A \cong \angle D$ and the including sides are proportional. The ratios of the sides show that vertex B corresponds to vertex F, and C to E. It then follows that the similar triangles are ABC and DFE. Furthermore, the constant ratio of 2 states that the length of each side of $\triangle ABC$ is twice the length of its corresponding side in $\triangle DFE$.

The authority for the first two illustrations is the definition of similarity and the fact that, in the case of triangles, either set of criteria implies the other. For the third illustration the authority is Theorem 8-13, and for the fourth illustration it is Postulate 8-12.

Example In $\triangle ABC$, $AB = 15$, $BC = 25$, $CA = 30$. In the similar $\triangle DEF$, $FD = 42$; find the lengths of DE and EF.

In Fig. 8-43, f represents the side $\overline{DE}$ and d represents $\overline{EF}$. The correspondence of vertices as given is $A \leftrightarrow D$, $B \leftrightarrow E$, and $C \leftrightarrow F$. The ratios of the sides are $\dfrac{f}{15} = \dfrac{d}{25} = \dfrac{42}{30} = \dfrac{7}{5}$. Since $25 = 5 \times 5$, $d = 5 \times 7 = 35$; $15 = 3 \times 5$, so $f = 3 \times 7 = 21$.

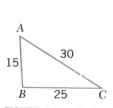
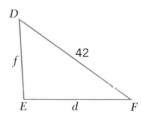

FIGURE 8-43 Similar triangles

EXERCISES

1 Draw figures similar to that of Fig. 8-30 to illustrate interior and exterior points for the following simple closed curves: (a) circle; (b) triangle; (c) rectangle; (d) hexagon.

2 Can a straight line segment joining an interior point of a simple closed curve to an exterior point have more than one point in common with the curve? Illustrate your answer.

3 A simple description of a simple closed curve in a plane is that it may be represented by a curve which can be drawn with a pencil on a sheet of paper without lifting the pencil and such that the starting point and end point coincide but the curve has no other points which coincide. Draw illustrations of simple closed curves in a plane using only curved lines; using only straight line segments; using combinations of the two.

4 Draw illustrations of closed curves in a plane which are not simple closed curves. Use curved lines only; use straight line segments only; use combinations of the two.

5 A famous German mathematician, Leonhard Euler (pronounced oil'er) (1707–1783), discovered a formula which expresses the relation existing between the number of faces (F), the number of vertices (V), and the number of edges (E) of any polyhedron. The formula is $F + V - E = 2$. Verify this formula for the polyhedrons of Figs. 8-37 to 8-39.

* 6 Prove Theorem 8-12.

* 7 Prove Theorem 8-13.

8 Which of the following conditions are sufficient to make the two given triangles similar? Name the similar triangles, when they exist, by ordering the vertices to indicate correspondence.

(a) $m \angle A = 35$, $m \angle B = 105$, $m \angle C = 40$; $m \angle D = 40$,
$m \angle E = 35$, $m \angle F = 105$

(b) $m \angle A = 50$, $m \angle B = 90$; $m \angle D = 50$, $m \angle E = 90$

(c) $m \angle A = 47$, $m \angle B = 95$, $m \angle C = 38$; $m \angle D = 47$,
$m \angle E = 113$, $m \angle F = 20$

(d) $m \angle B = 50$, $BA = 15$, $BC = 12$; $m \angle D = 50$, $DE = 24$,
$DF = 30$

(e) $AB = 40$, $BC = 10$, $CA = 35$; $EF = 20$, $FD = 5$, $DE = 25$

(f) $m \angle A = 45$, $m \angle B = 65$; $m \angle D = 65$, $m \angle E = 70$

(g) $AB = 24$, $BC = 32$, $CA = 40$; $EF = 25$, $FD = 20$, $DE = 15$

(h) $m \angle A = 72$, $AB = 75$, $AC = 48$; $m \angle D = 72$, $DE = 25$,
$DF = 24$

(i) $AB = 18$, $BC = 12$, $CA = 20$; $DE = 9$, $EF = 6$, $FD = 10$

* 9 Are these quadrilaterals similar?

(a) Rectangle $ABCD$ with $AB = DC = 12$, $AD = BC = 3$
Rectangle $EFGH$ with $EF = HG = 20$, $EH = FG = 5$

(b) Parallelogram $ABCD$ with $AB = DC = 75$, $AD = BC = 15$;
$m \angle A = m \angle C = 135$, $m \angle B = m \angle D = 45$

Parallelogram $EFGH$ with $EF = HG = 100$, $EH = FG = 20$; $m \angle E = m \angle G = 135$, $m \angle F = m \angle H = 45$

(c) Rectangle $ABCD$ with $AB = DC = 84$, $AD = BC = 63$
Rectangle $EFGH$ with $EF = HG = 125$, $EA = FD = 50$

10 The length of the sides of $\triangle ABC$ are as follows: $AB = 35$, $BC = 20$, and $CA = 40$. In the similar $\triangle DEF$, $DE = 42$. What are the lengths of DF and EF if the vertices correspond in this manner: $A \leftrightarrow D$, $B \leftrightarrow E$, $C \leftrightarrow F$.

8-9 THE CONCEPT OF SIZE

We have seen how any simple closed curve in a plane separates the plane into three disjoint sets of points: the set of points interior to the curve, the set of the curve, and the set exterior to the curve. The union of the set *Closed region* of interior points and the set of points of the curve is called a *closed region*, and the curve is the *boundary* of the region. If the simple closed curve is a circle, the region is called a *circular region;* if it is a polygon, the region is called a *polygonal region*. In particular, a closed region is called a *triangular region* or a *rectangular region* according to whether its boundary is a triangle or a rectangle (Fig. 8-44).

There is a positive real number associated with each line segment which tells the number of units of linear measure in the length of the line segment. This number is called the distance between the two end points of the segment. Similarly, the measure of an angle is a positive real number which tells the size of the angle in units of angular measure. Likewise, for every circular region or polygonal region there exists a positive real number, called the *area of the region*, which describes its size in units of area (square) measure. In a somewhat analogous fashion, spherical and polyhedral regions in space are defined. For each such region there exists a positive real number, the *volume of the region*, which describes its size in terms of units of volume (cubic) measure.

8-10 THE CONCEPTS OF CONGRUENCE AND SYMMETRY

Previously we have seen that two line segments are congruent when they have the same length. Also two angles, whether plane angles or dihedral angles, are congruent if they have the same measure. Since two line

Circular region   *Triangular region*   *Rectangular region*

FIGURE 8-44 Closed regions

segments, two plane angles, or two dihedral angles are always figures of the same shape, the condition that they be congruent is equivalent to stating that they must be figures of the same shape and size. This is in fact the basic condition for congruence.

Congruence of geometric figures   **Definition 8-24**   Two geometric figures are said to be *congruent* if they are the same in shape and size.

This definition, while satisfactory as the context for the present informal discussion, is more nearly a basis for intuitive acceptance than an attempt at rigorous statement. It is equivalent to saying that two geometric figures are congruent if they are so related that they would fit exactly if one were moved, without changing its shape or size, and placed on the other. For example, two circles whose radii are congruent are said to be *congruent circles*. Any polygonal region can be divided into a finite number of triangular regions such that any two of them either do not intersect or do intersect in a line segment or a point. Since this is the case, the congruency of polygons fundamentally can be based on the congruency of triangles. There are three basic conditions which are both necessary and sufficient for the congruency of triangles. These conditions are stated here as postulates.

Congruence of triangles   **Postulate 8-13 (SAS)**   If two triangles correspond to each other in such a way that two sides and the included angle of one triangle are congruent to the corresponding sides and included angle of the other, the two triangles are congruent.

Postulate 8-14 (ASA)   If two triangles correspond to each other in such a way that two angles and the included side of one triangle are congruent to the corresponding two angles and included side of the other, then the triangles are congruent.

Postulate 8-15 (SSS)   If two triangles correspond to each other in such a way that the three sides of one triangle are congruent to the corresponding three sides of the other triangle, then the triangles are congruent.

It should not be too difficult a conclusion to draw that corresponding parts of any two congruent figures are also congruent.

Symmetry   A plane region is *symmetric with respect to a point* if this point bisects every line segment through it with end points in the boundary of the region. The point is called the *center of symmetry*. The region is *symmetric with respect to a line* if the line forms with the boundary of the region two congruent figures such that if one were folded over on the line, it would

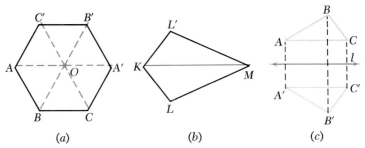

FIGURE 8-45 Symmetric plane regions

coincide with the other. The line is called an *axis of symmetry*. The regular hexagon (Fig. 8-45a) has each of the diagonals shown as axes of symmetry and their point of intersection as a center of symmetry. The hexagon has three other axes of symmetry. Can you describe them? The kite-shaped quadrilateral *KLML'* of Fig. 8-45b has *KM* as an axis of symmetry but has no center of symmetry. In this figure the triangles *KLM* and *KL'M* can be shown to be congruent. Two plane regions are said to be symmetric with respect to an axis of symmetry if their boundaries are congruent figures and the correspondence is such that if one were folded over on the axis it would coincide with the other. The triangles *ABC* and *A'B'C'* of Fig. 8-45c are symmetric with respect to the line *l*.

In space two polyhedrons are said to be congruent if their corresponding parts are, congruent and arranged in the same order. If the corresponding parts are arranged in the reverse order, the two polyhedrons are said to be symmetric. In Fig. 8-46 the polyhedrons *A-BCD* and *K-LMN* are congruent. They are each symmetric to *P-RST*. Two symmetric solids are somewhat like two gloves or two shoes of the same pair. They are sized alike but shaped in reverse order to each other.

EXERCISES

1 Given the following information about △*ABC* and *DEF*: △*ABC* ≅ △*DEF*; *A* ↔ *D*, *B* ↔ *E*, and *C* ↔ *F*. What can you say

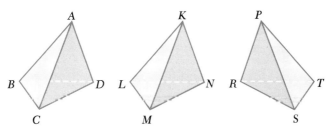

FIGURE 8-46 Congruent or symmetric solids

about the corresponding angles and sides of the two triangles? Give reasons to support your answer.

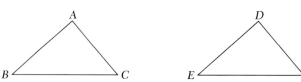

2   In the quadrilateral $ABCD$ shown, $\overline{AB} \cong \overline{AD}$ and $\overline{BC} \cong \overline{DC}$. (Note that the congruences are indicated by writing the same symbol on each segment of a congruent pair. This practice is followed generally for line segments and angles.) Is it a correct statement to say that the line segment $\overline{AC}$ bisects $\angle DAB$ and $\angle DCB$? Why or why not? (*Hint:* An angle is bisected by a line if the line divides it into two congruent angles.)

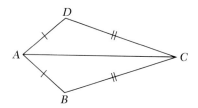

3   Referring to the figure below, prove $\triangle KLM \cong \triangle PLO$.

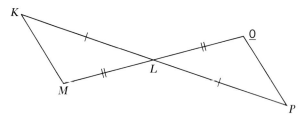

4   In the parallelogram $PQRS$ shown below, prove $\triangle PQR \cong \triangle RSP$. What angle is congruent to $\angle SRP$? To $\angle SPR$? Is the parallelogram symmetric with respect to the diagonal $\overline{PR}$? Why?

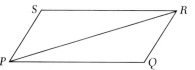

Prove the theorems in Exercises 5 to 8.

∗ 5   *Theorem 8-14.* Two triangles are congruent if two angles and a side of one are respectively congruent to two angles and a side of the other.

∗ 6   *Theorem 8-15.* Two right triangles are congruent if the two legs of one are congruent to the two legs of the other.

7 *Theorem 8-16.* In an isosceles triangle the angles opposite the congruent sides are congruent. (*Hint:* From the common point of the two congruent sides draw a line to the midpoint of the opposite side.)

8 *Theorem 8-17.* An equilateral triangle is also equiangular.

9 Describe three axes of symmetry of a regular hexagon other than those pictured in Fig. 8-45a.

10 For each of the given figures determine whether or not there exist one or more axes of symmetry and also a center of symmetry: (*a*) circle; (*b*) trapezoid; (*c*) isosceles triangle; (*d*) scalene triangle; (*e*) equilateral triangle; (*f*) quadrilateral with no two sides equal; (*g*) square; (*h*) rectangle; (*i*) parallelogram; (*j*) rhombus.

8-11 CONSTRUCTIONS WITH STRAIGHTEDGE AND COMPASS

Postulates 8-1 and 8-4 make it possible to draw a straight line connecting any two points in space, and assure us that this line will always lie in a plane. Postulate 8-11 makes it always possible to draw in a plane a circle with a given point as center and a given line segment as radius. The instruments used for these constructions are the *straightedge*, or *unmarked ruler*, for drawing a line segment to represent the straight line, and the *compass* for drawing a circle. Since these postulates are in essence included among the original postulates of Euclid, constructions with the straightedge

Euclidean constructions

and compass are frequently referred to as "euclidean constructions." Among such constructions there are a few that are considered the basic ones. These are the ones of interest in this discussion. As the different constructions are studied, it will become quite apparent that they depend upon the intersection properties of two lines, a line and a circle, and two circles. The intersection of two lines has been taken care of by Theorem 8-1, Postulate 8-6, and the discussion of page 281. The corresponding properties of the line and circle and of two circles will be taken care of here by Postulates 8-16 and 8-17, each of which is a provable theorem, but with proof beyond the scope of this treatment.

Intersection of line and circle

Postulate 8-16 Given a line and a circle in the same plane:

1 They have two points in common if the distance d from the center of the circle to the line is less than the radius r: $d < r$. (See Fig. 8-47a.)

2 They have one point in common if $d = r$. (See Fig. 8-47b.)

3 They have no points in common if $d > r$. (See Fig. 8-47c.)

When $d < r$, the portion of the line containing the two points of the line between P and Q form a chord of the circle. When $d = r$, the line which

Tangent

has one point A in common with the circle is called a *tangent* to the circle, the point A is called the *point of tangency*, and it can be shown that the radius $\overline{OA}$ forms a right angle with the tangent at the point of tangency.

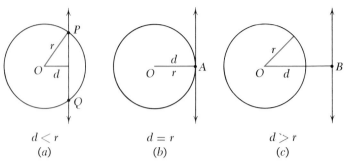

FIGURE 8-47 Intersection of a line and a circle

Postulate 8-17 Given two circles, with radii r_1 and r_2, in the same plane with distance d between their centers.

Intersection of two circles

1 They have two points of intersection if the sum of each two of r_1, r_2, and d is greater than the third. The two points will be on opposite sides of the line joining the centers. (See Fig. 8-48a.)

2 They have one point of intersection if the sum of any two is equal to the third. (See Fig. 8-48b.)

3 They have no points in common if the sum of any two is less than the third. (See Fig. 8-48c.)

Basic geometric constructions

Construction 8-1 At a point on a given line, construct an angle congruent to a given angle.

We are given the line l and $\angle ABC$ (Fig. 8-49).

Mark a point P on the line l. Select one of the half lines of which P is the boundary as the set of points which with P is to be one side of the desired angle. This is the ray $\overrightarrow{PT}$ of the figure. With the vertex B of $\angle ABC$ as a center and any convenient radius, draw the arc of a circle cutting the sides of the angle in the points M and N. With this same radius and P as a center, draw the arc of a circle cutting $\overrightarrow{PT}$ in Q. With Q as a center and MN as a radius, draw the arc of a circle. The two circles will intersect in the two points R and S. The rays $\overrightarrow{PR}$ and $\overrightarrow{PS}$ form with $\overrightarrow{PQ}$ the angles QPR and QPS, each of which is congruent to $\angle ABC$.

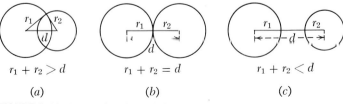

FIGURE 8-48 Intersection of two circles

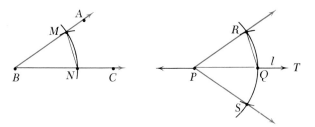

FIGURE 8-49   Construction of an angle congruent to a given angle

Proof

Statements   *Reasons*

Draw the line segments $\overline{RQ}$ and $\overline{MN}$ Postulate 8-1

In $\triangle QPR$ and NBM, $\overline{PQ} \cong \overline{BN}$, $\overline{PR} \cong \overline{BM}$,   They were so

$\overline{RQ} \cong \overline{MN}$ constructed

$\triangle QPR \cong \triangle NBM$ Postulate 8-15

$\angle QPR \cong \angle NBM$ Definition 8-24

A similar argument will prove $\angle QPS \cong \angle NBM$.

Construction 8-2   Construct a triangle whose sides are congruent to three given line segments.

Given the three line segments a, b, c (Fig. 8-50).

Draw any line l and on it mark the point A. With A as a center and c as a radius, draw the arc of a circle cutting l in B. Also with A as a center and b as a radius, draw the arc $\widehat{PQ}$. With B as a center and a as a radius, draw arc $\widehat{RS}$. These two arcs will intersect in two points C and C'. Draw $\overline{AC}$ and $\overline{BC}$, also $\overline{AC'}$ and $\overline{BC'}$. Either $\triangle ABC$ or $\triangle ABC'$ is the required triangle. Prove this statement.

What would happen in this construction if the sum of any two sides were not greater than the third?

Why could this construction be used to reproduce a given triangle?

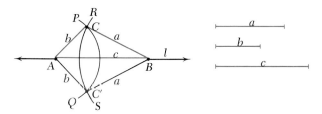

FIGURE 8-50   Construction of a triangle with sides congruent to three given line segments

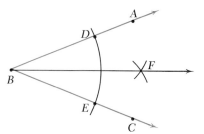

FIGURE 8-51 Bisection of a given angle

Construction 8-3 Bisect a given angle.

Given the angle ABC (Fig. 8-51).

With B as a center and any convenient radius, draw the arc of a circle cutting rays $\overrightarrow{BA}$ and $\overrightarrow{BC}$ in points D and E, respectively. With D as a center and a radius greater than one-half DE, draw a circle. With E as a center and the *same radius*, draw another circle. By Postulate 8-17 these two circles will intersect in two points, one of which will be on the opposite side of the line $\overleftrightarrow{DE}$ from B. Call this point F. Draw the ray $\overrightarrow{BF}$ which will bisect $\angle ABC$.

The proof of this construction is left to the reader.

Construction 8-4 Bisect a given line segment.

Given the line segment $\overline{PQ}$ (Fig. 8-52).

With P as a center and any radius which is greater than one-half PQ, draw a circle. With Q as a center and the *same radius*, draw another circle. By Postulate 8-17 these two circles will intersect in two points M and N, which are on opposite sides of $\overline{PQ}$. By Postulate 8-1 there exists a line joining these two points. Since the two points are on opposite sides of $\overline{PQ}$, the line $\overleftrightarrow{MN}$ will have a point in common with $\overline{PQ}$, call it O. This point O is the midpoint of $\overline{PQ}$.

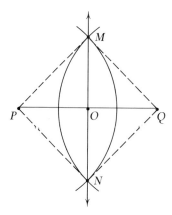

FIGURE 8-52 Bisection of a line segment

Proof

Statements	*Reasons*
Draw line segments $\overline{PM}$, $\overline{PN}$, $\overline{QM}$, and $\overline{QN}$	Postulate 8-1
$\overline{PM} \cong \overline{QM}$, $\overline{PN} \cong \overline{QN}$	Why?
$\triangle PMN \cong \triangle QMN$	Postulate 8-15
$\angle PMO \cong \angle QMO$	Definition 8-24
$\triangle PMO \cong \triangle QMO$	Postulate 8-13
$\overline{PO} \cong \overline{OQ}$	Definition 8-24
Therefore O is the midpoint of $\overline{PQ}$	Definition 8-8

Corollary The line $\overleftrightarrow{MN}$ forms right angles with $\overline{PQ}$.

Definition 8-25 The *perpendicular bisector* of a line segment is the line which passes through the midpoint of the segment and forms right angles with the segment. Any point on this perpendicular bisector is the same distance from each of the end points of the segment.

The line $\overleftrightarrow{MN}$ of Fig. 8-52 is the perpendicular bisector of the segment $\overline{PQ}$.

Construction 8-5 Draw a line through a given point so that it is perpendicular to a given line.

The construction is the same whether P is on the line (Fig. 8-53a) or is not on the line (Fig. 8-53b).

With P as a center and a convenient radius, draw a circle that will cut

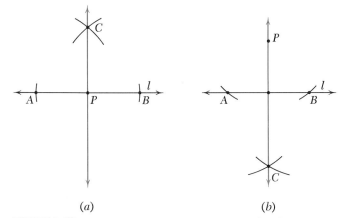

(a)   (b)

FIGURE 8-53 Construction of a line perpendicular to a given line

the line l in two points A and B. With any radius that is greater than PA or PB, use A and B each as a center to draw an arc of a circle. These two arcs will intersect in a point C. Draw $\overleftrightarrow{PC}$.

Why is $\overleftrightarrow{PC}$ the perpendicular bisector of $\overline{AB}$?

Since $\overleftrightarrow{PC}$ is the perpendicular bisector of $\overline{AB}$, it follows that it is perpendicular to the line l of which the segment $\overline{AB}$ is a subset.

EXERCISES

1   *Construction 8-6.* Referring to the figure, construct a triangle which will have line segments a and b as sides and $\angle C$ as the included angle.

EXERCISE 1

2   *Construction 8-7.* Referring to the figure, construct a triangle with angles B and C as two of its angles and line segment a as the included side.

EXERCISE 2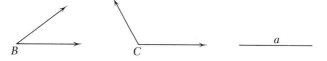

3   Construct a triangle which will have as sides line segments whose lengths are 3, 4, and 2 inches, respectively.

4   Draw three line segments and then use them as sides to construct a triangle. What precaution do you have to take in selecting the line segments?

5   Draw two angles and a line segment. Construct a triangle which will have the two given angles as two of its angles and the given segment as a side opposite one of these angles. What precaution must you remember in determining the sizes of the two given angles? Discuss the different possibilities of this construction.

* 6   Prove that Construction 8-2 is a valid construction.

* 7   Prove that Construction 8-3 is a valid construction.

* 8   Prove the corollary to Construction 8-4.

9   (*a*) Construct an isosceles triangle.

   (*b*) Construct the perpendicular bisector of the base of the isosceles triangle.

   (*c*) Why should this perpendicular bisector pass through the point common to the two equal sides?

10   (*a*) Construct an equilateral triangle.

   (*b*) Construct the perpendicular bisector of each side.

(c) Why should the perpendicular bisector of each side pass through the point of intersection of the other two sides?

(d) If you have made a careful construction, the three perpendicular bisectors will seem to have a point in common. Why should they have a point in common, and therefore intersect?

11   Use a line segment of 1-inch length as a radius to draw a circle. If you mark any point on this circle and use it as a center with the same 1-inch radius to mark another point on the circle and then repeat this until you have gone around the entire circle, you should have returned to the same point from which you started and have six points marked. Join these six points. The figure you have constructed is a regular hexagon. Construct the six axes of symmetry of this regular hexagon.

12   Construct a regular hexagon. Do not erase the circle used in drawing the hexagon. By selective shading you can get a simple architectural design.

13   Construct a regular hexagon. Draw the line segments joining alternate vertices. You should now have an equilateral triangle. Shading will again produce an architectural design.

14   Draw a circle with any convenient radius. Now draw one of its diameters and then construct the perpendicular bisector of this diameter. Mark the points where this perpendicular bisector cuts the circle. Join these points to the end points of the given diameter. You should have a square.

*15   Draw a circle with any convenient radius. Draw two diameters so that they form an angle of 30°. Use Exercise 14 to construct two squares each with one of these diameters as a diagonal. By shading you can get an architectural design. (*Hint:* Use a protractor as an aid in getting an angle of 30°.)

*8-12   THE STRUCTURE OF A GEOMETRY

From the historian T. L. Heath† we learn that Euclid, when he compiled the 13 books of his *Elements*, listed as first principles 23 definitions, 5 postulates, and 5 common notions. The common notions were statements accepted as true assertions related to all areas of quantitative subject matter rather than to geometry in particular. The postulates were specifically related to geometry. We are concerned here with the 5 postulates.

Euclidean postulates

1   A straight line can be drawn from any point to any other point.

2   A finite straight line can be produced continuously in a straight line.

3   A circle may be described with any center and distance.

4   All right angles are equal to one another

---

† T. L. Heath, "The Thirteen Books of Euclid," 2d ed., vol. I, pp. 153–155, Dover Publications, Inc., New York, 1956.

5 If a straight line falling on two straight lines makes the interior angles on the same side together less than two right angles, the two straight lines, if produced indefinitely, meet on that side on which the angles are together less than two right angles.

This fifth postulate is the famous *parallel postulate*, which has been referred to by some writers as one of the greatest single statements in the history of science. It has been stated in this chapter as Postulate 8-6 in the equivalent, but more readily comprehensible, form known as Playfair's postulate. The other four postulates also have been stated in the same or equivalent form to that used by Euclid. For these reasons the geometry of this chapter represents a portion of what is familiarly known as *euclidean geometry*. This is in contrast to the noneuclidean geometries of Bolyai (1802–1860), Lobachevski (1793–1856), and Riemann (1826–1866). Each of these geometries was arrived at through a distinct contradiction of the parallel postulate.

In the Bolyai-Lobachevski geometry, the first four euclidean postulates remain intact, while the parallel postulate is replaced by:

Replacements of the parallel postulate

Given a line l and a point P not contained in l. In the plane determined by l and P there exist more than one line through P and parallel to l.

In the riemannian geometry the parallel postulate is replaced by:

Given a line l and a point P not contained in l. In the plane determined by l and P there exists no line through P and parallel to l.

In this geometry there is also a basic difference in the statement of the second postulate.

Noneuclidean geometries

As might be expected, there are differences in the theorems of euclidean geometry and those of either noneuclidean geometry. These will occur, however, only in those cases where the contradictory postulates are appealed to as support for the argument needed to establish a particular theorem. For example, the congruency theorems, stated here as Postulates 8-13 to 8-15, are provable in euclidean geometry without any appeal to the parallel postulate. They are equally valid in both noneuclidean geometries. On the other hand, appeal to this postulate is necessary to prove the euclidean theorem:

The sum of the angles of a triangle is equal to two right angles.

The corresponding noneuclidean theorems are:

Bolyai-Lobachevski: The sum of the angles of a triangle is less than two right angles.

Riemann: The sum of the angles of a triangle is greater than two right angles.

The question thus arises as to which theorem is to be accepted. The answer to this question is that the decision is determined not by examining the theorems but by examining the postulates. Each theorem is a valid theorem in its respective context. Also, mathematicians have been able to establish that each noneuclidean geometry is as consistent as is euclidean geometry. No internal inconsistencies have been discovered throughout the centuries of study of euclidean geometry. While this is not proof of the fact, it does lend strong support to the feeling that no internal inconsistencies ever will be discovered. Thus, in geometry as in the study of number, a change in the basic postulates produces a change in the subject matter of the derivable system. The choice of postulates will be controlled primarily by the context of study. Euclidean geometry will continue to furnish the geometrical model for the solution of the more general problems of engineering, science, and environmental experience. On the other hand, the two-dimensional noneuclidean geometry of Riemann can be represented on the surface of a sphere and thus can serve as a model for the solution of certain types of navigational problems. Furthermore, Riemann created a large set of noneuclidean geometries, one of which (and distinct from the one suggested here) Einstein used as the model for his general theory of relativity. Also, it has been suggested that probably the best model for the study of binocular vision is to be found in the geometry of Bolyai and Lobachevski.†

The procedure followed in the development of this chapter has been to use patterns of deductive reasoning to derive from first principles (postulates and definitions) specific properties (theorems) of the system. Such a technique is known as *synthetic* in contrast to the *analytic* technique which, through recourse to cartesian coordinates, employs equations as an aid in making the interpretations and derivations from first principles. The analytic technique, which was introduced in Secs. 8-5 and 8-6, will be pursued in somewhat more detail in Chap. 10. A geometry may be either synthetic or analytic, depending upon the technique of investigation used.

Another important characteristic of a geometry is the space in which it is oriented. The development of this chapter has been restricted to spaces of one, two, and three dimensions. In the discussion of points on a line the orientation was in one dimension; in the study of points and lines in

---

† Howard Eves and Carroll V. Newsom, "An Introduction to the Foundations and Fundamental Concepts of Mathematics," rev. ed., p. 78, Holt, Rinehart and Winston, Inc., New York, 1965.

a plane it was in two dimensions; and in the study of points, lines, and planes in space it was in three dimensions. In a cartesian frame of reference one coordinate is required to locate a point in one dimension; two coordinates to locate a point in two dimensions; and three coordinates to locate a point in three dimensions. One of the great advantages of the use of cartesian coordinates and the analytic technique is the facility afforded for the extension of geometric concepts to many dimensions. For example, the geometry of the relativity theory is four-dimensional in that *time* is one of the coordinates. A point is oriented in this space by the ordered quadruple (x,y,z,t) where x, y, and z are the ordinary cartesian coordinates of a three-dimensional space, and t is the time coordinate.

Translation and rotation

In our brief study of geometry we have employed such techniques as "moving along a line" or "turning through an angle." In more elegant language these acts are known, respectively, as *transformations* of *translation* and *rotation*. If you move this book along a table or turn it over you will not affect the size of the book; its measurements will remain unchanged, or *invariant*. Measurements of length, angle size, area, and volume are among the more important invariants under the transformations of translation and rotation. Along with all other invariable quantities and relations, these invariants become the topics of study in the geometry which employs these transformations.

Projection

Another familiar type of transformation is *projection*. Any visit to a movie theater will furnish many illustrations of the use of this transformation. It is not difficult to comprehend that measurements, such as those mentioned previously, are no longer invariants under this transformation. As an illustration, one has but to reflect upon the relation of the image on the screen to the image on the movie film. The relative positions of point of light, film image, and screen can change length, angle size, area, and volume almost at will. There are other types of invariants which become the subject matter for study in the geometry of projection. A few examples are the relations of points on a line, of lines passing through a point, and certain types of ratios. It is of significance to note that these are also invariants under translation and rotation.

There are still other types of transformations which lead to other forms of geometric study.

In conclusion we might summarize the steps necessary to structure an area of geometric study in the following manner. (The order of listing is not necessarily the only sequence of selection.)

Steps for structuring an area of geometric study

1 Select the space in which the study is to be oriented.

2 List the fundamental postulates, undefined elements, and defined terms upon which the study is to be based.

3 Determine whether the synthetic method, the analytic method, or some combination of the two is to be used.

4 Build the configurations to be studied.

5 Select the transformations to be used.

6 Identify the invariants to be studied.

8-13 REVIEW EXERCISES

1–38 Write a carefully constructed answer to each of the guideline questions of this chapter.

39 In the accompanying figure, *LMN-PQR* represents a *triangular prism*, and if *BCDE* is a square, *A-BCDE* represents a *square pyramid*.

(*a*) In each figure count the points, lines, and planes represented. (*Hint:* There are also lines represented but not drawn.)

(*b*) In each figure count and name the plane angles as drawn.

(*c*) In each figure count and name the dihedral angles.

(*d*) If $\overleftrightarrow{MQ}$ is perpendicular to plane *PQR*, is there a relation between ∠*PQR* and the dihedral angle *P-MQ-R*?

(*e*) How many planes are represented as passing through each line of each figure?

(*f*) What is the minimum number of points necessary to determine a line in space?

(*g*) If $\overleftrightarrow{LM}$ and $\overleftrightarrow{PQ}$ are parallel lines, what relation exists between $\overleftrightarrow{LM}$ and $\overleftrightarrow{RQ}$?

(*h*) How many planes are represented as passing through each point of each figure?

(*i*) What is the minimum number of planes necessary to determine a point in space?

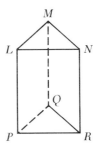

40 Given five points *A*, *B*, *C*, *D*, and *E*, no three of which are collinear and no four of which are coplanar:

(*a*) How many distinct lines and planes do these five points determine?

(*b*) Identify each line and each plane by naming the points which determine it.

*41 Follow the instructions of Exercise 40 for the points *P*, *Q*, *R*, *S*, *T*, *U*, no three of which are collinear and no four of which are coplanar.

42 Angle POR is the complement of $\angle POS$ and the supplement of $\angle POQ$. If $m \angle POR = 30$, what is the measure of each of the other angles?

43 Angles A and C are supplementary while angles B and C are complementary. If $m \angle C$ is 20 degrees more than $m \angle B$, what is the measure of each angle?

*44 Use the concept of half plane to describe each of the four quadrants of a rectangular frame of reference in a plane.

*45 (a) List ordered triples of numbers which will locate points in each of the eight octants of space.

(b) How many lines are determined by these eight distinct points?

(c) Letter each of the points and identify each of the lines of (b) by the points which determine it.

(d) If no four of the points are in the same plane, how many distinct planes will be determined by the eight points?

(e) Identify each plane by the letters naming the points which determine it.

46 The table lists, correct to the nearest minute, the longitude and latitude of several cities. Answer the following questions based on the table.

(a) Arrange the cities of the northern hemisphere in order from the northernmost city to the southernmost city.

CITY	LONGITUDE	LATITUDE
Atlanta, Ga.	84°24′W	33°45′N
Camden, N.J.	75°7′W	39°57′N
Chicago, Ill.	87°38′W	41°52′N
Columbia, S.C.	81°2′W	34°0′N
Denver, Col.	104°59′W	39°45′N
Dutch Harbor, Alaska	166°33′W	53°33′N
Fresno, Calif.	119°47′W	36°44′N
Honolulu, Hawaii	157°52′W	21°18′N
Los Angeles, Calif.	118°14′W	34°3′N
Nashville, Tenn.	86°47′W	36°10′N
Natal, Brazil	35°13′W	5°47′S
New Orleans, La.	90°4′W	29°57′N
Rome, Italy	12°30′E	41°54′N
St. Louis, Mo.	90°12′W	38°38′N
San Francisco, Calif.	122°25′W	37°47′N
Sydney, Australia	151°12′E	33°52′S
Yokohama, Japan	139°40′E	35°27′N

(b) Which city is most nearly due east of Los Angeles? Of Chicago?

(c) When it is 12 noon on Sunday in Honolulu, Hawaii, what time is it in Sydney, Australia?

(d) When it is 10 A.M. in Los Angeles, what time is it in Nashville? In Denver? In Dutch Harbor, Alaska? In Honolulu? In Camden, N.J.? In Rome? In Sydney? In Natal, Brazil? In Yokohama, Japan?

(e) Which city is farther east, Camden, N.J., or Natal, Brazil?

(f) Which of the listed cities is most nearly on the same latitude circle with Yokohama, Japan?

47 Distinguish between congruence, similarity, and symmetry in plane figures.

48 Given the right triangles ABC and DEF. Use two different methods to find the length of $\overline{AC}$.

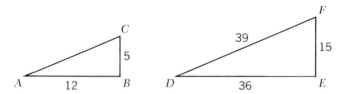

EXERCISE 48

49 Given the line segment $\overline{AC}$ and $\angle C$. Construct a right triangle with $\overline{AC}$ as its hypotenuse and $\angle C$ at the point C.

EXERCISE 49

*50 Use a line segment $1\frac{1}{2}$ inches long as a radius to construct a circle.

(a) Inscribe a regular hexagon in the circle.

(b) Erect the perpendicular bisector of each side of the hexagon.

(c) Why should these bisectors all pass through the center of the circle?

(d) Produce each bisector until it intersects the minor arc determined by the side of the hexagon.

(e) Connect each of these points with the end points of the corresponding side of the hexagon. You now have a *regular dodecagon* (12 sides) inscribed in the circle.

*INVITATIONS TO EXTENDED STUDY

1 Describe the relative positions of three distinct lines in space.

2 Use the concept of half planes to define the interior and exterior of an angle.

3 *Prove this theorem:* If a plane intersects two parallel planes, then it intersects them in parallel lines.

4 Polar coordinates are used to establish a coordinate system in the plane which is distinct from a rectangular system. Investigation of such a system can be both challenging and interesting.

5 Describe the eight octants of a left-handed cartesian frame of reference in space as the intersection of half spaces.

6 Discuss the application of the concept and principles of similarity in the enlargement of photographs.

7 Use Postulate 8-12 as an accepted truth and then prove as theorems Postulates 8-13 and 8-14.

8 Use the pythagorean theorem to prove that the distance d between the two points $P_1(x_1, y_1)$ and $P_2(x_2, y_2)$ is given by the formula

$$d = \sqrt{(x_2 - x_1)^2 + (y_2 - y_1)^2}$$

Prove the theorems of Exercises 9 to 14.

9 The three bisectors of the angles of a triangle intersect in a common point.

10 The point of intersection of the bisectors of the angles of a triangle is the center of the circle inscribed in the triangle. (*Hint:* A circle is *inscribed* in a triangle if it is tangent to each side of the triangle. A line is *tangent* to a circle if it is perpendicular to a radius of the circle at a point of the circle.)

11 The three perpendicular bisectors of the sides of a triangle intersect in a common point.

12 The point of intersection of the perpendicular bisectors of the sides of a triangle is the center of a circle circumscribed about the triangle. (*Hint:* A *circle* is *circumscribed* about a triangle if the three vertices of the triangle are points of the circle.)

13 The three medians of an equilateral triangle are also altitudes of the triangle. (*Hint:* A median of a triangle is a line drawn from a vertex to the midpoint of the opposite side. An *altitude* of a triangle is a line drawn from a vertex perpendicular to the opposite side.)

14 The sum of the measures of the interior angles of a quadrilateral is twice the measure of a straight angle.

15 Draw a triangle and inscribe a circle in it.

16 Draw a triangle and circumscribe a circle about it.

17 Find illustrations in nature, in art, and in architecture of the use of the concept and principles of symmetry.

18 Use the basic constructions given in Chap 8 to create and construct architectural designs.

THE CONCEPT OF MEASUREMENT

GUIDELINES FOR CAREFUL STUDY

In Chap. 7 the concept of the real number line was introduced. By selecting one point on a line, which was designated as the reference point, and then a second point to establish the distance between the two points as a "unit distance," it was possible to set up a one-to-one correspondence between the points of the line and the set of all real numbers.

In Chap. 8 this unit distance was identified as a measurable line segment in terms of which the length of any line segment, or the distance between two points, could be determined. The existence of a positive real number as the measure of any straight line segment was postulated, and the technique for finding the distance between any two points on the number line was defined. Similarly, the existence of positive real numbers as measures for angles, areas, and volumes, respectively, was postulated. The investigation and discussion of techniques for finding and using such measures, in the main, was reserved until later. It is the purpose of this chapter to pursue this study further and to examine the difficulties that are present along with the precautions necessary in making and using the more familiar measurements essential to everyday living.

The following questions will serve as helpful guidelines to careful study of Chap. 9:

1　What is the real number line?
2　What is meant by the measure of a line segment? An angle? An area? A volume?
3　What is the technique for finding the distance between two points?
4　When are two line segments said to be congruent? Two angles? Two triangles?
5　When are two triangles said to be similar?
6　What is meant by the ratio of the lengths of two line segments?
7　What is meant by a measurement?
8　What is meant by the process of measuring?
9　Why are all measurements obtained through the use of a measuring scale necessarily approximate?
10　What is the basic distinction between exact and approximate numbers?
11　What is the basic distinction between direct and indirect measurement?
12　What are some of the reasons for the existence of standard units of measure?
13　What is the metric system of measurements and what are its basic units? How are other units of the various scales of this system formed and named?
14　What is meant by a denominate number? How are denominate numbers used in computation?
15　What is the basic distinction between precision and accuracy in measurement? How do they affect computation with measurement numbers?

333

16   What are basic distinctions between apparent error, permissible error, tolerance, and relative error?

17   What is scientific (standard) notation and how may it be used?

18   What is meant by the sine, cosine, and tangent of an angle? How may they be used in making measurements?

19   What is meant by a statistical measure?

20   What are some of the distinctions between the three measures of central tendency of a distribution?

21   What are some of the implications and precautions to be realized in the use of each measure of central tendency of a distribution?

22   What is the distinction between the variance and the standard deviation of a distribution? For what purposes are these measures used?

INTRODUCTION

Measurement has two meanings

The word "measurement" carries with it a certain ambiguity of meaning. In one sense it is synonymous with "mensuration," which is the subject matter of that branch of geometry which deals with the finding of lengths, areas, volumes, and other forms of quantifying size. In a second sense it means the result of any such quantification of size. It is in the first sense that the word is used in the title of this chapter. Within the chapter it will be used in the connotation of the second meaning, as set forth in Definition 9-1. In either context we are confronted with a further extension of the concept of number which we now have the responsibility to investigate.

This new extension is somewhat different from those of previous chapters. There is no need here for the invention of new numbers but rather for a more flexible concept of the quantifying characteristics of real numbers to provide for a contrast between exact and approximate quantification. Also there is no necessity for defining new operational procedures and remodeling definitions of recognized well-defined operations. Rather, the demand now is for an adaptation of the familiar techniques of addition, subtraction, multiplication, and division used in operating with numbers resulting from exact quantification to patterns for computing with numbers resulting from approximate quantification. Also, there is need for the establishment of guidelines for the use and interpretation of the results obtained from such computation.

9-1   COUNTING, ESTIMATING, MEASURING

Counting

In Chap. 2 *counting* was identified as the process of determining the cardinal number of a set. The numbers which result from such a process can be verified because of the discrete and unitary character of the elements being enumerated. Furthermore, the counting process, in essence, establishes a one-to-one correspondence between the elements of some chosen set of elements and a subset of the set of natural numbers. The numbers which result from counting are thus records of exact results, as they establish a *measurement* which answers the question: How many?

Definition of measurement

Definition 9-1   A *measurement* is the quantitative characteristic of a set, object, or entity of any description which is subject to correlation with some subset of the set of real numbers.

Example   How many chairs are needed to accommodate the members of the Glee Club?

By actual count of the members of the Glee Club the measurement of the "size" of the group would be determined. The count could be verified and established as an exact result which would answer the question which posed the problem, namely: How many?

Estimating

A second significant technique used in determining measurements is that of *estimating*. Such measures as "the distance from the earth to the sun is 93,000,000 miles," "the mass of the earth is 6,000,000,000,000,000,-000,000,000,000 (six octillion) grams," and a molecule of water weighs 0.0000000000000000000000296 (296 ten-septillionths) gram" are not wild guesses but are estimates based on well-established scientific procedures. Less scientific, yet significant, measurements are estimates such as "2,000,000 people witnessed the Tournament of Roses Parade." An interesting measurement which results from a rather carefully planned pattern of counting, yet is an estimate, is the census figure. The Bureau of the Census announced that "according to the final count of the returns of the Census of 1970, the population of the United States is 200,251,326." Although this number, which is the official measurement of the "size" of the population of the United States, has the appearance of an exact number and was derived from very refined techniques for counting large collections of objects, it is necessarily an approximation to the true population figure. This is due, of course, to the fact that change in the population figure, caused by birth, death, immigration, and emigration, was taking place even during the time the count was in process.

In the above discussion a measurement has been exhibited as a number, obtained as the result of some pattern of counting or estimating to serve as a quantifier of size. There is still a third means of arriving at a measurement of a quantity, and that is the process of *measuring*. Quantities such

Measuring

as length, width, area, volume, and capacity are characterized by continuity, rather than discreteness, of composition. Any effort to determine the measurement of such quantities must, of necessity, resort to a combination of counting and estimating. It is accomplished through the use of accepted scales on which countable units have been marked. Thus, in *measuring* the length of a tabletop, one would use a yardstick or tape and count the number of units (feet or inches) and estimate the fractional part of the chosen unit. A combination of the results of these two processes would give the measurement for the length of the table. Such measurements, once obtained, are not capable of verification in the same sense as a measurement

Results from measuring are approximate

resulting from the count of the size of a small gathering of people, for example. The qualities of discreteness and unitariness are absent. Although the inch may have been used, and a number of inches counted, in the process of arriving at the measurement, the unit is divisible and one cannot be assured that separate divisions will agree. Furthermore, the second application of the measuring instrument, although it may be the same instrument applied by the same individual, is subject to variation. Any measurement resulting from an act of measuring is, by the very nature of the act, an approximation.

9-2 EXACT AND APPROXIMATE NUMBERS†

Exact numbers have been the primary concern in the development of this text up to this point. From a study of the natural numbers and their properties we proceeded to a consideration of the integers, the rational numbers, and the real numbers as they evolved as desirable extensions of the natural number system. At each point of the development the need for extension grew out of a desire to maintain the property of closure with respect not only to the established operations but also to the new operations that were defined to meet computational needs.

In contrast, approximate numbers are real numbers obtained from estimating and measuring. They are also obtained as the result of rounding numbers or giving decimal expression, for computational purposes, to irrational numbers and certain types of rational numbers.

If a square of any dimension is constructed, the length of its diagonal will be $\sqrt{2}$ times the length of a side. Here $\sqrt{2}$ is used as an exact number. It is obtained as the result of the extraction of the positive square root of the exact number 2. Any rational approximation to this numerical value would be an approximate number. The quotient of $5 \div 3$ gives the rational

---

†Attention should be called to the fact that the phrases "exact numbers" and "approximate numbers" are used here with explicitly defined connotations. There are writers who feel that the adjectives "exact" and "approximate" should not be used to qualify the numbers, but rather the ideas they are attempting to express. Thus, 12 would carry an exact connotation, whether it be "a number expressing an exact count," as when used to enumerate the number of inches in 1 foot, or "a number expressing an approximation," as when used to record the result from an act of measuring or estimating. While it is agreed that the essence of number is exactness, it seems that the effort to avoid the phrases "exact numbers" and "approximate numbers" introduces a great deal of undesirable circumlocution into discussions where needed distinctions are made between "counting results" and "measuring results." For the benefits of simplicity of expression and reference, the two terms will be used in this chapter within the context described in this section.

number $\frac{5}{3}$. This quotient may also be expressed in decimal form as $1.\overline{6}$. In either the common fraction or the repeating decimal fraction form, this number is an exact number. If, however, it were used in a computational situation which required decimal expression, an approximation such as 1.7, 1.67, or 1.667 would have to be used. If a person visits a grocery store and makes a purchase of one can of peas, priced at three cans for 64 cents, he pays the approximate price of 22 cents and not the exact price of $21\frac{1}{3}$ cents. Furthermore, even the context of use will at times determine whether a number is exact or approximate. This is the case when a person speaks of 2 pounds of apples. If he is enumerating the number of pounds, then 2 is exact, but if he is telling how much the apples actually weigh, then 2 is approximate. Also, if "4 quarts" is used to express a measure of capacity, then 4 is used as an approximate number, while in the tabular definition of 1 gallon (4 quarts = 1 gallon), it is used as an exact number.

EXERCISES

1   Distinguish between exact and approximate numbers.

State whether you think each number used in these exercises is exact or approximate and give reasons to support your answer.

2   Light travels with a speed of 186,000 miles per second.
3   There are 5,280 feet in 1 mile.
4   The auditorium will seat 1,200 people.
5   The school library has 14,400 books.
6   Forty-three boys reported for football practice.
7   $\sqrt{10} = 3.162$; $\sqrt[3]{125} = 5$.
8   We drove 485 miles yesterday.
9   The fastest recorded time for the 100-yard dash is 9.1 seconds.
10   There are 86,400 seconds in 24 hours.
11   The speed record for automobiles is 622.407 miles per hour.
12   The score of the baseball game was 5 to 0.
13   The 1970 population figure for the city of Los Angeles was 2,781,829.
14   Harry is 5 feet 11 inches tall and weighs 168 pounds.
15   Mrs. Byrd bought 2 chickens. They weighed 4 pounds 8 ounces and she paid $2.29 for them.
16   The Declaration of Independence was signed on July 4, 1776.
17   The postal rate for first-class mail is "8 cents per ounce or any fraction thereof." The cost for sending a 6-ounce letter will be 48 cents.

9-3   THE APPROXIMATE NATURE OF MEASUREMENT

It should be quite evident that a measurement resulting from an estimation is intrinsically approximate. It is not always so immediately evident that a measurement which results from the application of a measuring scale is

FIGURE 9-1   Ruler graduated to tenths of an inch

just as essentially approximate, regardless of how refined the scale graduation might be. The approximate nature of such a measurement is illustrated in Fig. 9-1. Here a ruler, which has been graduated to tenths of an inch, is being used to measure the length of the line segment $\overline{AB}$. The point A is at the zero point of the scale of the ruler. The point B lies between the 4- and 5-inch marks, and so it is determined by counting that the total number of complete inches in the length is 4. The number of tenths of an inch beyond the 4-inch mark can also be counted and discovered to be 3. In other words, the number of complete tenths of an inch that B is removed from A is 43. Close examination justifies the estimate that B is a little more than halfway between the 4.3-inch mark and the 4.4-inch mark. Thus, we use both counting and estimating to arrive at the measurement of 4.36 inches for the line segment $\overline{AB}$. The 6 appearing in the measurement is, of course, questionable. It has resulted from an estimate, and the care of making the estimate is a very significant factor in determining just *Sources of* what number should be used. Carelessness is not the only source of error *error in* in making this reading. Two people looking at the picture in the figure might *measuring* relate B to slightly different positions on the ruler. If another person were to take the same or another ruler and use it, there could be a difference in the way he would fit the zero point at A. Also, the rulers themselves might not be perfectly made. Changes in temperature or other weather conditions, carelessness, poor eyesight, inexperience, and many other error sources combine to make it impossible for us to obtain anything but an approximate measurement. This would be the case even if the end point B *appeared* to be exactly at the 4-inch mark of the ruler. Thus the two readings, in this latter case, of "$\overline{AB}$ is 4 inches long" and "$\overline{AB}$ is 4.0 inches long" are significantly different statements. Because of the approximate nature of measurement, the first measure can only mean that the measurer is implying that the length of $\overline{AB}$ lies somewhere between 3.5 and 4.5 inches. The second statement denotes that the length of $\overline{AB}$ lies somewhere between 3.95 and 4.05 inches.

9-4 DIRECT AND INDIRECT MEASUREMENT

Direct The illustration in the preceding section is that of a *direct measurement.*
measurement It pictures a situation in which a graduated scale, a measuring instrument, was applied directly to the object whose measurement was to be deter-

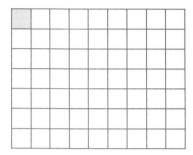

FIGURE 9-2 Determining the area of a rectangle

mined. Not all measurements can be obtained from such direct procedures. The rectangle of Fig. 9-2 is 7 units wide by 9 units long. What is the area of this rectangle? We have seen how, in order to determine the length or width of such a rectangle, we can apply a measuring instrument and read the appropriate measurement. We have no such device to apply in order to determine the area, and so we have to resort to indirect methods. Any

Indirect measurement

measurement obtained by such methods is called an *indirect measurement.* By observation, we realize that each small square in the diagram represents a unit of square measure (unit of area) of the type corresponding to the linear unit used in measuring the length and width. Therefore, it is only necessary to contrive a scheme for counting these small squares. Geometry provides us with a method that is intuitively acceptable. We can observe that in each horizontal row there are 9 such small squares and that there are 7 such rows. It is evident then that there must be $7 \times 9 = 63$ square units in the area of the rectangle. Of course, similar reasoning, or the commutative property of multiplication, tells us that 9×7 gives the correct number of square units also. Another form of indirect measure is provided by the use of the pythagorean theorem (Theorem 8-11) to find the length of the hypotenuse of a right triangle when the lengths of the two legs are known, or the length of the diagonal of a square when we know the length of a side.

There are many important methods of making indirect measurements. When a person steps on a scale he reads from a graduation a number which he uses to complete the sentence: My weight is ____ pounds. The scale compares the mass of the body of the person being weighed with that of a standard unit of weight. Temperature is another important indirect measurement. The measuring instrument, a thermometer, consists essentially of some confined substance which has at least one measurable property that changes with temperature variation and a scale for measuring this change.

Other significant forms of indirect measurement will be discussed later in this chapter.

EXERCISES

1 Explain how measurements of weight are determined by means of counting and estimating.

2 Why are measurements of length, area, volume, and capacity only approximations?

3 In the definition "one foot is 12 inches," is the number 12 used as an exact number or as an approximate number?

4 The following statement is made: There are 750 seats in the school auditorium.

(*a*) Explain how this number might be an approximate number.

(*b*) Explain how it might be an exact number.

Which of the measurements in Exercises 5 to 18 are direct and which are indirect? Give reasons to support your answer.

5 The size of an angle, by means of a protractor

6 Time, by means of a watch or clock

7 Distance, by means of the odometer in an automobile

8 Speed, by means of the speedometer in an automobile

9 The height of an airplane, by means of an altimeter

10 The height of a child, by means of a tape measure

11 Atmospheric pressure, by means of a barometer

12 Amount of gas or electricity used, by means of a meter

13 The velocity of the wind, by means of an anemometer

14 The revolutions of an airplane engine, by means of a tachometer

15 The amount of dress goods purchased, by means of a yardstick

16 The contents of a measuring cup, by means of graduations etched on its side

17 The area of a triangle, by means of the formula $A = \frac{1}{2}bh$ where b represents the linear units in the base and h the linear units in the altitude

18 The contents of a grain bin, by means of a bushel measure

9-5 THE NEED FOR MEASUREMENT UNITS

We have seen how the natural numbers came into existence in the human effort to give quantitative characterization to a set of objects. They provide us with a means of measuring the quantitative content of discrete sets of objects, i.e., of answering the question: How many? When a person states that a room contains 65 chairs, he has, in a very true sense of the word, measured a certain type of content of the room. The unit used in making the measurement is "chair." Indeed, if he has counted each chair in the room, he has secured an exact measure of this particular room content,

one that is unchanging and verifiable. As has been pointed out previously, such exactness of measurement is always possible when the content to be measured is composed entirely of discrete elements. This is not the case, however, when the content is of fluid, or continuous, composition. For example, a statement that a pitcher contains 4 pints of water can be exact *only* if the individual means that he filled a pint container four times and emptied it each time into the pitcher. If the "4 pints" is used in the sense of a measure of capacity, then it must be considered as approximate because of the various elements of error present, such as judgment as to fullness of the container each time, calibration of the container, and care in pouring. Thus, there would be an exact answer to the question: How many? Only an approximate answer would be possible to the question: How much? The length of a distance and the weight of a mass are also measurements of "objects" of fluid, or continuous, content. The numbers used *Denominate* to express such measurements are called *denominate numbers* to distin-
numbers guish them from the exact numbers used to give the measurement in situations of discrete content.

Sets of discrete elements provide natural units, such as the "chair" mentioned above, for measuring the content of a set. When there are no such identifiable discrete elements, we have to invent ways and means of providing a suitable unit of measure which can be counted in order to obtain an appropriate measurement. The "pint" of the preceding paragraph, which is a subunit of the official gallon; the yard with its subunits foot and inch; and the bushel with its subunits peck, quart, and pint are such inventions which have been created in an effort to systematize and standardize measurement in those situations characterized by continuity of structure.

Primitive human beings in their efforts to provide suitable units, took *Primitive* recourse in convenience. Their principal concern was to have a unit that
units would enable them to establish a measure that would be personally satisfying. They were not confronted with concern for their neighbor, either far or near, nor were they involved in problems of commerce, industry, or travel. It was quite natural that they used units which were personally convenient (see Fig. 9-3). The digit (width of a finger), palm (width of hand palm), span (distance between tips of thumb and little finger in outstretched hand), cubit (distance from elbow to tip of outstretched hand), furlong (length of furrow one could plow before resting oxen), pace (distance between heel of one foot and heel of the same foot when next in contact with ground—a double step), and others were in frequent use. Similar convenient designations of units were practiced in other forms of making measurements.

It is of course evident that there would be wide differences in measurements resulting from the use of such personally oriented units. Thus, the demands of commerce and industry led naturally and persistently to the desirability and necessity for standard units of measure—units not only of national but also of international identity.

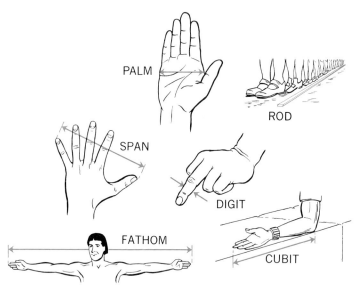

FIGURE 9-3 Early units of measure

9-6 STANDARD UNITS OF MEASURE

International units of measure

In 1893 the *international meter* and *kilogram* were adopted as the basic standards for measuring length and mass in the United States. Under these prescriptions the *yard* was defined to be $\frac{3,600}{3,937}$ meter, or 36.00 units (inches) of which there are 39.37 in a meter. The official *avoirdupois pound* was defined to be 7,000 grains, which made it approximately equivalent to 0.454 kilogram. In 1959 these equivalences were modified to define the *international yard* as 0.9144 meter and the *international avoirdupois pound* as 0.45359237 kilogram. The *international meter*, first defined (1889) to be one ten-millionth of the meridional distance from the pole to the equator, now (since 1960) is defined to be 1,650,763.73 wavelengths of the orange-red radiation in vacuum of krypton 86. The *international kilogram* is defined to be 15,432.356 grains, and the *international liter* as the volume of a kilogram of water under pressure of 76 centimeters of mercury at 4° centigrade.

The metric system

The metric system of weights and measures is entirely decimal in nature. The basic units are meter, gram, and liter. In each table prefixes are then used to indicate both higher and lower units on each scale. Greek derivatives are used as prefixes to indicate higher points on the scale, while Latin derivatives are used to indicate points of lower division. The table of linear measure is given in Fig. 9-4. The tables for weight and capacity may be constructed in exactly the same manner, using *gram* as the basic unit for weight and *liter* for capacity.

---

10 millimeters (mm) = 1 centimeter (cm)
10 centimeters = 1 decimeter (dm)
10 decimeters = 1 meter (m)
10 meters = 1 dekameter (dkm)
10 dekameters = 1 hectometer (hm)
10 hectometers = 1 kilometer (km)
10 kilometers = 1 myriameter (M)

FIGURE 9-4   Metric linear measure

The English system   The English system of weights and measures is nondecimal in structure. Furthermore, there is no simple pattern, as in the metric system, that may be followed in building the several tables of linear, area, volume, capacity, and weight measure. For proper use each such table must be learned as a separate entity. There are, of course, tables of equivalents which relate each basic English unit to its corresponding international standard. This confusion in table structure, as compared with the simplicity of structure of the metric system, is one of the arguments used by those who have proposed that the metric system of weights and measures be made the universally accepted system. The basic decimal structure of the metric system is another of the supporting arguments used by such groups.

9-7 DENOMINATE NUMBERS

Despite the recognized fact that all measurement of continuous content is by nature approximate, human society has made a constant effort to provide better instruments and more appropriate units for the purpose of obtaining the closest possible agreement between the observed measurement of a quantity and its true measurement. This has led to the prescription and definition of units and subunits, with their concomitant interrelationships, all of which are summarized for our ready use in familiar tables of weights and measures.†

Mach numbers   In these modern days of jet propulsion, satellite projection, and atomic fission we have seen both macroscopic expansion and microscopic division of these units. For example, the *Mach numbers* have been defined for better measurement of supersonic speeds and the light-year for astronomic distances. A plane traveling at Mach 2 would be traveling at a speed twice the speed of sound in the surrounding atmosphere or, under certain conditions, at approximately 1,500 miles per hour. The announced speed of one of the early man-made satellites was Mach 24. This meant that it traveled at a speed of approximately 18,000 miles per hour.

Light-year   The astronomical unit of 1 *light-year* is defined to be the distance light will travel in 1 year's time, or a distance of 5,870,000,000,000 miles. One

---

† See inside back cover.

of the purposes of this unit is to express the distances of different stars and planets from the earth. The North Star, for example, is a distance of 47 light-years from the earth. This means that when you look at this star, the light you see was emitted from the star 47 years ago. Alpha Centauri is the closest star to the earth, at a distance of 4.3 light-years. To appreciate something of the significance and convenience of the light-year as a unit for measuring astronomical distances, you might express in miles each of the two mentioned distances, 47 light-years and 4.3 light-years. In contrast, *Atomic mass unit* the *atomic mass unit* is used in discussions involving atomic weights. It is defined to be 0.0000000000000000000001660 gram, or 1,660 octillionths gram. This is equivalent to 0.00000000000000000000000000366 pound, or 366 hundred-octillionths pound. Atomic nuclei are made up of two kinds of primary particles: *protons* and *neutrons*. The proton has a mass of 1.00758 atomic mass units and carries a single unit of positive electric charge. The neutron has a mass of 1.00897 atomic mass units and carries no electric charge. This nucleus is surrounded by negatively charged particles called *electrons*. The mass of an electron is 0.00055 atomic mass unit.

The above illustrations are given simply to emphasize the tremendous flexibility and sensitivity of both the English and the metric systems of measurement. The numbers used to enable us to describe the results of such measurements, no matter how large or how small they may be, are recognized as belonging to the denominate numbers introduced earlier.

The introduction of denominate numbers is not an extension of the real number system and does not create a demand for definition of operational processes in order that we might use them. Rather, it presents a new context in which previously defined operations with the real numbers may be used, thus calling for pertinent adaptations and interpretations. Consider the two following examples.

Example Jack walked a distance of 300 yards from his house to the store. He then walked from the store to school, a distance of 400 yards. (1) What was the total distance he walked? (2) What is the distance d from Jack's home to the school? Note carefully the two questions.

The answer to the first question is simple. Jack walked a total distance of 700 yards. Three illustrations are given in Fig. 9-5 to indicate that the answer to the second question is impossible until more is known about the relative locations of home, store, and school. Only three of the many possible relative positions of home, store, and school are shown. In each case the actual distance from Jack's home to the school is shown in color. With the orientation as indicated in each respective illustration of Fig. 9-5, a different computation (indicated below) is necessary to find the answer to the question.

1   *Addition* $d = 300$ yd $+ 400$ yd $= 700$ yd

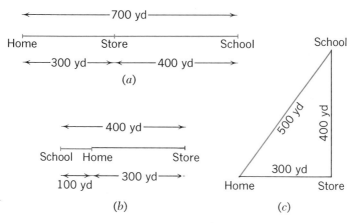

FIGURE 9-5 Three interpretations of distance data

2 *Subtraction* $d = 400$ yd $- 300$ yd $= 100$ yd
3 *Pythagorean theorem* $d = \sqrt{(400)^2 + (300)^2} = 500$ yd

This example has illustrated the necessity of thorough recognition and careful consideration of all the physical characteristics of a problem situation before the nature of the operational procedure needed to solve the problem may become apparent.

Example Consider the two following methods of finding the length of a line segment $\overline{AC}$.

1 A yardstick is used to find that one part $\overline{AB}$ of the line segment is 2 feet long, and the remaining portion $\overline{BC}$ is 4 inches long. There would be no difficulty presented in determining that the length of $\overline{AC}$ is the sum of the determined lengths; that is, $AC = 2$ ft $+ 4$ in., or, better, 2 ft 4 in.
2 A yardstick is used to find that $\overline{AB}$ is 2 feet long and a meterstick to find that $\overline{BC}$ is 4 decimeters long. Although the basic technique is the same as in the previous case, there would be no inclination to find the sum 2 ft $+ 4$ dm and give the length of $\overline{AC}$ as 2 ft 4 dm. It would be recognized that before any combination could be made, it would be necessary either to express both the measurements in the English system or to express both in the metric system.

This second example has illustrated the fact that two measurements of like character and in the same system may be combined into one measurement by addition, but that the numerical coefficients are not added unless the units are like units. When the units are not like units, the addition is indicated by juxtaposition. The result is a mixed measurement number. In fact, the measurement 2 ft 4 in. could have been written $2\frac{1}{3}$ ft; the use of the unit "inch" provides a means of avoiding fractions. Under certain

5 yd	2 ft	8 in.
8 yd	2 ft	11 in.
5 yd	1 ft	9 in.
6 yd	0 ft	5 in.
24 yd	5 ft	33 in.
26 yd	1 ft	9 in.

FIGURE 9-6

circumstances 28 in. would be the preferable form for expressing this measurement.

In carrying out the fundamental operations with denominate numbers not only does one have the ordinary conversion problems incident to place value in the structure of the numbers used, but also there are problems of conversion from one denomination to the other. These are made, of course, in terms of the pertinent tables of weight or measure.

Computation with denominate numbers

Example Find the sum of the following measurements: 5 yd 2 ft 8 in.; 8 yd 2 ft 11 in.; 5 yd 1 ft 9 in.; 6 yd 5 in. Notice (Fig. 9-6) that the columns align measurements expressed in the same units. It then follows that the numerical coefficients in each column can be added. The first sum is 24 yd 5 ft 33 in., but it is the accepted practice that the simplification of this result calls for the modification of any given denominate number to a form such that no denomination of a lower order contains any integral multiples of units of higher order. Thus 33 in. = 2 ft 9 in., and so 9 in. is recorded in the column for this denomination while the 2 ft is combined with the 5 ft already obtained in the sum to give 7 ft, which converts to 2 yd 1 ft. Hence the accepted answer for this sum is 26 yd 1 ft 9 in.

Example What is the product of 5 × 6 hr 36 min 30 sec? The first product (Fig. 9-7) is the usual multiplication.

$$150 \text{ sec} = 2 \text{ min } 30 \text{ sec}$$
$$180 \text{ min} + 2 \text{ min} = 182 \text{ min}$$
$$= 3 \text{ hr } 2 \text{ min}$$
$$30 \text{ hr} + 3 \text{ hr} = 33 \text{ hr} = 1 \text{ day } 9 \text{ hr}$$

A result of this kind might be left in either of the last two forms indicated, depending upon the emphasis in a particular problem situation.

	6 hr	36 min	30 sec
			5 sec
	30 hr	180 min	150 sec
	33 hr	2 min	30 sec
1 day	9 hr	2 min	30 sec

FIGURE 9-7

8 gal	1 qt	1 pt
5 gal	3 qt	2 pt
7 gal	4 qt	3 pt
5 gal	3 qt	2 pt
2 gal	1 qt	1 pt

FIGURE 9-8

Example Subtract 5 gal 3 qt 2 pt from 8 gal 1 qt 1 pt. In this exercise there is the problem of converting higher denominations to equivalent lower denominations in order that the subtraction will be possible. The exercise is rewritten (Fig. 9-8) to exhibit what takes place. This, of course, is not necessary in the working of the problem, but it can be helpful.

$$1 \text{ qt} = 2 \text{ pt} \qquad 1 \text{ pt} + 2 \text{ pt} = 3 \text{ pt} \qquad 3 \text{ pt} - 2 \text{ pt} = 1 \text{ pt}$$
$$1 \text{ gal} = 4 \text{ qt} \qquad 4 \text{ qt} + 0 \text{ qt} = 4 \text{ qt} \qquad 4 \text{ qt} - 3 \text{ qt} = 1 \text{ qt}$$
$$7 \text{ gal} - 5 \text{ gal} = 2 \text{ gal}$$

There are two types of problems that arise in division, depending on whether the emphasis is on partition or on measurement.

Example

1 What is $\frac{1}{3}$ of 11 lb 3 oz? This is a partition division. 11 lb $\div$ 3 = 3 lb with 2 lb as a remainder (Fig. 9-9a).

$$2 \text{ lb} = 32 \text{ oz}$$
$$32 \text{ oz} + 3 \text{ oz} = 35 \text{ oz}$$
$$35 \text{ oz} \div 3 = 11\tfrac{2}{3} \text{ oz}$$
$$\frac{1}{3} \text{ of } 11 \text{ lb } 3 \text{ oz} = 3 \text{ lb } 11\tfrac{2}{3} \text{ oz}$$

2 What is the ratio of 1 bu 2 pk 2 qt to 3 pk 6 qt? This is a measurement division. Before the division can be accomplished the two measures must be converted to the same denomination. This is usually accomplished best by changing to lower denominations (Fig. 9-9b). The ratio of 1 bu 2 pk 2 qt to 3 pk 6 qt is the same as the ratio of 50 qt to 30 qt or 5 to 3.

EXERCISES

1 Construct the tables for weight and capacity in the metric system.
2 Under the equivalence established by the definition of 1 yard, express 1 inch in terms of 1 meter. Make your computation correct to four decimal places.
3 Use the result of Exercise 2 to express 1 inch in terms of centimeters.
4 The new official definition of 1 meter is 1,650,763.73 wavelengths of

(a)

$$
\begin{array}{r}
3 \text{ lb} \quad 11\frac{2}{3} \text{ oz} \\
3 \overline{\smash{\big)}\ 11 \text{ lb} \quad 3 \text{ oz}} \\
9 \text{ lb} \\
\hline
2 \text{ lb} \quad 3 \text{ oz} = 35 \text{ oz} \\
33 \\
\hline
2 \\
2 \\
\hline
0
\end{array}
$$

(b)

$$
\begin{array}{llll}
1 \text{ bu} = 4 \text{ pk} & & = 32 \text{ qt} \\
& 2 \text{ pk} & & = 16 \text{ qt} \\
& & & \quad 2 \text{ qt} \\
\hline
1 \text{ bu} & 2 \text{ pk} & 2 \text{ qt} = 50 \text{ qt} \\
& 3 \text{ pk} & & = 24 \text{ qt} \\
& & & \quad 6 \text{ qt} \\
\hline
& 3 \text{ pk} & 6 \text{ qt} = 30 \text{ qt}
\end{array}
$$

$$\frac{50 \text{ qt}}{30 \text{ qt}} = \frac{5}{3}$$

FIGURE 9-9

the orange-red light of krypton 86. How many such wavelengths are there in 1 inch?

5 If 1 avoirdupois ounce = 437.5 grains, express 1 grain in terms of the avoirdupois ounce. Make your computation correct to five decimal places.

Using the data given on page 344, determine the answers to the questions in Exercises 6 to 10.

6 What is the distance to the North Star in miles?
7 What is the distance to the star Alpha Centauri in miles?
8 What is the weight in grams of one proton?
9 What is the weight in grams of one electron?
10 What is the weight in grams of one neutron?

In Exercises 11 to 13, find the sums.

11 4 yd 2 ft 6 in.   12 6 hr 17 min 26 sec   13 1 bu 2 pk 3 qt
     5 yd   5 in.   8 hr 45 min   2 bu 3 pk 5 qt
     2 yd 1 ft 8 in.   9 hr 20 min 56 sec   4 bu   7 qt
     3 yd 2 ft   10 hr 6 min 8 sec   2 bu 1 pk 6 qt

In Exercises 14 to 17, find the differences.

14 6 sq yd 3 sq ft 32 sq in.   15 year month day
     1 sq yd 8 sq ft 84 sq in.   1963   5   9
                                         1928   8   25

16   5 m   9 cm 6 mm
       2 m 2 dm 8 cm 4 mm

17   3 g 2 dg 5 cg 2 mg
      1 g   8 cg 9 mg

18 Use only decimal fractions to solve Exercises 16 and 17.

19 Find this sum in two different ways: (a) without the use of decimal fractions; (b) using only decimal fractions.

              2 l 5 dl 4 cl 1 ml;   8 dl 5 ml;   4 l 7 cl

In Exercises 20 to 23, find the products.

20 6 ft 7 in.
     ×8

21 4 bu 1 pk 5 qt 1 pt
             ×7

22 5 gal 3 qt 1 pt
        ×3

23 3 days 16 hr 35 min 45 sec
                   ×6

In Exercises 24 to 27, find the quotients. What is the remainder in each case?

24 6⟌8 sq yd 2 sq ft 100 sq in.

25 5⟌12 gal 1 pt

26 8⟌12 m 8 cm

27 3⟌4 g 9 dg 5 cg

In Exercises 28 to 31, find the ratios.

28 3 yd 8 ft to 4 ft 3 in.   29 1 bu 2 pk 4 qt to 3 bu 1 pk
30 4 m 7 dm 3 cm 2 mm to 23 m 6 dm 6 cm
31 5 mg to 1 kg
∗32–35 Work Exercises 26, 27, 30, and 31 using only decimal fractions.

9-8 PRECISION AND ACCURACY

As has been pointed out previously, the fact that all measurement is by nature approximate does not prevent us from creating units and constructing instruments to enable us to secure observed values that most nearly represent true values of any desired measurements. In the manufacturing industry, for example, the Johansson gauge blocks are used to make the refined measures necessary in such engineering problems as cylinder grinding and piston fitting. These blocks are so highly polished and possess such adhesive properties that when fitted together they may be used as a standard for making measurements with an error of less than one hundred-thousandth of an inch (0.00001 in.). Such quality of refinement in a measuring instrument is called *precision*. Thus, the more precise of two carefully made measurements is the one that has been made with the scale showing the greater degree of refinement in unit subdivision.

Precision

FIGURE 9-10 Precision in measurement

The picture in Fig. 9-10 illustrates a 6-in. ruler graduated to sixteenths of an inch. It is being used to measure the length of the line segment $\overline{AB}$. The point A is flush with the zero mark of the ruler, and the identification of the position of the point B will give the desired length of $\overline{AB}$. How long is $\overline{AB}$? There are six possible answers, depending on how precisely we wish to make the measurement.

1   *To the nearest inch,* $\overline{AB}$ is 4 in. long. This means that B lies nearer to the 4-in. mark than to any other inch mark on the scale. In other words, it lies somewhere between $3\frac{1}{2}$ and $4\frac{1}{2}$ in. Thus, in this case there is a maximum possible error of $\frac{1}{2}$ in. Another way of giving the length is 4 in. $\pm \frac{1}{2}$ in.

2   *To the nearest half-inch,* $\overline{AB}$ is also 4 in. long. This means that B lies nearer to the 4-in. mark (or the mark which indicates eight half-inches from A) than to any other half-inch mark on the scale. In other words, it lies somewhere between $3\frac{3}{4}$ and $4\frac{1}{4}$ in. This time the maximum possible error is $\frac{1}{4}$ in. Another way of giving the length is 4 in. $\pm \frac{1}{4}$ in.

3   *To the nearest quarter-inch,* $\overline{AB}$ is $4\frac{1}{4}$ in. long. This means that B lies nearer to the $4\frac{1}{4}$-in. mark (or the 17-quarter-inch mark from A) than to any other half-inch mark on the scale. In other words, it lies somewhere between $4\frac{1}{8}$ and $4\frac{3}{8}$ in. Another way of giving the length is $4\frac{1}{4}$ in. $\pm \frac{1}{8}$ in.

4   *To the nearest eighth-inch,* $\overline{AB}$ is $4\frac{2}{8}$ in., or $4\frac{1}{4}$ in., long. Explain what this means. Remember that, in this case, $4\frac{1}{4}$ in. is simply a way we have of writing 18 eighth-inches. Why can this measure be written as $4\frac{1}{4}$ in. $\pm \frac{1}{16}$ in.? If this measure should be given merely by stating the mixed number, why does $4\frac{1}{4}$ in. not give the true picture?

5   *To the nearest sixteenth-inch,* what is the length of $\overline{AB}$? What is the maximum possible error? What would be used to fill the blanks in the statement $\overline{AB}$ is ____ in. $\pm$ ____ in.?

6   One could even make an estimate of the length of $\overline{AB}$ correct *to the nearest thirty-second of an inch.* What estimate would you make? What is the maximum possible error of the most justifiable reading you can give the measurement?

For the line segment $\overline{AB}$ of Fig. 9-10 there exists an exact measurement. The measurements resulting from the six illustrated techniques are all approximations to this unobtainable true measurement. Connected with each approximation, therefore, there is an error which is, of course, unobtainable. As has been pointed out, in each reading of the measurement

Apparent (absolute) error

there is a maximum possible error which is $\frac{1}{2}$ of the smallest unit of measure used. This maximum possible error is called the *apparent,* or *absolute, error.* Thus, it follows that *the more precise of two measures is that one which has the smaller apparent error.* For example, in the above illustration the measurement $4\frac{1}{4}$ in. $\pm \frac{1}{16}$ in. is more precise than $4\frac{1}{4}$ in. $\pm \frac{1}{8}$ in. The apparent error of the first measurement ($\frac{1}{16}$ in.) is smaller than the apparent error of the second measurement ($\frac{1}{8}$ in.). It should be evident that twice the apparent error of a measurement gives the interval about the observed measurement within which the actual measurement (which is unobtainable) must lie. Similarly, where measurements are recorded in decimal fractions the degree of precision is indicated by the number of decimal places involved. A measure recorded as 6.3 in. would be less precise than one recorded as 6.30 in. In the first case, the measurement is made to the nearest tenth of an inch; the apparent error is 0.05 in.; the measurement lies between 6.25 and 6.35 in. In the second case, the measurement is made to the nearest hundredth of an inch; the apparent error is 0.005 in.; the measurement lies between 6.295 and 6.305 in.

This last illustration emphasizes the falsity of a statement that is sometimes heard, namely, to annex a zero to the right of a decimal fraction does not affect the fraction. This is manifestly incorrect. As illustrated above, the significance of the measure 6.3 in. was changed materially by annexing the zero, since 6.30 in. means something entirely different. The true and usable statement is as follows: To annex a zero to the right of the decimal does not change *the numerical value* of the decimal. The two decimals 6.3 and 6.30 have the same numerical value for computational purposes. The results derived from the computation, however, are subject to significantly different interpretations depending upon which measure was used. Similarly, it should be quite obvious that in measurement numbers, $7\frac{0}{8}$ in. is quite different from 7 in. The first measurement, $7\frac{0}{8}$ in., means that the unit of precision is $\frac{1}{8}$ in. and the measurement is 7 in. $\pm \frac{1}{16}$ in., while 7 in. can mean only that the unit of precision is 1 in. and the measurement is 7 in. $\pm \frac{1}{2}$ in., or possibly 7 in. $\pm$ 0.5 in. What is the difference in meaning of $7\frac{1}{2}$ in. and 7.5 in.? Which is the more precise?

Tolerance

Another important type of error that figures very significantly in industry is the *permissible error,* better known as *tolerance.* It is simply a value defining the amount of the maximum allowable error or departure from true value or performance. For example, the specified tolerance of $\pm$0.004 in. on a shaft whose dimension is specified as 2.000 in. $\pm$ 0.004 in. means that all shafts measuring anywhere between 1.996 and 2.004 in. will be considered as satisfactory.

Precision in language usage is thus seen to be essential for specification of precision in measurement. If, for example, the length of the line segment $\overline{AB}$ on page 350 is stated to be 4 in., what are you, the reader, to think is meant? No prescription has been made as to the unit of precision. All that you can do is to assume that the smallest unit specified indicates the desired precision. Therefore, all you can interpret the statement to mean is that the length is given to the nearest inch; i.e., the length of $\overline{AB}$ is 4 in. $\pm\, \frac{1}{2}$ in., or merely somewhere between $3\frac{1}{2}$ and $4\frac{1}{2}$ in. If the person making the statement wants you to know that to the nearest half-inch, the line segment is 4 in. long, or 4 in. $\pm\, \frac{1}{4}$ in., it is his responsibility so to state. What would the statement have to be to carry the meaning that the length of a given line segment is 4 in. correct to the nearest sixteenth of an inch?

In dealing with the errors of measurement we have concern not only for their size but also for their significance. For example, an error of $\frac{1}{2}$ in. in a measurement of 6 ft 4 in. (76 in.) is far less serious than is an error of $\frac{1}{2}$ in. in 4 in. This significance is measured by the ratio of the error to the measurement, which is known as *relative error*.

Relative error

Accuracy The more accurate of two measurements is that one which has the smaller relative error.

Percent of error The relative error of a given measure may also be expressed as a *percent of error* by converting the ratio into a decimal fraction and multiplying by 100. As will be seen later, there is no point in carrying such a division out to more than the first nonzero digit.

While the criterion of the relative error is a very valuable one in the determination of the usefulness of any given measure, it should be kept in mind that the intended use of the measure also enters into the determination of the seriousness of the accompanying error. An error of $\frac{1}{2}$ in. in the wrong direction where a person is walking along a 6 ft 4 in. ledge of a high mountain cliff could be far more serious than an error of $\frac{1}{2}$ in. in the wrong direction when he is rolling a marble down a 4-inch track.

Example Consider these two measurements: (1) The distance from the earth to the moon is 239,000 miles. (2) The estimated diameter of the average atom is 0.0000000079 in. (79 ten-billionths of an inch).

1 What is the apparent error of each measurement?
2 What is the relative error of each measurement?
3 What is the percent of error of each measurement?
4 Which measure is the more precise?
5 Which measure is the more accurate?

1 The apparent error in 239,000 miles is 500 miles; in 0.0000000079 in. it is 0.00000000005 in.

2 The relative error in 239,000 miles is $\dfrac{500}{239,000} = \dfrac{5}{2,390} = \dfrac{1}{478} =$ 0.002; in 0.0000000079 in. the relative error is $\dfrac{0.00000000005}{0.0000000079} = \dfrac{5}{790} = \dfrac{1}{158} = 0.006.$

3   The percent of error in 239,000 is 0.2%; in 0.0000000079 it is 0.6%.

4   The measurement 0.0000000079 in. is the more precise of the two.

5   The measurement 239,000 miles is the more accurate of the two.

Significant digits

The comparisons of this example emphasize the importance of attention to *significant digits*, or *significant figures*, in a number expressing a measurement or an approximation of any kind. The significant digits in any number may be determined in accordance with the following criteria:

1   Any nonzero digit is significant.

2   Zeros used merely for the purpose of placing the decimal point are not significant.

3   All other zeros are significant.

Each of the numbers 632, 25.0, 0.250, 304, and 0.000123 has three significant figures. There should be no question about 632. In 25.0 the zero is not needed for the placing of the decimal point; therefore, since it is used, it is significant. In 0.000123, all zeros are merely for placing the decimal point, and so they are not significant.

In the example the distance to the moon is given to three significant figures, while the estimated diameter of the average atom is given to two significant figures. Thus the more accurate measurement was that one given to the greater number of significant figures. This is in general true. For example, a measurement of 3,256 miles is far less precise but much more accurate than 7.6 in.; 0.5 mile is much larger than 0.05 in., but $\frac{0.5}{3,256} = \frac{5}{32,560}$ is much smaller than $\frac{0.05}{7.6} = \frac{5}{760}$. The more accurate of two numbers having the same number of significant digits would be that one which has the larger digit in the extreme left significant-digit position of the number. The measurement 0.00934 in. is both more precise and more accurate than 39.4 in. Why?

Example   How many significant digits are there in each of these measurements: (1) 16 ft 2 in.; (2) 6 lb $5\frac{1}{2}$ oz; (3) 6 gal 2 qt 1 pt; and (4) 6°32′14″?

1   The smallest unit of measure in 16 ft 2 in. is 1 inch. The 16 ft 2 in. is merely an accepted form for recording the measurement of 194 in. There are three significant digits.

2   The unit of precision in 6 lb $5\frac{1}{2}$ oz is $\frac{1}{2}$ ounce. The measurement is 203 half-ounces, and there are three significant digits.

3   The unit of precision is 1 pint in 6 gal 2 qt 1 pt. The measurement is 53 pints, and there are two significant digits.

4   The unit of precision is 1 second. The measurement is 23,534 seconds, and there are five significant digits.

9-9 SCIENTIFIC NOTATION

A bit of reflection should make it rather obvious that nonsignificant zeros can contribute no significant figures to the result of any computation. They must be taken into consideration only in order to ensure that the significant figures of the result are assigned their proper positional values. In other words, nonsignificant figures can serve only as aids in the proper placing of the decimal point in a given or computed approximate number. Scientific notation provides a very effective method for simplifying any computation involving numerals burdened with nonsignificant digits, whether many or few. The concept of "significant digits" provides a more convenient language than that of Sec. 6-11 for stating the rule to be used in converting from either form of notation, positional or scientific, to the other.

The approximate distance from the earth to the sun has been given as 93,000,000 miles. Written in scientific notation this distance is given as 9.3×10^7. Observe how this is accomplished:

Converting to scientific notation

1   The significant digits of the number are recorded in their positions relative to each other (in this number they are 9 and 3).

2   These digits are then written so that there is only one digit to the left of the decimal point (9.3).

3   This number is then multiplied by the power of 10 necessary to restore its digits to their proper positional value in the original number. (The 9 is in the ones place in 9.3 and in the ten-millions place in 93,000,000. The ten-millions place is seven positional places to the left of the ones place. The necessary multiplier is, therefore, 10^7. Of course, the same multiplier will also restore 3 to its proper positional value.)

The value of the atomic mass unit has been established scientifically as 0.000000000000000000000001660 gram; this number written in scientific notation is 1.660×10^{-24} gram. Remember that the negative exponent means that the decimal point must be moved to the left in translating from scientific notation to positional notation. Notice how scientific notation removes all questions as to significance or nonsignificance of zeros. This fact is emphasized a bit more clearly by considering the established figure for the velocity of light, 3.00×10^{10} centimeters per second. There is no question here concerning the significance of the two zeros, and it is readily recognized that this figure is correct to three significant figures. This same number written in positional notation would be 30,000,000,000 centimeters per second, and a means for indicating the significance of the two zeros must be invented. Here the second zero from the left has been underscored to indicate that it is significant, thus automatically giving significance to the zero between it and the 3. Other devices are used to indicate significant zeros in such numbers. There is no generally accepted practice. Scientific notation removes the necessity for the standardization of any such symbol.

Explain the scientific notation form in each of these examples:

Positional notation	*Scientific notation*
239,000	2.39×10^5
0.00118	1.18×10^{-3}
34.00	3.400×10^1
150,000,000	1.5000×10^8
0.0000000079	7.9×10^{-9}

9-10 ADDITION AND SUBTRACTION

If two measurements are combined by the process of addition, it should be evident that the error in the computed measurement will be the algebraic sum of the errors in the given measurements (see Fig. 9-11). Similarly, if one measurement is subtracted from another, the error in the computed measurement will be the algebraic difference between the errors in the given measurements. Thus, if all measurements used in addition or subtraction were given with their respective errors, it would be a simple matter to arrive at a resultant measurement and its pertinent error. In actuality, however, measurements are given most frequently with no specified error, so that the only error known in each case is the maximum possible error. It is safe to assume, though, that these errors will tend to balance against each other in that some will be negative and some positive. It should be evident that if the measurements are given to different degrees of precision, the error in the least precise measure will always be larger than all other errors. For this reason, in any addition or subtraction, the result cannot be any more precise than the least precise measurement used in the computation.

Example Assume that each of four teams, composed of two individuals each, has been given a tape measure graduated so that measurements to the nearest thousandth of a foot can be made. Each team is asked to measure the length of one side of a four-sided plot of ground, and no specifications are given as to how the measurements are to be made. It is conceivable that the four measurements shown in Fig. 9-12 might be the result. What is the most justified measurement for the perimeter of the

$$
\begin{array}{cc}
M + E & M + E \\
\underline{+\quad m + e} & \underline{-\ (m + e)} \\
(M + m) + (E + e) & (M - m) + (E - e)
\end{array}
$$

FIGURE 9-11 Computation of errors

	MINIMUM SUM	MAXIMUM SUM	COMPUTED SUM
	11.5	12.5	12
	14.25	14.35	14.3
	20.515	20.525	20.5
	16.8665	16.8675	16.9
	63.1315	64.2425	63.7

The most justified result is 64.

(*a*) (*b*) (*c*)

FIGURE 9-12 Finding the perimeter of a polygon

figure? The *perimeter* of a polygon is the distance around it. In this case it is the sum of the four measurements that can be obtained from the obtained data.

Consider each of the three sums given here. The sum in column (*a*) gives the minimum possible value for the sum, and the one in column (*b*) gives its maximum possible value. So the true value lies somewhere between these extremes. In column (*c*) each measurement was rounded to within one position of the same degree of precision as that of the least precise, namely, 12 ft. The sum was then rounded to the lowest degree of precision. It will be observed that the value so obtained is well within the limits found in columns (*a*) and (*b*).

For these reasons the standard rule for the addition or subtraction of approximate numbers is:

Addition and subtraction with approximate numbers In addition and subtraction of approximate numbers, when they are expressed decimally, first round all numbers to within one position of the same degree of precision as the least precise number, then compute. The result should then be rounded to the lowest degree of precision.

The above illustration consists of measurements which are expressed in decimal-fraction fashion. The procedure for addition and subtraction of measures expressed in terms of common fractions is illustrated in the next example.

Example What is the perimeter of a four-sided figure whose sides have the following measurements: $14\frac{1}{2}$ in.; $21\frac{1}{4}$ in.; $9\frac{3}{8}$ in.; $18\frac{5}{16}$ in.?

The $14\frac{1}{2}$ in. is the least precise of the four measurements. It thus controls the precision of the final answer. In order to accomplish the addition, equivalent fractions with false precision must be introduced (see Fig. 9-13). The correction for this has to be taken care of by the form in which the answer is given. This answer must be qualified because it is not

---

$14\frac{1}{2}$ $=$ $14\frac{8}{16}$ (an incorrect degree of precision)
$21\frac{1}{4}$ $=$ $21\frac{4}{16}$ (an incorrect degree of precision)
$9\frac{3}{8}$ $=$ $9\frac{6}{16}$ (an incorrect degree of precision)
$18\frac{5}{16}$ $=$ $18\frac{5}{16}$ (correct precision)
$\overline{\qquad\quad 63\frac{7}{16}}$ (an incorrect degree of precision)
The correct answer is $63\frac{7}{16}$ in. $\pm\frac{1}{4}$ in.

FIGURE 9-13 Precision in English units

correct to sixteenths of an inch but only to one half-inch. Therefore, the perimeter is $63\frac{7}{16}$ in. correct to the nearest half-inch, or $63\frac{7}{16}$ in. $\pm\frac{1}{4}$ in. This means that the limits of the computed perimeter are $63\frac{3}{16}$ in. and $63\frac{11}{16}$ in. rather than $63\frac{13}{32}$ in. and $63\frac{15}{32}$ in., or $63\frac{7}{16}$ in. $\pm\frac{1}{32}$ in.

9-11 MULTIPLICATION AND DIVISION

Just as precision is the guide to satisfactory computation in the addition and subtraction of approximate numbers, so accuracy serves as the guide in multiplication and division. The detail of analysis is quite a bit more technical than in addition and will be dispensed with here. However, an example involving limits can again serve as an indication of the procedure to use.

Example How many square feet are there in the area of a rectangle whose dimensions are 14.82 by 26.3 ft?

The minimum dimensions that could be represented by the given measurements are 14.815 by 26.25 ft, thus giving a minimum area of 388.89375 sq ft. Similarly, the maximum dimensions are 14.825 by 26.35 ft, giving a maximum area of 390.63875 sq ft.

When the more accurate number 14.82 is rounded to the same number of significant digits as 26.3 and the product obtained, we have 14.8 $\times$ 26.3 $=$ 389.24. This is seen to be well within the limits of the extreme values for the product. Since the computation could not be expected to yield a result more accurate than the accuracy of the least accurate number used, the product of the measurement would be given as 389 sq ft, a measurement still well within the computed limits.

Example The product of two measurements is 389 sq ft. If one measurement is 14.82 ft, what is the other?

The limiting values can be obtained here by dividing the minimum possible value of the product by the maximum possible value of the given measurement, and the maximum possible value of the product by the minimum possible value of the measurement. These results are shown in columns

MINIMUM VALUE		MAXIMUM VALUE		COMPUTED VALUE	
	26.20 . . . ft		26.29 . . . ft		26.28 ft
14.825	388.50000	14.815	389.50000	14.8	389.000
	296 50		296 30		296
	92 000		93 200		930
	88 950		88 890		888
	3 0500		4 3100		420
	2 9650		2 9630		296
	8500		1 34700		1240
			1 33335		1184
			1365		56
	(*a*)		(*b*)		(*c*)

FIGURE 9-14 Significant digits in division

(*a*) and (*b*), respectively, of Fig. 9-14. Also, in column (*c*) is shown the computed value obtained by rounding the more accurate number to the smaller number of significant figures. The quotient is within the extreme limits, but the justified result obtained when rounding this quotient to three significant figures is 26.3 ft. The fact that this is outside the extremes is not disturbing since the figure in the extreme right position of any approximate number is always in question.

These two examples suffice to illustrate the rule commonly given for the multiplication and division of approximate numbers:

Multiplication and division with approximate numbers In the multiplication and division of two approximate numbers, express both numbers to the same degree of accuracy (same number of significant figures) as the one of lesser accuracy. The result will have this degree of accuracy.

In a series of such computations the same rule applies, but it usually is safer and productive of better results to carry one extra digit along until the final result is obtained. Thus, for example, when an approximate number is raised to an integral power, the final result can be given to the same number of significant figures as contained in the original number.

Example A cube each of whose sides is 1.81 in. will have a volume = $(1.81 \text{ in.})^3 = 5.93$ cu in.

Any time an exact number enters any such computation with approximate numbers, it has no effect at all on the degree of approximativeness. An exact number may be thought of as having as many significant digits as one desires.

Example The circumference of a circle with radius 4.16 in. is $2\pi(4.16$ in.$)$. In this formula 2 is an exact number and π has been computed to many significant figures, and so 4.16 is the number that controls the accuracy of the result. The answer can be given to three significant figures. It is 26.1 in.

Finally, the process of extracting square roots enters into a sufficient number of the computations of elementary mathematics to justify mentioning the rule governing the results of extracting square roots of approximate numbers.

Square root The *square root* of an approximate number can be computed correct to the same number of significant digits as the given number.

Example What is the radius of a circle whose area is 7.64 sq in.?
The formula is $A = \pi r^2 = 3.14r^2$ where the value of π is taken correct to three significant digits.
Therefore we have

$$7.64 = 3.14r^2$$

$$r^2 = \frac{7.64}{3.14} = 2.43$$

$$r = \sqrt{2.43} = 1.56 \text{ in.}$$

EXERCISES

1 In any given approximate number, what digits are considered to be significant digits?
2 Explain the difference between these three measurements: 8 ft; 8.0 ft; 8.00 ft. Give the limits in each case.
3 Explain the difference between each pair of measurements: 4 in. and $4\frac{0}{8}$ in.; 15 ft 4 in. and $15\frac{1}{3}$ ft. Give the limits in each case.
4 Scientists tell us that a speed of approximately Mach 33.4 is required to hurl an object out of the earth's atmosphere. Use 750 miles per hour as the approximate equivalent of Mach 1 to express this required speed in miles per hour.
5 What is the unit of precision in each of these measures?

(a) 8 ft 3 in.	(b) 5 bu 3 pk	(c) 10 lb 5 oz
(d) 58 min 15 sec	(e) 6 pk	(f) 1 gal 2 qt
(g) 0.00020 in.	(h) 117.000 miles	(i) 2 min 15.2 sec
(j) 221,000 miles	(k) 10,000 dollars	(l) 5 yd 1 ft 10.0 in.

6 What are the maximum apparent error, the relative error, and the percent of error in each measure of Exercise 5?

7 Which is the more precise and which the more accurate in each of the following pairs of measurements?

(*a*) 123.6 in. or 24.3 ft   (*b*) 36 min or 360 sec
(*c*) 30 ft or 3 yd   (*d*) 58 ft or 572 miles
(*e*) 195 lb or 80 oz   (*f*) 5 gal 3 qt or 3 qt 1 pt

8 Find the perimeter of each of these polygons.
(*a*) Sides: 8 ft 3 in.; 6 ft 9 in.; 4 ft 6 in.
(*b*) Sides: 10 ft 8 in.; 12 ft 9 in.; 18 ft 7 in.
(*c*) An equilateral triangle each of whose sides is 16.00 in.
(*d*) A square each of whose sides is 5 yd 2 ft 8 in.
(*e*) A rectangle with sides 6 ft 10 in. and 4 ft 8 in.
(*f*) Sides: 216.3 ft; 400.0 ft; 320.26 ft; 315.780 ft; 293.504 ft
(*g*) Sides: 16 in.; 20.00 in.; 17.3 in.; 8 in.; 11.2 in.; 7.001 in.
(*h*) A regular pentagon each of whose sides is 40.0 in.
(*i*) A regular hexagon each of whose sides is 32.06 in.
(*j*) A regular octagon each of whose sides is 2.5 in.

9 Write formulas for finding the perimeter of each of these polygons: (*a*) triangle; (*b*) equilateral triangle; (*c*) rectangle; (*d*) square; (*e*) regular pentagon; (*f*) regular hexagon.

10 Find the area and perimeter of a rectangular field which is 46.00 by 3.64 ft. Why should the area be given only to the nearest square foot while the perimeter should be given to the nearest one-hundredth of a foot? (*Hint:* The formula for the area is $A = lw$ where l and w represent the two dimensions.)

11 Find the area and perimeter of a square each of whose sides is 56.4 in. Why can the perimeter be given correct to the nearest tenth of an inch and the area only to the nearest 10 square inches? (*Hint:* The formula for the area is $A = s^2$ where s represents the length of a side.)

12 Find the side of a square with each of the given areas:

(*a*) 2.25 sq in.   (*b*) 144 sq in.   (*c*) 323 sq in.

13 Find the circumference and area of a circle whose radius is as given:

(*a*) $r = 2.5$ in.   (*b*) $r = 0.134$ ft

(*Hint:* Use $\pi = 3.14$)

14 Find the circumference of a circle whose area is 445 sq in.

15 Find the area of a circle whose circumference is 7.62 in.

16 Write in scientific notation each of the numbers in the following statements:
(*a*) The diameter of the smallest visible particle is approximately 0.005 centimeter.
(*b*) There are approximately 800,000,000,000,000,000,000,000 molecules in 1 pound of cane sugar.

(c) The diameter of an average red blood corpuscle is 0.0000316 in.

(d) The wavelength of visible light varies from 0.000072 to 0.000040 cm.

(e) The speed of sound through a steel rod is 16,410.0 ft per sec.

(f) The average distance of the planet Jupiter from the earth is 483,900,000 miles.

17 Write in positional notation each of the numbers in the following statements:

(a) The average length of time a movie image remains on the screen is 6.4×10^{-2} sec.

(b) The mass of the earth is approximately 6.0×10^{27} grams.

(c) The nearest known star is approximately 2.5×10^{13} miles from the earth.

(d) The mass of one molecule of water is approximately 3×10^{-23} gram.

(e) The speed of light is approximately 1.86×10^5 miles per sec.

18 One light-year is defined as the distance light will travel in one year's time. What is the length of a light-year if the speed of light is 186,282 miles per sec?

19 The distance from the sun to the earth is 92,956,524 miles. Correct to the nearest second, how long does it take a ray of sunlight to reach the earth?

Use scientific notation to work Exercises 20 to 25.

*20 Compute the distance in miles from the earth to the North Star; to Alpha Centauri. (See page 344 for distances in light-years. Compare answers and technique with Exercises 6 and 7 of Sec. 9-7.)

*21 Compute the mass in grams of a proton, a neutron, and an electron. (See page 344 for data. Compare answers and technique with Exercises 8 to 10 of Sec. 9-7.)

*22 The star Arcturus is 223,700,000,000,000 miles from the earth. How many years does it take light from Arcturus to reach the earth?

*23 The approximate velocity of sound in water is 4,720 feet per sec. How deep is a lake at a point where it takes sound 0.030 sec to travel from the surface to the bottom of the lake and back?

*24 The diameter of the moon is 2,160 miles and it is 239,000 miles from the earth. The diameter of the sun is 864,000 miles, and correct to four significant digits, its distance from the earth is 92,960,000 miles. Check the truth of this statement: The ratio of the diameter of the sun to the diameter of the moon is essentially the same as the ratio of their respective distances from the earth.

*25 "One inch of rainfall" means that a surface area has been covered uniformly to the depth of one inch. How many gallons of water would there be in an 11-foot-square container holding one inch of rain? One cubic foot of water weighs about 62.4 pounds and there are about 8.345 pounds to one gallon.

9-12   THE TANGENT AND COTANGENT RATIOS

History has it that the Greek mathematician Thales (ca. 600 B.C.) used a familiar property of similar triangles to find the heights of pyramids. One version is that he noted the length of the shadow of the pyramid at a time when the length of his shadow was the same as his height, thus making the height of the pyramid the same as the length of its shadow. Another version of the story is that he equated the ratio between the height of a stake and the length of its shadow to the ratio between the unknown height of a pyramid and the length of its shadow. From this proportion he was able to determine the height of the pyramid. These ratios are at times referred to as "shadow ratios."

The shadow ratio

Example   Assume that the distance from the point of the shadow of a pyramid to the center of its base is 16$\underline{0}$ feet at the same time the shadow of a stake, 16.0 feet high, is 8.0 feet long (see Fig. 9-15). What is the height of the pyramid?

The line segments $\overline{PS}$ and $\overline{AB}$, which represent sun rays, are parallel. From this fact it can be established that the angles of triangle PRS and ACB are congruent. It then follows from Theorem 8-13 and the implications of Definition 8-23 to triangles (pages 310 and 309) that the two triangles are similar and that ratios of corresponding sides are equal. Therefore we have

$$\frac{h}{16\underline{0}} = \frac{16.0}{8.0}$$

from which $h = 320$ feet. Why is the zero not significant?

This use of the equality of ratios of corresponding sides of similar triangles provides a very effective technique for making indirect measurements not only of inaccessible heights but also of other types of length.

Example   To measure the distance across a stream, surveyor's instruments were used to set up the diagram of Fig. 9-16. The angles BCA and AED

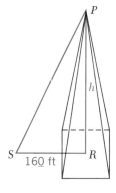

FIGURE 9-15   Using similar triangles to find heights

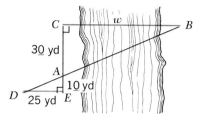

FIGURE 9-16   Using similar triangles to find distances

are right angles. The points C, A, and E are collinear as are the points B, A, and D. $CA = 30$ yd, $AE = 10$ yd, and $DE = 25$ yd. Find CB.

Why are $\angle CAB$ and DAE congruent?
Why are $\triangle CAB$ and DAE similar?
Why is $\dfrac{w}{30} = \dfrac{25}{10} = \dfrac{5}{2}$?

From this proportion it follows that $w = 75$ yd.

The two preceding examples illustrate two simple applications of the similarity relationship to right triangles. Further significance of this relationship may be discovered through an investigation of the similar right triangles of Fig. 9-17. The central angle of the circle with center at O is the common angle of the right triangles DAO, EBO, and FCO. The angles DAO, EBO, and FCO are congruent angles since each is a right angle. It thus follows (Theorem 8-13) that the three triangles are similar. Also the angles ODA, OEB, and OFC are congruent angles, each of which is a complement of the angle O. From these similar triangles there is a set of equal ratios which correspond to the "shadow ratio" used previously. They are

$$\frac{DA}{OA} = \frac{EB}{OB} = \frac{FC}{OC}$$

It should be evident from the property of similar triangles that if a point

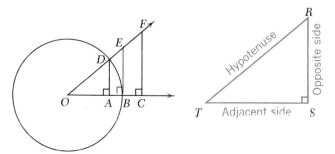

FIGURE 9-17   The tangent ratio

P were taken anywhere along the ray $\overrightarrow{OF}$ and a line segment $\overline{PM}$ were drawn perpendicular to the ray $\overrightarrow{OC}$, the triangle PMO would be similar to each of the triangles of the figure. Then

$$\frac{PM}{OM} = \frac{DA}{OA} = \frac{EB}{OB} = \frac{FC}{OC}$$

Furthermore, if $\angle STR$ in the right triangle RST is congruent to $\angle O$, then $\triangle RST$ is similar to each of the triangles DAC, EBO, and FCO. From these similar triangles we have

$$\frac{RS}{TS} = \frac{DA}{OA} = \frac{EB}{OB} = \frac{FC}{OC}$$

Each of the illustrations has been selected to make the point that, for any given angle, the ratio

Measure of the length of the side opposite the angle
Measure of the length of the side adjacent to the angle

or, more briefly, the ratio

Opposite side
Adjacent side

is a fixed ratio. In other words, there is a specific value of this ratio associated with each acute angle; and, conversely, with each value of the ratio there is associated a specific acute angle. For this reason we say that the

Tangent ratio

ratio is a *function* of the angle, and we give it the name *tangent ratio*. Symbolically, we write tangent $\angle O$ or tan $\angle O$. In other words,

$$\tan \angle O = \frac{\text{opposite side}}{\text{adjacent side}} = \frac{DA}{OA} = \frac{EB}{OB} = \frac{FC}{OC}$$

The reason this ratio is given the name tangent may be discovered from a further study of $\triangle EBO$ of Fig. 9-17. If the circle is considered as a unit circle, that is, a circle whose radius is one unit, then $OB = 1$, and $\tan O = \frac{EB}{OB} = \frac{EB}{1} = EB$. An examination of the figure exhibits $\overline{EB}$ as a line segment of the line tangent to the circle with B as the point of tangency.

In $\triangle RST$ of Fig. 9-17, angle TRS is the complement of angle STR. The side opposite $\angle TRS$ is $\overline{TS}$, and the adjacent side is $\overline{RS}$; therefore tan $\angle TRS = TS/RS$. This ratio may be written in terms of $\angle STR$ as follows:

$$\text{Tangent of complementary angle of } \angle STR = \frac{\text{adjacent side}}{\text{opposite side}}$$

Cotangent ratio

The symbol cot $\angle STR$, abbreviation for *"cotangent of STR,"* is used to represent this ratio:

$$\cot \angle STR = \frac{\text{adjacent side}}{\text{opposite side}}$$

If A is used to represent any angle, we have

$$\tan A = \frac{\text{opposite side}}{\text{adjacent side}} \tag{9-1}$$

$$\cot A = \frac{\text{adjacent side}}{\text{opposite side}} \tag{9-2}$$

A comparison of Eqs. (9-1) and (9-2) brings to light a very important relation which exists between the tangent and cotangent of any angle, namely,

$$\cot A = \frac{1}{\tan A} \tag{9-3}$$

or, in equivalent form,

$$\tan A \cot A = 1 \tag{9-4}$$

We have seen that as long as $\angle A$ remains fixed in size the tangent ratio remains constant in value. What happens as $\angle A$ changes in size? The three angles AVB, AVC, and AVD of Fig. 9-18 have the same vertex V and the same initial side $\overline{VA}$. The tangent ratios are

$$\tan \angle AVB = \frac{AB}{VA} \qquad \tan \angle AVC = \frac{AC}{VA} \qquad \tan \angle AVD = \frac{AD}{VA}$$

In the ratios the length of the adjacent side $\overline{VA}$ remains fixed. A comparison of the change in the angle with the corresponding change in the length of the opposite side reveals that the value of the ratio increases as the angle increases and decreases as the angle decreases.

From Eq. (9-3) we have that the tangent and cotangent are reciprocal functions. This means that as the value of the tangent function increases, the value of the cotangent function decreases, and as the value of the tangent function decreases, the value of the cotangent function increases. Furthermore, from Fig. 9-18, it should be evident that for an angle of $0°$, the length of the opposite side is 0. From this fact we have $\tan 0° = 0$. It therefore follows from the reciprocal relation (9-3) that $\cot 0°$ *is undefined*, since division by zero is an undefined operation. Similarly, as we shall be able to establish more clearly in Sec. 9-15, $\cot 90° = 0$ and $\tan 90°$ *is undefined*.

9-13 THE SINE AND COSINE RATIOS

Another set of equal ratios between corresponding sides of the similar triangles of Fig. 9-17 is

$$\frac{DA}{OD} = \frac{EB}{OE} = \frac{FC}{OF} = \frac{RS}{TR} = \frac{\text{opposite side}}{\text{hypotenuse}}$$

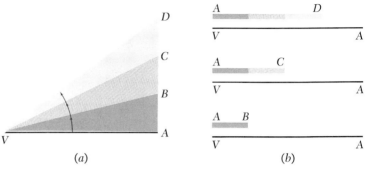

FIGURE 9-18 (*a*) The tangent ratio increases as the angle increases. (*b*) Ratios compared.

This ratio, which also remains constant for any fixed angle, is called the *sine* ratio, abbreviated to *sin*. Thus, if A represents any given angle, we have

Sine ratio

$$\sin A = \frac{\text{opposite side}}{\text{hypotenuse}} \tag{9-5}$$

Still another set of equal ratios from Fig. 9-17 is

$$\frac{OA}{OD} = \frac{OB}{OE} = \frac{OC}{OF}.$$

Closer examination of each right triangle involved in these ratios will reveal that each ratio is the sine of the angle which is the complement of the central angle O. Each of these ratios is in turn equal to TS/TR of triangle RST, which is the sine of the complement of $\angle STR$. If the symbol A is used to represent the angle at O and its equal angle STR, we have

$$\text{Sine of the complement of } A = \frac{TS}{TR} = \frac{\text{adjacent side}}{\text{hypotenuse}}$$

Following the pattern of Sec. 9-12, we define "sine of the complement of A" as the *cosine* A, which is abbreviated to read *cos* A. Thus

Cosine ratio

$$\cos A = \frac{\text{adjacent side}}{\text{hypotenuse}} \tag{9-6}$$

From a study of Fig. 9-19 we can obtain background for the labels sine and cosine attached to these functions and also an idea as to how the value of the ratio changes with change in the size of the angle. The etymology of the word "sine" has it derived from an Arabic word meaning "half chord." If, in Fig. 9-19, we consider the circle as a unit circle, we have

$$\sin \angle BOP = PB \qquad \sin \angle NOC = NC \qquad \sin \angle MOD = MD$$

In each case the function is indeed a half chord. Also it is to be observed that as the size of the angle increases the sine of the angle increases in value.

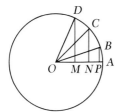

FIGURE 9-19 The sine function and the cosine function

Similarly, $\cos \angle POB = OP$; $\cos \angle NOC = ON$; and $\cos \angle MOD = OM$, from which it is clear that as the size of the angle increases, the value of the cosine decreases.

When the angle is a $\underline{0}°$ angle, the terminal side of the angle coincides with the initial side. This means, in Fig. 9-19, that the point B, for example, would coincide with point A. Thus, for a $\underline{0}°$ angle the measure of the length of the opposite side is $\underline{0}$, while the measure of the length of the adjacent side is the same as that of the hypotenuse. From these facts it follows that $\sin \underline{0}° = \underline{0}$ and $\cos \underline{0}° = 1$. Similarly, when the angle is a $90°$ angle, the measure of the length of the adjacent side is $\underline{0}$ while the measure of the length of the opposite side is the same as the measure of the hypotenuse. From these facts we have $\sin 9\underline{0}° = 1$ and $\cos 90° = \underline{0}$.

From the preceding discussion it should be rather evident that there exists a definite relationship between $\sin A$ and $\cos A$. However, it is not a reciprocal relationship such as that existing between $\tan A$ and $\cot A$ [see Eq. (9-10)].

Frequently it is helpful and convenient to use a right triangle such as ACB of Fig. 9-20, with right angle at C and its sides lettered to correspond to the angles opposite, as a triangle of reference. With this reference we have

$$\sin A = \frac{\text{opposite side}}{\text{hypotenuse}} = \frac{a}{c}$$

$$\cos A = \frac{\text{adjacent side}}{\text{hypotenuse}} = \frac{b}{c}$$

$$\tan A = \frac{\text{opposite side}}{\text{adjacent side}} = \frac{a}{b}$$

$$\cot A = \frac{\text{adjacent side}}{\text{opposite side}} = \frac{b}{a}$$

(9-7)

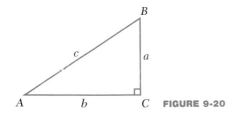

FIGURE 9-20

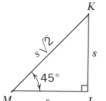

M   s   L   **FIGURE 9-21**

Trigonometric
functions   These functions are called *trigonometric functions*. The word "trigo-
nometry" literally means "triangle measure," and it is from the use of these
functions in measurement problems associated with triangles that they have
derived their name. Actually, they have many significant interpretations and
uses other than those associated with triangles. It is beyond the scope of
this book to pursue their study beyond their association with the right
triangle.

9-14 TABLE OF NATURAL FUNCTIONS

In the isosceles right triangle KLM of Fig. 9-21 with $ML = LK = s$, we
have $MK = s\sqrt{2}$ (Theorem 8-11). From these facts it follows that

$$\sin 45° = \frac{s}{s\sqrt{2}} = \frac{1}{\sqrt{2}} = \frac{\sqrt{2}}{2} = 0.7071 \qquad \text{correct to four places}$$

$$\cos 45° = \frac{s}{s\sqrt{2}} = \frac{1}{\sqrt{2}} = \frac{\sqrt{2}}{2} = 0.7071$$

$$\tan 45° = \frac{s}{s} = 1$$

$$\cot 45° = \frac{s}{s} = 1$$

The triangle PQR of Fig. 9-22 is an equilateral triangle. The line segment
$\overline{PS}$ is perpendicular to the base at its midpoint. By the pythagorean theorem
we have

$$(PS)^2 + (QS)^2 = (PQ)^2$$

or
$$(PS)^2 + (\tfrac{1}{2}s)^2 = s^2$$

$$(PS)^2 = \tfrac{3}{4}s^2$$

$$PS = \frac{s\sqrt{3}}{2}$$

Hence

$$\sin 60° = \frac{s\sqrt{3}/2}{s} = \frac{\sqrt{3}}{2} = 0.8660 = \cos 30°$$

$$\cos 60° = \frac{\frac{1}{2}s}{s} = \frac{1}{2} = 0.5000 = \sin 30°$$

$$\tan 60° = \frac{s\sqrt{3}/2}{s/2} = \sqrt{3} = 1.732 = \cot 30°$$

$$\cot 60° = \frac{\frac{1}{2}s}{s\sqrt{3}/2} = \frac{1}{\sqrt{3}} = \frac{\sqrt{3}}{3} = 0.5774 = \tan 30°$$

Table of natural functions

The values computed for the functions of 0°, 30°, 45°, 60°, and 90° are merely illustrations of what can be done for angles of any size. These values are frequently referred to as *natural functions*. Tables of such values have been constructed for use in making various types of indirect measurements. Figure 9-23 is such a table. This table is very brief and is given primarily for illustration purposes. In steps of 1° the approximate values of the four functions are listed for all angles from 0° to 90°. Notice that there are no values given either for tan 90° or for cot 0°. These function values do not exist since the respective functions are not defined for these angles. Also notice that as the angle-measure values increase so do the values of the sine and tangent functions, but the values of the cosine and cotangent functions decrease.

The next two examples illustrate how the table may be used (1) to find the function value for a given angle and (2) to find the acute angle which corresponds to a given function value.

Example (1) What are the values of sin 25°, cos 38°, tan 12°, cot 45°? (2) What are the values of sin 48°, cos 87°, tan 62°, cot 75°?

1 The entries in the extreme left column of the table are angle measures, in intervals of 1°, from 0° to 45°. For angle-measure values of the left column the function label is at the top of each respective column. In the column under the label "sin" and opposite 25° the entry 0.4226 is the value of sin 25°. Similarly, cos 38° = 0.7880, tan 12° = 0.2126, and cot 45° = 1.000.

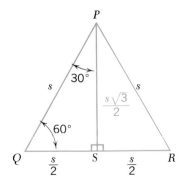

FIGURE 9-22

DEGREES	sin	cos	tan	cot	DEGREES
0°	.0000	1.000	.0000	. . . . .	90°
1°	.0175	.9998	.0175	57.29	89°
2°	.0349	.9994	.0349	28.64	88°
3°	.0523	.9986	.0524	19.08	87°
4°	.0698	.9976	.0699	14.30	86°
5°	.0872	.9962	.0875	11.43	85°
6°	.1045	.9945	.1051	9.514	84°
7°	.1219	.9925	.1228	8.144	83°
8°	.1392	.9903	.1405	7.115	82°
9°	.1564	.9877	.1584	6.314	81°
10°	.1736	.9848	.1763	5.671	80°
11°	.1908	.9816	.1944	5.145	79°
12°	.2079	.9781	.2126	4.705	78°
13°	.2250	.9744	.2309	4.331	77°
14°	.2419	.9703	.2493	4.011	76°
15°	.2588	.9659	.2679	3.732	75°
16°	.2756	.9613	.2867	3.487	74°
17°	.2924	.9563	.3057	3.271	73°
18°	.3090	.9511	.3249	3.078	72°
19°	.3256	.9455	.3443	2.904	71°
20°	.3420	.9397	.3640	2.747	70°
21°	.3584	.9336	.3839	2.605	69°
22°	.3746	.9272	.4040	2.475	68°
23°	.3907	.9205	.4245	2.356	67°
24°	.4067	.9135	.4452	2.246	66°
25°	.4226	.9063	.4663	2.145	65°
26°	.4384	.8988	.4877	2.050	64°
27°	.4540	.8910	.5095	1.963	63°
28°	.4695	.8829	.5317	1.881	62°
29°	.4848	.8746	.5543	1.804	61°
30°	.5000	.8660	.5774	1.732	60°
31°	.5150	.8572	.6009	1.664	59°
32°	.5299	.8480	.6249	1.600	58°
33°	.5446	.8387	.6494	1.540	57°
34°	.5592	.8290	.6745	1.483	56°
35°	.5736	.8192	.7002	1.428	55°
36°	.5878	.8090	.7265	1.376	54°
37°	.6018	.7986	.7536	1.327	53°
38°	.6157	.7880	.7813	1.280	52°
39°	.6293	.7771	.8098	1.235	51°
40°	.6428	.7660	.8391	1.192	50°
41°	.6561	.7547	.8693	1.150	49°
42°	.6691	.7431	.9004	1.111	48°
43°	.6820	.7314	.9325	1.072	47°
44°	.6947	.7193	.9657	1.036	46°
45°	.7071	.7071	1.000	1.000	45°
DEGREES	cos	sin	cot	tan	DEGREES

FIGURE 9-23   Natural functions of acute angles

2   In the extreme right column the entries are, in intervals of 1°, angle measures from 45° to 9$\underline{0}$°. For angle-measure values of the right column the function label is at the bottom of each respective column. In the column over the label "sin" and opposite 48° the entry 0.7431 is the value of sin 48°. Similarly, cos 87° = 0.0523, tan 62° = 1.881, and cot 75° = 0.2679.

The explanation of the organization of the table of natural functions lies in the fact that in such an arrangement of angle-measure values the entry in the left column is the complement of the entry directly opposite it in the right column. This means that the same function value serves for two angles. For example, sin 25° = 0.4226 = cos 65°; cos 38° = 0.7880 = sin 52°; tan 62° = 1.881 = cot 28°; cot 45° = 1.000 = tan 45°.

Example   (1) Given sin A = 0.8746, what is the measure of $\angle A$? (2) For what measure of A is cos A = 0.8746? (3) Given tan A = 0.1944, how many degrees are there in $\angle A$? (4) For what measure of A is cot A = 0.1944?

1   For sin A = 0.8746 the table shows 0.8746 over "sin" and opposite 61° *in the right column;* therefore $m \angle A$ = 61°.
2   For cos A = 0.8746, the table shows 0.8746 under "cos" and opposite 29° *in the left column;* therefore $m \angle A$ = 29°.
3   For tan A = 0.1944, the number is in the column under "tan" and opposite 11°; therefore $m \angle A$ = 11°.
4   For cot A = 0.1944, the number is in the column over "cot" and opposite 79°; therefore $m \angle A$ = 79°.

If Fig. 9-23 were restricted merely to use for the entries found therein, its value would be extremely limited. An established principle, followed as a guide in the use of tables such as this one, is:

Use of table   Two significant figures in function values correspond to angle measurement correct to the nearest degree; three significant figures in function values correspond to angle measures correct to the nearest ten minutes; and four significant figures in function values correspond to angle measures correct to the nearest minute.

Interpolation   Such fractional parts of degrees can be found by a process called *interpolation,* which is illustrated in the next two examples.

Example

1   In Fig. 9-24, assume the numbers 5, 12, and 13 to be exact numbers. Use sin B = $\frac{5}{13}$ to find $m \angle B$ correct to the nearest minute.

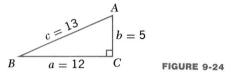

FIGURE 9-24

$$\sin B = \frac{5}{13} = 0.3846 \qquad \text{correct to four significant figures}$$

The number 0.3846 is not to be found in the table, and so we look for the two numbers nearest to it and then build this auxiliary table by using these values with their corresponding angle measures.

$$1° = 60' \begin{bmatrix} -23° & 0.3907 \\ x \begin{bmatrix} -\angle B & 0.3846 \\ -22° & 0.3746 \end{bmatrix} 0.0100 \end{bmatrix} 0.0161$$

This table shows that the measure of $\angle B$ lies between 22° and 23°. Remember that the sine of any angle increases in value as the measure of the angle increases. This means that $m \angle B$ is larger than 22° but smaller than 23°. The auxiliary table shows that a difference of 60' in the angle corresponds to a difference of 0.0161 in the sine. Our problem is to find how much difference in the angle will correspond to the difference of 0.0100 in the sine. This can be found from the proportion

$$\frac{x}{60'} = \frac{0.0100}{0.0161} = \frac{100}{161}$$

$$x = \frac{100}{161} \times 60' = 37' \qquad \text{correct to the nearest minute}$$

Since the angle increases with the sine, this is to be added to 22° to get the measure of $\angle B$.

$$m \angle B = 22°37'$$

2 Use $\tan B = \frac{5}{12}$ to find the measure of $\angle B$.

$$\tan B = \frac{5}{12} = 0.4167$$

The auxiliary table is

$$1° = 60' \begin{bmatrix} -23° & 0.4245 \\ x \begin{bmatrix} -\angle B & 0.4167 \\ -22° & 0.4040 \end{bmatrix} 0.0127 \end{bmatrix} 0.0205$$

$$\frac{x}{60'} = \frac{0.0127}{0.0205} = \frac{127}{205}$$

$$x = \frac{127}{205} \times 60'$$

$$x = 37'$$

The tangent function also increases as the angle increases, and so we have

$$m \angle B = 22°37'$$

Example Find the measure of the hypotenuse $\overline{RT}$ of the right triangle RST of Fig. 9-25.

$$\cos 62°15' = \frac{0.8562}{d} \quad \text{or} \quad d = \frac{0.8562}{\cos 62°15'}$$

The first problem is that of finding cos 62°15′, and the angle is not listed as an entry in the table. This means that we have to interpolate between values for 62° and 63°. In order to do this we must remember that the function value for the cosine of an angle decreases as the size of the angle increases. Since this is true, we shall have to *subtract* the interpolated value from that for the cos 62°. The auxiliary table is

$$
60' \left[15' \left[\begin{array}{ll} 63° & 0.4540 \\ 62°15' & \\ 62° & 0.4695 \end{array} \right. x \right. \right] 0.0155
$$

$$\frac{x}{0.0155} = \frac{15'}{60'} = \frac{1}{4}$$

$$x = \frac{0.0155}{4} = 0.0039$$

$$\cos 62°15' = 0.4695 - 0.0039 = 0.4656$$

Therefore

$$d = \frac{0.8562}{0.4656} = 1.839 \qquad \text{correct to four significant figures}$$

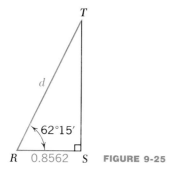

T

d

62°15′

R 0.8562 S **FIGURE 9-25**

EXERCISES

1 Given the two similar triangles ACB and RST of the figure. Find the measure of $\overline{AB}$ and $\overline{RS}$.

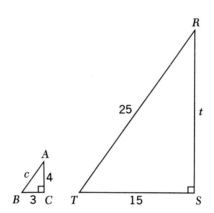

EXERCISE 1

2 If the length of the shadow of a flagpole is 65 feet at the same time the length of a person's shadow is the same as his height, what is the height of the flagpole?

3 In order to measure the length of a lake, surveying instruments were used to set up the diagram of the figure. The angles ADP and BCP are right angles. Using the data of the figure, find the length of the lake.

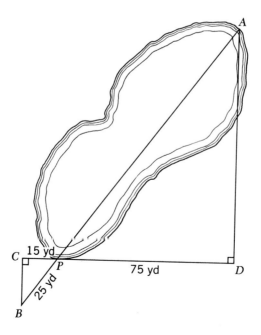

EXERCISE 3

In Exercises 4 to 10, we use $\triangle ABC$ of the figure as a guide to find values, expressed as common fractions, for sin A, cos A, tan A, and cot A.

4 $a = 3, b = 4, c = 5$
5 $a = 20, b = 21, c = 29$
6 $a = 51, b = 140, c = 149$
7 $a = 9, b = 40, c = 41$
8 $a = 15, b = 36, c = 39$
9 How does $\angle A$ of Exercise 8 compare with $\angle B$ of Fig. 9-24? Give reasons to support your answer.
10 $a = 105, b = 140, c = 175$

EXERCISES
4 to 10

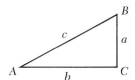

11 Is $\angle A$ of Exercise 10 congruent with $\angle A$ of any of the previous exercises?
12 Select a set of values for a, b, and c such that the corresponding $\angle A$ will be congruent to $\angle A$ of Exercise 6.
13 In Exercises 4 to 8 and 10 express in common fraction form sin B, cos B, tan B, and cot B.
14 In Exercise 2 what is the *angle of elevation* of the top of the flagpole? (*Hint:* The angle of elevation would be the angle made with the horizontal by the line of sight from the tip of the shadow to the top of the flagpole.)
15 What would have been the angle of elevation of the top of the flagpole if the ratio of its height to the length of its shadow had been $\frac{5}{12}$? (*Hint:* Use the example of Fig. 9-24.)
16 Express as a decimal fraction, correct to four decimal places, the value of each trigonometric function found in Exercises 4 to 8 and 10.
17 Is there any value of $m \angle A$ in the interval $0° \le m \angle A < 180°$ for which either of these statements is not true?

   (*a*) cos $A \cdot$ tan $A =$ sin A (*b*) sin $A \cdot$ cot $A =$ cos A

18 Find each of these function values: sin 45°, tan 75°, cos 58°, cot 16°, sin 67°12′, cos 23°50′, tan 24°45′, cot 68°19′.
19 Find the measure of the angle A in each of these cases: sin $A =$ 0.8090; cos $A =$ 0.7771; tan $A =$ 0.4040; cot $A =$ 2.246.
20 Find the measure of the angle x correct to the nearest minute in each of these cases: sin $x =$ 0.5540; cos $x =$ 0.6556; tan $x =$ 1.590; cot $x =$ 0.6480.
21 To reach a bridge which is 35 feet above the level of the road an incline was started at a distance of 525 feet from the bridge. Correct to the nearest 10′, what was the angle at which the approach had to rise?

22 Find the height of a telephone pole if at a point 125 feet from the base of the pole the angle of elevation to the top of the pole is 21°30′.

23 A guy wire is to be run from a point 25.00 feet up on a pole to a point on the ground at a distance of 18.00 feet from the base of the pole. What angle will this guy wire make with the ground?

24 What will be the length of the guy wire of Exercise 23?

25 What is the measure of the central angle whose chord is 25.0 inches if the radius of the circle is 35.0 inches?

*26 A hiker determines that, from his position, the angle of elevation of the top of a television tower is 37°30′. The height of the tower is known to be 2,000 ft. Which measure gives the closest approximation to the hiker's straight line distance from the base of the tower: $\frac{1}{4}$ mile; $\frac{1}{2}$ mile; $\frac{3}{4}$ mile; 1 mile?

*27 The original height of the largest Egyptian pyramid was 481 ft. The length of each side of its square base was 755 feet. At what angle to the horizontal was each face inclined? (See figure.)

*28 Assume that on one face of the pyramid of Exercise 27 a line was drawn from the apex (the highest point A of the pyramid) to the midpoint B of the base. Correct to the nearest foot, what would be the length of the segment $\overline{AB}$? (See figure.)

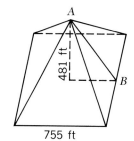

EXERCISES 27 and 28

*29 The Leaning Tower of Pisa originally stood upright. Today it stands approximately as shown in the figure. Correct to the nearest 10 minutes, what is the measure of its angle of inclination ($\angle A$ of the figure)?

EXERCISE 29

*9-15 TRIGONOMETRIC IDENTITIES

An examination of Eqs. (9-7) brings to light certain fundamental relations that exist between the four functions. For example,

$$\frac{\sin A}{\cos A} = (\sin A) \div (\cos A) = \frac{a}{c} \div \frac{b}{c}$$

$$= \frac{a}{c} \times \frac{c}{b} = \frac{a}{b}$$

$$= \tan A$$

Hence we have
$$\tan A = \frac{\sin A}{\cos A} \qquad (9\text{-}8)$$

Similarly,
$$\cot A = \frac{\cos A}{\sin A} \qquad (9\text{-}9)$$

Equations (9-8) and (9-9) may be used to get a clearer picture of why $\tan 90°$ and $\cot 0°$ are undefined. Since $\cos 90° = 0$, it follows from (9-8) that $\tan 90°$ is undefined. Also, since $\sin 0° = 0$, it follows from (9-9) that $\cot 0°$ is undefined.

A fundamental relation existing between $\sin A$ and $\cos A$ is an immediate consequence of the pythagorean theorem (Theorem 8-11). An application of this theorem to $\triangle ACB$ of Fig. 9-20 gives

$$c^2 = a^2 + b^2$$

If both sides of this equality are multiplied by $1/c^2$, we get

$$\frac{1}{c^2} \times c^2 = \frac{1}{c^2} \times (a^2 + b^2)$$

which by the property of the multiplicative inverse and the distributive property gives

$$1 = \frac{a^2}{c^2} + \frac{b^2}{c^2}$$

Comparison with Eqs. (9-7) now yields

$$\sin^2 A + \cos^2 A = 1 \qquad (9\text{-}10)$$

Equations (9-8) to (9-10) express relationships between the respective functions which are true equalities for all values of the angle for which the functions are defined. Such equalities are called *identities*.

Trigonometric identity **Definition 9-2** A *trigonometric identity* is a stated equality between expressions composed of trigonometric functions which is true for all values of the angle for which the functions are defined.

The fact that this definition is restricted to trigonometric functions of angles and is not stated as generally as it might be is not of concern here. This discussion is limited to the acute angles of a right triangle, with the exception that angles of $0°$ and $90°$ have been included. It must be remembered that $\cot A$ is undefined when $m \angle A = 0°$ and $\tan A$ is undefined when $m \angle A = 90°$.

Equations (9-8) to (9-10) and also (9-3) and (9-4) may now be stated in the form of the identities which they are. The symbol $\equiv$ is used to indicate such an identity and is read "is identically equal to."†

Fundamental
identities

$$\tan A \equiv \frac{\sin A}{\cos A} \tag{9-11}$$

$$\cot A \equiv \frac{\cos A}{\sin A} \tag{9-12}$$

$$\tan A \equiv \frac{1}{\cot A} \tag{9-13}$$

$$\tan A \cot A \equiv 1 \tag{9-14}$$

$$\sin^2 A + \cos^2 A \equiv 1 \tag{9-15}$$

Identity (9-11) holds for all values of A except $90°$ since $\tan 90°$ is not defined; identity (9-12) holds for all values of A except $0°$ since $\cot 0°$ is not defined. Identities (9-13) and (9-14) hold for all values of A except $0°$ and $90°$, and identity (9-15) holds for all values of A.

Identities between expressions composed of trigonometric functions may be established in any of three ways.

1 Start from a known identity and derive the desired relation.
2 Transform the expression on either side of the stated equality into the expression of the other side, or into an expression which is identically equal to it.
3 Transform each of the expressions of the stated equality into the same expression or into two expressions which are known to be identically equal. When this technique is used one must be sure that the steps can be reversed on either side of the equality (significant only when dealing with identities more intricate than those of interest here).

Example

1 Derive Eq. (9-14) from Eq. (9-13).
Equation (9-13) is an identity as a consequence of the definitions of $\tan A$ and $\cot A$. Multiply both sides of this known identity by $\cot A$ to get $\tan A \cot A \equiv 1$.

---

† Attention is called to the fact that the symbol $\equiv$ was used in Chap. 5 with an entirely different connotation. Since the contexts of the two uses are always clearly distinct from each other, there is no confusion in this seeming ambiguity.

2 Prove $\cos A \equiv \sqrt{1 - \sin^2 A}$ for $0° \leq m\angle A \leq 90°$.
Equation (9-15) is a known identity, whence

$$\sin^2 A + \cos^2 A \equiv 1 \quad \text{or} \quad \cos^2 A \equiv 1 - \sin^2 A$$

Extracting the square root, we get

$$\cos A \equiv \pm\sqrt{1 - \sin^2 A}$$

By definition $\cos A \geq 0$, whence

$$\cos A \equiv \sqrt{1 - \sin^2 A}$$

Example Prove that $\cos^4 A - \sin^4 A \equiv \cos^2 A - \sin^2 A$.

We shall attempt to convert the expression on the left side of this equality into the one on the right.

$\cos^4 A - \sin^4 A$ may be written as the difference of two squares in the form

$$\cos^4 A - \sin^4 A \equiv (\cos^2 A)^2 - (\sin^2 A)^2$$

Factoring this difference of two squares, we get

$$\cos^4 A - \sin^4 A \equiv (\cos^2 A + \sin^2 A)(\cos^2 A - \sin^2 A)$$

From Eq. (9-15) we have

$$\cos^2 A + \sin^2 A \equiv 1$$

whence
$$\cos^4 A - \sin^4 A \equiv 1 \cdot (\cos^2 A - \sin^2 A)$$
$$\equiv \cos^2 A - \sin^2 A$$

EXERCISES

In Exercises 1 and 2 use Theorem 8-11 and Eqs. (9-8) to (9-10) to find the remaining functions of angle A of the right $\triangle ABC$. Express each result as a common fraction.

1 $\sin A = \dfrac{7}{25}$ 　　 2 $\tan A = \dfrac{12}{35}$

In Exercises 3 to 26 prove each identity for $\angle A$ in the interval $0° \leq m\angle A \leq 90°$. Indicate any value of A for which an identity is not true.

3 $\sin A \cot A \equiv \cos A$

4 $\cos A \tan A \equiv \sin A$

5 $\cos A + \sin A \tan A \equiv \dfrac{1}{\cos A}$

6 $\sin A + \cot A \cos A \equiv \dfrac{1}{\sin A}$

7 $\cos^2 A - \sin^2 A \equiv 2\cos^2 A - 1$

8 $\quad \tan A + \cot A \equiv \dfrac{1}{\sin A \cos A}$

9 $\quad 1 + \cot^2 A \equiv \dfrac{1}{\sin^2 A}$

10 $\quad \tan^2 A + 1 \equiv \dfrac{1}{\cos^2 A}$

11 $\quad \sin(90° - A) \equiv \cos A$

$\qquad \cos(90° - A) \equiv \sin A$

$\qquad \tan(90° - A) \equiv \cot A$

$\qquad \cot(90° - A) \equiv \tan A$

12 $\quad \sin A \equiv \sqrt{1 - \cos^2 A}$

13 $\quad \cos^4 A - \sin^4 A \equiv 1 - 2\sin^2 A$

14 $\quad \dfrac{1 + \sin A}{\cos A} \equiv \dfrac{\cos A}{1 - \sin A}$

15 $\quad \tan^4 A - 1 \equiv \dfrac{1 - 2\cos^2 A}{\cos^4 A}$

16 $\quad \dfrac{\sin^2 A}{1 + \cos A} \equiv 1 - \cos A$

17 $\quad (\tan A - \cot A)\sin A \cos A + \cos^2 A \equiv \sin^2 A$

18 $\quad (\tan A + \cot A)\sin A \cos A - \sin^2 A \equiv \cos^2 A$

∗19 $\quad \cos^4 A - \sin^4 A + 1 \equiv 2\cos^2 A$

∗20 $\quad \cos^2 A + (\tan A - \cot A)\sin A \cos A \equiv \sin^2 A$

∗21 $\quad \dfrac{\sin A + \tan A}{1 + \cos A} \equiv \tan A$

∗22 $\quad \dfrac{1 + \tan A}{1 + \cot A} \equiv \tan A$

∗23 $\quad \dfrac{\cos A}{1 + \sin A} + \tan A \equiv \dfrac{1}{\cos A}$

∗24 $\quad \tan^2 A - \sin^2 A \equiv \tan^2 A \sin^2 A$

∗25 $\quad (\tan A + \cot A)^2 \equiv \dfrac{1}{\sin^2 A \cos^2 A}$

∗26 $\quad \dfrac{\cot A \tan^2 A}{\cos A + \tan A \sin A} \equiv \sin A$

9-16 STATISTICAL MEASURES

Whether in school, in the community, or in business, it frequently is not only convenient but necessary to rely on certain established techniques for detecting, analyzing, and interpreting information from large masses of data. For example, it is much more convenient, and frequently gives much more significant information, to speak in terms of the mean test score of a large group of students than to try to think in terms of the scores of all the

Distribution individual students. Such a set of measures or values of the same, or similar, things is called a *distribution*. The *mean*, as used here, is a *statistical measure of central tendency* of a distribution, or a score which indicates a central point about which all other scores are distributed. There are two other measures of central tendency, the *mode* and the *median*. The distinctions between these three measures are clarified by the three following definitions and the accompanying example.

Mean   **Definition 9-3**   The *mean* of a distribution is that term which results from dividing the sum of all the terms by the total number of terms.

Mode   **Definition 9-4**   The *mode* of a distribution is that term which occurs most frequently.

Median   **Definition 9-5**   The *median* of a distribution is that term of such size that of all the remaining terms, there are just as many greater as there are smaller.

Example   Compute the mean, median, and mode for the distribution of test scores given in Fig. 9-26.

1   The total number of scores is 35. The total sum of all the terms is 210. Mean $= \frac{210}{35} = 6$.

2   There are 35 different scores. The mid-score is therefore the eighteenth one. Starting with either the highest score, 10, or the lowest score, 2, the median is 7.

SCORE	FREQUENCY	SUM OF TERMS
10	1	10
9	7	63
8	1	8
7	9	63
6	4	24
5	2	10
4	3	12
3	4	12
2	4	8
1	0	0
Total	35	210

Mean score $= \dfrac{210}{35} = 6$ Median score $= 7$ Mode $= 9$ and 7

FIGURE 9-26   Distribution of test scores

*Bimodal
distribution*

3 The mode is also 7 since this is the score with the greatest frequency. Actually, there are two scores, 9 and 7, which occur with much more frequency than any of the others. For this reason this distribution might be said to be *bimodal*, with 9 and 7 as the two modes.

A purely descriptive way of thinking about these three measures is:

1 *The mean is a computed measure.* It is obtained by finding the arithmetic average of all terms of the distribution.

2 *The median is a counted measure.* It is obtained by merely arranging the terms in sequence of size and then counting from either the largest or the smallest term to that one which is the middle term. When there is an odd number of terms, as in the example, there is no ambiguity as to which is the middle one. When there is an even number of terms there exists an ambiguity as to what the middle term is, since there are two instead of one. If, in the example, the total number of scores were increased to 36 by adding 1 to the frequency of any score from 1 through 6, then such an ambiguity would exist. The two middle terms would be the eighteenth and nineteenth. Counting from the lowest score upward, the eighteenth score would be 6, and 7 would be the nineteenth. Counting from the largest score downward, 7 would be the eighteenth and 6 the nineteenth. In such cases as this an acceptable interpretation is to take the average of the two middle terms as the median. In this modified example the median score would be 6.5.

If, in the example, the total frequency of 36 were brought about by increasing by 1 the frequency of any of the scores 7 through 10, both the eighteenth and nineteenth terms would be 7, counting from either end. There would be no ambiguity in this case, and the median score would be 7.

3 *The mode is an observed measure.* It is obtained by merely observing that it is the term of the distribution which occurs the greatest number of times. There can be distributions of only one mode or of more than one, such as the bimodal distribution of the example. While they seldom occur in distributions of any significant size, it is conceivable that there could be situations in which no mode exists.

Generally speaking, the mean is considered the best measure of central tendency to use for purposes of describing a distribution as a whole. This is due to the fact that it is sensitive to any change whatsoever in the pattern of a distribution while the median and the mode are not. This sensitivity, however, at times serves as a distinct disadvantage; in particular, this is true if the distribution is characterized by measures concentrated at one of the extremes. This limitation is caused by the fact that one of the properties of the mean is that there is a balance between the distances of the scores from the mean. This is illustrated in Fig. 9-27, where the distribution of the example has been repeated. In this figure the column headed "Deviations" indicates how much each score deviates, or differs, from the

Deviation

SCORE	FREQUENCY	DEVIATIONS (SCORE − MEAN)	TOTAL DEVIATIONS (FREQ. × DEVIATION)	SQUARES OF DEVIATIONS†
10	1	+4	+4	16
9	7	+3	+21	63
8	1	+2	+2	4
7	9	+1	+9	9
6	4	0	+36	
5	2	−1	−2	2
4	3	−2	−6	12
3	4	−3	−12	36
2	4	−4	−16	64
1	0	−5	0	0
	35		−36	206

Mean score $= 6$

Variance $= \dfrac{206}{35} = 5.89$

Standard deviation $= \sqrt{5.89} = 2.43$

†Each entry in this column may be obtained by squaring the deviation and multiplying the result by the corresponding frequency; for example, $63 = (3 \times 3) \times 7 = 9 \times 7$.

FIGURE 9-27   Computation of variance and standard deviation

mean score 6. The deviation, or difference, for any given score is found in this manner: If the mean of the distribution is M, then the deviation d for any score S is given by $d = S - M$. For example, the deviation which corresponds to 4 is -2 since $4 - 6 = -2$, and the deviation which corresponds to 9 is $+3$ since $9 - 6 = 3$. Thus all scores smaller than 6 have a negative deviation and those larger than 6 have a positive deviation. The column headed "Total Deviations" shows that the total of the negative deviations exactly balances the total of the positive deviations. This property is one which is characteristic of the mean and not of either the median or the mode.

Each measure of central tendency performs a significantly intrinsic function. As we have seen, the mean is sensitive to all characteristics of a distribution and makes a good measure to indicate the size of *all* terms of a set of data. There are times when the better measure is the median, since it is not affected by extremes of the distribution. For example, it would be a far better indicator of the salary status of a community composed entirely of salaried employees of a given large business or industrial company having only two or three highly salaried executives. Also, there are times when the mode is the only measure of interest. This would be the case if a retail merchant or a manufacturer were interested in the most popular model or product. The mode would show the concentration of frequencies to indicate popularity.

In spite of the effectiveness of the mean for studying relative sizes of distribution of data, it has distinct limitations. For example, a set of scores on the test of the previous example could be as follows:

$$\{6,6,6,6,6,6,6,6,6,6\}$$

In this case 10 students took the test and each student had a score of 6. The mean, as well as the median and mode, of this distribution is 6. But the distribution has an element of concentration not possessed by the distribution of the example discussed. It is quite evident that the *scatter*, or *variability*, of a set of scores carries as great a significance in characterizing a distribution as does its central tendency. A much more intelligent comparison of two distinct distributions of scores can be obtained through combining an examination of their respective measures of variability with a study of their respective means.

Variability

The problem of determining a measure of the variability of a set of scores involves finding some number, determined from the scores, with these properties: (1) it should be independent of the mean of the distribution; (2) it should not be affected by the signs of the deviations from the mean; (3) it should be small when the scores are clustered closely together and large when they are widely scattered; and (4) it should be more sensitive to the intrinsic nature of the deviations than to their total frequency.

There are two related measures which meet these requirements and which serve, in combination with the mean, to give a rather complete picture of any distribution of scores. They are called *variance* and *standard deviation*.

Variance

Definition 9-6 The *variance* of a distribution is the quotient of the sum of the squares of the deviations from the mean divided by the total frequency.

Standard deviation

Definition 9-7 The *standard deviation* of a distribution is the positive square root of the variance.

Either of these two measures can be used to describe the variability of a distribution. No effectiveness is lost or gained in choosing one over the other. It should be evident that when one is small the other is small, and when one is large the other is large. The symbol for the variance is σ^2 (read "sigma squared"), and for the standard deviation it is σ ("sigma"). The standard deviation possesses certain mathematical properties not possessed by the variance, but these are of no concern to this discussion. It is of significance here to see just how these measures meet the four conditions previously specified: (1) they are independent of the mean since their computation uses only the deviations from the mean and in no way involves the mean; (2) they are not affected by the signs of the deviations since each deviation is squared; (3) they are affected by the scatter of the scores since the squares of large deviations will be large and the squares of small deviations will be small; and (4) they are more sensitive to the intrinsic nature of the deviations than to their total frequency since this is a fundamental characteristic of the mean of a set of measures.

The computation and use of these two measures of variability are illustrated in the next example.

Example Compute the mean, variance, and standard deviation for the data of Figs. 9-27 and 9-28.

Note that the total frequency and the mean of the two distributions are exactly the same. In Fig. 9-28, however, the scores are clustered more closely about the mean than are those of Fig. 9-27. This fact is reflected in the smaller variance and standard deviation of this distribution as compared with the same measures of the other, as shown in this summary.

	Figure 9-27	*Figure 9-28*
Mean	6	6
Variance	5.89	1.94
Standard deviation	2.43	1.39

SCORE	FREQUENCY	DEVIATION (SCORE − MEAN)	TOTAL DEVIATIONS (FREQ. × DEVIATION)	SQUARES OF DEVIATIONS
10	0	+4	0	0
9	0	+3	0	0
8	5	+2	+10	20
7	9	+1	+9	9
6	10	0	+19	
5	5	−1	−5	5
4	4	−2	−8	16
3	2	−3	−6	18
2	0	−4	0	0
1	0	−5	0	0
	35		−19	68

Mean score = 6

Variance = $\dfrac{68}{35}$ = 1.94

Standard deviation = $\sqrt{1.94}$ = 1.39

FIGURE 9-28 Computation of variance and standard deviation

When the mean and variance or standard deviation for a given distribution are known, we have two very effective measures for comparing the distribution with any other similar distributions.

EXERCISES

Compute the three measures of central tendency for the distributions in Exercises 1 to 3.

1 36, 33, 38, 48, 46, 35, 44, 37, 42, 34, 41, 47, 32, 45, 40, 39, 43
2 3, 4, 5, 5, 5, 9, 8, 7, 7, 4, 5, 5, 6, 7, 7, 7, 8, 3, 6, 5, 5, 7, 6, 8, 7, 5, 5, 7, 6, 5, 9

   Hint: First make a frequency distribution similar to that of Fig. 9-26.

3

Score	Frequency	Score	Frequency
110	2	96	17
108	1	94	8
106	4	92	4
104	4	90	5
102	4	88	2
100	6	86	3
98	8		

4 In each of the Exercises 1 to 3, show that the total of all negative deviations from the mean is the same as the total of all positive deviations.
5 Find the variance and standard deviation of each of the distributions of Exercises 1 to 3.
6 The mean of one set of IQ scores is 98 with a standard deviation of 6, and the mean of a second set is 110 with a standard deviation of 15. How do the two distributions compare?
7 If you are told that the mean of a set of scores is 10 and the variance is 0, what can you say about the scores?

9-17 REVIEW EXERCISES

1–22 Write a carefully constructed answer to each of the guideline questions of this chapter.
23 If evaporated milk were priced to sell at 2 cans for 39 cents, the price charged for 1 can would be 20 cents. Explain how the number 20 can be considered either as an exact number or as an approximate number.
24 Which use of 12 is exact and which is approximate in these sentences: (*a*) The stick is 12 in. long; (*b*) There are 12 in. in 1 ft. Explain the difference.
25 Explain the difference between each pair of measurements.
   (*a*) $2\frac{3}{4}$ ft $\pm \frac{1}{8}$ ft and $2\frac{3}{4}$ ft $\pm \frac{1}{16}$ ft
   (*b*) 8.6 in. and 8.60 in.

26 In each case which measurement is the more precise and which is the more accurate? Give reasons to support your answers.
   (*a*) 48 in. and 4 ft
   (*b*) 195 lb and 2 lb 3 oz

27 Which measures are direct and which are indirect?
   (*a*) The length of a room by use of a tape measure
   (*b*) The height of a cliff by trigonometric formula
   (*c*) The distance a person walks by means of a pedometer
   (*d*) A person's temperature by means of a thermometer

28 There are 1,650,763.73 wavelengths of the orange-red light of krypton 86. If 1 yd = 0.9144 m, how many such wavelengths are there in one yard? Compute your answer correct to the nearest hundredth of one wavelength.

29 Find these sums.

   (*a*) 8 hr 6 min 37 sec   (*b*) 8 gal 3 qt 1 pt
         7 hr 50 min 43 sec   7 gal 2 qt
        10 hr 27 min 20 sec   3 gal 2 qt 1 pt
         7 hr 45 min   4 gal   1 pt

   (*c*) 5 yd 2 ft 8 in.   (*d*) 8 m 9 dm 5 cm
             6 ft 7 in.   6 m 7 dm   4 mm
         1 yd   9 in.   4 m   6 cm 8 mm

30 Express these ratios as fractions in their lowest terms.
   (*a*) 3 ft 4 in. to 2 yd 2 ft
   (*b*) 5 dm 4 mm to 1 m 8 cm

31 Find these quotients.

   (*a*) 4 | 10 gal 3 qt   (*b*) 5 | 19 yd 2 ft 2 in.

32 Which figure expresses correctly the area of a rectangle whose length is 9.24 cm and whose width is 7.3 cm?

   (*a*) 67.452 sq cm   (*b*) 67.45 sq cm
   (*c*) 67.5 sq cm   (*d*) 67 sq cm

   Explain your answer.

33 What is the perimeter of a triangle whose sides are 8 ft $6\frac{1}{2}$ in., 7 ft $4\frac{3}{4}$ in., and 5 ft $3\frac{1}{8}$ in.? Explain the correct precision in your answer.

34 Use scientific notation to carry out this computation.

$$\frac{32,100 \times 0.004156}{876,420 \times 0.000000123}$$

35 What is the height of the Eiffel Tower if the length of its shadow is 738 ft when its angle of elevation is 53°7′?

36 To measure the width of a river a person used the measurements of the figure. Correct to the nearest foot, what was the width of the river?

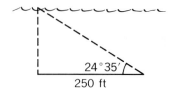

EXERCISE 36

37 The entrance to a bridge is 26 ft above the road level. The approach to the bridge makes an angle of 13°25′ with the horizontal. Correct to the nearest foot, what is the length of this approach?

*38 Prove these trigonometric identities for $0° \leq m \angle x \leq 90°$. Indicate any value for $m \angle x$ for which any identity is not defined.

(a) $\dfrac{1 + \sin x}{\cos x} + \dfrac{\cos x}{1 + \sin x} \equiv \dfrac{2}{\cos x}$

(b) $\sin^2 x + \cos^2 x(1 - \tan^2 x) \equiv \cos^2 x$

(c) $\dfrac{\sin x}{1 - \cos x} - \dfrac{1 - \cos x}{\sin x} \equiv 2 \cot x$

(d) $\dfrac{\tan x - \cot x}{\tan x + \cot x} \equiv 2 \sin^2 x - 1$

39 First construct a frequency distribution of these scores, and from this distribution, compute the mode, median, mean, variance, and standard deviation: 90, 87, 84, 79, 77, 89, 83, 77, 75, 86, 84, 87, 79, 78, 76, 75, 89, 84, 79, 77, 74, 74, 74, 73, 71, 70, 80, 78, 77, 75, 75, 73, 68, 72, 89, 83, 79, 78, 76, 69, 64, 62, 87, 84, 83, 78, 74, 69, 61, 90, 83, 77, 76, 74, 68, 87, 72, 84, 81, 80, 79, 76, 72, 69, 62, 84, 78, 77, 69, 80, 77, 65, 61, 83, 80, 78, 76, 73, 67.

*INVITATIONS TO EXTENDED STUDY

1 Make a catalog of those uses of number which give rise to approximate numbers rather than exact numbers.

2 Make a catalog of primitive units of measure along with their limitations.

3 Make a comparative study of the characteristic properties of measurement as practiced in selected leading countries of the world.

4 Make a study of the historical background of the various standard units of measure.

Prove these three trigonometric identities.

5 $\dfrac{\sin^3 A + \cos^3 A}{\sin A + \cos A} \equiv 1 - \sin A \cos A$

6   $\sin A(1 - \tan^2 A)\left(\dfrac{1}{\cos A - \sin A} + \dfrac{1}{\cos A + \sin A}\right) \equiv 2 \tan A$

7   $\dfrac{\sin A \cos A}{\sin^2 A - \cos^2 A} \equiv \dfrac{\tan A}{\tan^2 A - 1}$

8   Show that in any acute-angle triangle ABC

$$\frac{a}{\sin A} = \frac{b}{\sin B} = \frac{c}{\sin C}$$

where a, b, and c are the sides opposite the respective angles A, B, and C.

9   Extend the definitions of the trigonometric functions of angles to those of the second, third, and fourth quadrants.

10   Extend the definitions of the trigonometric functions of angles to negative angles.

11   Look up the use of the *radian*, *mil*, and *grad* as units of angle measure.

12   Investigate the use of *percentiles* and *percentile ranks* to describe the position of a term in a distribution.

13   Investigate the use of *quartiles* and *deciles* to describe the position of a term in a distribution.

14   Investigate the use of the concepts of *range* and *class interval* in the construction of a frequency distribution.

15   Investigate the use of the class interval and the *guessed mean* in computing the mean, variance, and standard deviation of a given distribution.

THE CONCEPTS OF
RELATION AND
FUNCTION

GUIDELINES FOR CAREFUL STUDY

In the early chapters of this book we were concerned with the concept of number and with sets of numbers. Operations were defined, and basic postulates became the laws to govern the use of these operations on certain specified sets of numbers. Within each system, in addition to other properties derivable from those postulated, certain fundamental relations were discovered to exist. Such relations as equal, greater than, and less than, each identifiable by certain intrinsic characteristics, were used to establish a correspondence which compared one number or a set of numbers with another number or set of numbers.

Similarly, in Chaps. 8 and 9, points and sets of points, identified as various types of geometric configurations, were compared by means of such relations as concurrency, collinearity, coplanarity, congruency, and similarity as well as longer than, shorter than, larger than, and smaller than. These relations were discovered to be of great significance in the contemplation of position, shape, and size.

Not only is the concept of relation fundamental to the intelligent study of the structure of number systems and geometric configurations, but it is indeed a cornerstone of all mathematical endeavor, whether elementary or advanced. It is the purpose of this chapter to examine this concept more closely and under cloak of greater generality. We shall be particularly interested in a specific type of relation that is called *function*, and in its tremendous importance in elementary mathematics.

The following questions will serve as helpful guidelines to careful study of Chap. 10:

1 What is meant by the union or intersection of two sets?
2 What is the law of trichotomy? In what different forms may it be stated?
3 What is meant by the absolute value of a number?
4 What are the properties of equality? Of inequality?
5 What is the number line?
6 What is a number field?
7 What is the meaning of each of these concepts: concurrency, collinearity, and coplanarity?
8 When are two geometric figures said to be congruent? Similar?
9 What is an ordered pair of elements?
10 What is a rectangular frame of reference?
11 What is a constant? A variable? A parameter?
12 What is a relation? A function?
13 What are the domain and range of a relation or function?
14 What is the meaning of the symbol $f(x)$?
15 What is the formula for the linear function, and why is it called the linear function?
16 What is meant by the slope of a straight line?

393

17   What are the different forms for the equation of a straight line?

18   What is meant by a zero of a function?

19   What is a system of equations? What is its solution set?

20   What is meant by a consistent system of equations?

21   Why is an inequality of the form $y > ax + b$ not a function?

22   What is meant by a strict inequality? A mixed inequality?

23   What is meant by direct variation? Inverse variation? Joint variation?

24   What is a constant of variation?

25   What is a statistical graph?

26   What are the different types of statistical graphs and what are some of their distinguishing characteristics?

27   What parts do intuition, induction, and deduction play in an intelligent approach to problem solving?

28   What different types of difficulty does one encounter in the effort to solve verbal problems?

INTRODUCTION

*Conjunction,
equivalence,
order*

The relations emphasized in the previous chapters are those which can be classified as relations of *conjunction, equivalence,* or *order. Union* and *intersection* relate the elements of two or more sets; *concurrency* relates lines, planes, or lines and planes which have a point in common; *collinearity* relates points, planes, or points and planes which have a line in common; and *coplanarity* relates points, lines, or points and lines which have a plane in common. *Equality, congruence,* and *similarity* have in common the three *characteristic properties of equivalence* (Fig. 10-1). The order (inequality) relations *greater than* ($>$) and *less than* ($<$) do not satisfy the reflexive and symmetric properties. They do, however, satisfy the transitive property, and, through this property, they serve to order the natural numbers and the integers as well as the numbers of either the field of rational numbers or the field of real numbers. They order the points on the number line, and, since positive real numbers serve as measures of size, these two relations also order geometric configurations when comparisons of size are made.

PROPERTIES OF EQUIVALENCE	ILLUSTRATION
1 *Reflexive:* $a = a$	1 5 dollars $=$ 5 dollars
2 *Symmetric:* If $a = b$, then $b = a$.	2 If 5 dollars $=$ 10 half-dollars, then 10 half-dollars $=$ 5 dollars.
3 *Transitive:* If $a = b$ and $b = c$, then $a = c$.	3 If 5 dollars $=$ 10 half-dollars and 10 half-dollars $=$ 500 pennies, then 5 dollars $=$ 500 pennies.

FIGURE 10-1

It now becomes desirable to examine the general concept of relation, which lends great power and facility to mathematical investigation. The previously discussed relations will be seen to be special cases of this more general concept. In actuality our principal concern will be with the concept of *function*, a particular type of relation, but one that has sufficient generality to be of great significance in the study of mathematics.

10-1 VARIABLE AND CONSTANT

The concepts of *variable* and *constant* are fundamental to any discussion of relation and function. In Chap. 3 the letter n was used to represent a natural number, that is, it was used to represent any element of the set of natural numbers $N = \{1,2,3,4, \ldots\}$. In Chap. 4 the letter z was used to represent any element of the set of integers

$$Z = \{0,\pm 1,\pm 2,\pm 3, \ldots\}$$

and in Chap. 6 the letter q was used to represent any number of the form $\frac{a}{b}$ where a and b are integers and $b \neq 0$, that is, any element of the set of all rational numbers. The letters n, z, and q thus were used as symbols to represent any element of a set of elements. They were, therefore, used as *variables* in accordance with this definition:

Variable and its domain
Definition 10-1 A *variable* is a symbol used to represent any arbitrary element of a given set containing two or more elements. The set is called the *domain* or *replacement set* of the variable.

Constant
If the replacement set for a symbol contains only one element, then the symbol is called a *constant*. The numeral 2 is a symbol which represents the quantitative characterization of each of the following sets: the eyes of a normal person, the hands of a normal person, the feet of a normal person, the half-dollars in a dollar, the nickels in a dime. Since the replacement set for this symbol contains only one element, it is a constant. The Greek letter π (pi) is a symbol used to represent any element of a set of ratios, each that of the circumference of any given circle to its respective diameter. Since it can be proved that these ratios are all equal, the set contains one and only one element and, therefore, the symbol is a constant.

There is still another use of symbols in this context which is of importance. It is, in fact, something of a hybrid of the concepts of constant and variable. Consider, for example, the familiar formula $d = rt$, which gives the distance traveled in any given amount of time at any specified rate. In this formula each of the symbols d, r, and t is a variable. The symbol d represents any element of a set of numbers used as measures of distance; r represents any element from a set of measures of rate or speed of travel; and t represents any element from a set of measures of time. Let us change the emphasis slightly and set the formula in the context to give the distance traveled at a *constant rate of speed* in a given time. In this interpretation we have not changed the character of representation for the symbols d and t—they still are variables with the same domains as specified previously—but we have changed the character of representation for r. It is now used to represent a constant rate of speed. It is a constant representing one fixed value. But what fixed value is it? The formula may be used to find the distance traveled in any given time at 40 miles per hour, 50 miles per hour, 186,000 miles per second, or any arbitrary speed one might choose to use. *Arbitrary constant (parameter)* Such a symbol is called an *arbitrary constant*, or a *parameter.* In other words, a parameter is a symbol which, by its arbitrary selection, characterizes an existing relationship between the variables of a given discussion. Attention will be called in subsequent pages to other specific occurrences of the use of parameters.

10-2　RELATION AND FUNCTION

In Sec. 4-2 attention was called to the fact that the definition of "positive integers" is such as to establish a one-to-one correspondence between the set of all positive integers and the set of all natural numbers. The definition thus sets up a relation that exists between these two sets such that to a given natural number there corresponds one and only one positive integer. Later it was established that Theorem 8-13 can be used to set up a relation of similarity between a triangle such as $\triangle ABC$ of Fig. 10-2 and any triangle which has two of its angles respectively congruent to two of the angles of $\triangle ABC$. This is a one-to-many correspondence, since there exist many distinct triangles which meet the conditions of the theorem. Triangles DEF, DLM, and DRS of Fig. 10-2 are three such triangles. Similarly, any one of the three postulates 8-13 to 8-15 can be used to establish a one-to-many correspondence of congruency with $\triangle ABC$.

The three relations described here, though quite distinct in nature, have one very basic characteristic in common, namely, each relation sets up a correspondence between an element of one set and one or more elements

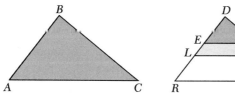

FIGURE 10-2　Similarity of triangles

n	p
1	$+1$
2	$+2$
3	$+3$
4	$+4$
5	$+5$
6	$+6$
7	$+7$

FIGURE 10-3 Selected values for $p = {}^{+}n$

of a second set. If, in the first example, we use the symbol p as a variable to represent any element of the set of all positive integers, then the relation may be symbolized by the set of pairs of elements $\{(n,p)\,|\,p = +n$ and n is a natural number$\}$. The relation is thus seen to be one of equality. Similarly, if t is used to represent $\triangle ABC$, each of the other two relations also may be represented by a set of pairs of elements. The relation of similarity may be expressed in the form $\{(t,s)\,|\,s \sim t$ and t is a triangle$\}$, and the relation of congruency in the form $\{(t,c)\,|\,c \cong t$ and t is a triangle$\}$.

Whether the relation was established by verbal description, as in the first paragraph of this section, by formula, or by set selector, as in the second paragraph, it is true in each case that two sets of elements are involved. This is the essential common characteristic that makes it possible to give a clear-cut definition of the concept of relation as used in mathematics. Before attempting the definition, however, it is desirable to clarify the language and symbolism used.

Ordered pair   Recall that a symbol of the form (x,y) is called an *ordered pair* simply because a pair of symbols x and y is used and one of them, x, is written first and the other, y, is written second. Thus (x,y) is a symbol for a set of ordered pairs in which x is the *first element* and y is the *second element*. For example, the symbol (n,p) of the preceding illustration represents the set of all ordered pairs of which the first element, n, is a natural number and the second element, p, is a positive integer. We shall list a few selected members of this set of ordered pairs:

$$\{(1,+1),\ (2,+2),\ (3,+3),\ (4,+4),\ \ldots\}$$

In this relation, to each first element there corresponds one and only one second element. This correspondence is also shown very explicitly by the table of Fig. 10-3, in which a few selected values are paired.

The relations of similarity and congruence of the previous paragraphs do not lend themselves very well to table representation. A simple illustration, which has their common characteristic of a one-to-many correspondence yet does submit to table representation, is that of the relation "is the square root of." This relation may be expressed as the ordered pair of real numbers (x,y) where x is any nonnegative real number, y represents the elements

x	y
0	0
1	± 1
2	$\pm \sqrt{2}$
4	± 2
$\dfrac{64}{9}$	$\pm \dfrac{8}{3}$
9	± 3

FIGURE 10-4   Selected values for $y = \pm \sqrt{x}$

of the set of all real numbers, and the formula defining the relation between x and y is $y = \pm \sqrt{x}$. A few selected elements from this set of ordered pairs are $(0,0)$, $(1,+1)$, $(1,-1)$, $(2,+\sqrt{2})$, $(2,-\sqrt{2})$, $(\frac{64}{9},+\frac{8}{3})$, $(\frac{64}{9},-\frac{8}{3})$, $(9,+3)$, $(9,-3)$. Here again this correspondence is expressed very clearly in the table of selected values of Fig. 10-4. The symbol $(x,\pm\sqrt{x})$ might be used to represent the same set of ordered pairs.

Whether the correspondence is one-to-one or one-to-many, a relation is clearly and explicitly characterized by a set of ordered pairs.

Relation, its domain and range   **Definition 10-2**   A *relation* is a set of ordered pairs in which there is associated with each first element at least one second element. The set of all possible first elements is called the *domain* of the relation, and the set of all possible second elements is called the *range*.

Dependent and independent variables   In the symbol (x,y) representing a set of ordered pairs, the variable x representing elements from the domain of the relation is called the *independent variable;* the variable y representing elements from the range is called the *dependent variable*. Thus in the relation (x,y) described by the formula $y = \pm\sqrt{x}$, the independent variable x represents any element to be selected from the domain of the relation, the set of all nonnegative real numbers; and the dependent variable y represents any element from the range of the relation, the set of all real numbers. Similarly, in the relation (n,p) described by the formula $p = +n$, the independent variable n selects elements from the domain of the relation, the set of all natural numbers; and the dependent variable p selects elements from the range of the relation, the set of all positive integers.

Each of the relations represented by the respective symbols (n,p) and (x,y) has been depicted by verbal description, set selector, formula, ordered pairs, and a table of values. Another very helpful technique for portraying the distinctive characteristics of any particular relation is the graph. Figure 10-5 presents the graph of each of the relations (n,p) and (x,y) of the present discussion. Each graph is an incomplete graph, since only a few of the related pairs are pictured. Since the domain and range of each relation are both infinite sets of points, it is impossible to draw a complete graph in either case. The graph of $p = +n$ (Fig. 10-5a) consists only of a set of discrete points. This is due to the fact that each variable can be replaced only by integral values. No fractions or irrational numbers can be used. The graph of $y = \pm\sqrt{x}$ (Fig. 10-5b) is a curve with no breaks in it. There are points whose coordinates involve fractions, such as $C(\frac{64}{9},\frac{8}{3})$ and $D(\frac{64}{9},-\frac{8}{3})$, and irrational numbers, such as $A(2,\sqrt{2})$ and $B(2,-\sqrt{2})$, as well as the points whose coordinates are ordered pairs of integers, $(0,0)$, $(1,1)$, $(1,-1)$, $(4,2)$, $(4,-2)$, $(9,3)$, and $(9,-3)$. It can be proved that for any arbitrarily selected positive real number, there exists in the field of real numbers two numbers, equal numerically but opposite in sign, each of which when squared will give the arbitrarily selected number. This is illustrated graphically in Fig. 10-5b by the fact that if, at any arbitrarily selected point on the positive x axis, a line is drawn perpendicular to the x axis, it will cut the graph of the curve $y = \pm\sqrt{x}$ in two points, each of which represents a square root of the real number identified with the point.† In Fig. 10-5b, let $P(5,0)$ represent the arbitrary point on the positive x axis; then the two points on the curve determined by the line perpendicular to $\overrightarrow{OX}$ at P are L, which is the point $(5,\sqrt{5})$, and M, which is the point $(5,-\sqrt{5})$. The line segments $\overline{PL}$ and $\overline{PM}$ are congruent segments whose common length is $\sqrt{5}$ units. Since this construction is possible whatever position the point P takes on the positive x axis, the graph of the relation $y = \pm\sqrt{x}$ has no

Continuous breaks in it and is thus said to be a *continuous curve*. This particular curve
curve   is called a *parabola*.

† Where $x = 0$ the two points are coincident.

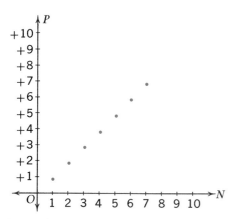

Incomplete graph of the relation $p = +n$ whose domain is the set of all natural numbers and range is the set of all positive integers

(a)

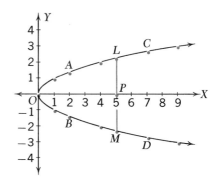

Incomplete graph of the relation $y = \pm\sqrt{x}$ whose domain is the set of all nonnegative real numbers and range is the set of all real numbers

(b)

FIGURE 10-5

Up to this point in our discussion the similarities of the two relations (n,p), where $p = +n$, and (x,y), where $y = \pm\sqrt{x}$, have been emphasized. They both can be depicted by verbal description, formula, ordered pairs, table of values, and graph. Second, both are characterized by a clearly delineated domain and range from which values are selected by an independent variable and a dependent variable, respectively. Third, both can be represented by an ordered pair of variables, one of which, for convenience, may be called the first element and the other the second element. In spite of all these very important characteristics of similarity, the two relations possess one characteristic of dissimilarity which is of great significance. This distinguishing characteristic is most vividly portrayed by the formula, the table of values, or the graph. From the graphs of Fig. 10-5 it should be quite evident that the relation (x,y), where $y = \pm\sqrt{x}$, is such that to each nonzero value of the first element, or independent variable, there correspond *more than one* value for the second element, or dependent variable. On the other hand, the relation (n,p), where $p = +n$, is such that to each value of the first element there corresponds *one and only one* value of the second element. This is the important element of dissimilarity between the two relations. All relations which have in common with (n,p) this important characteristic are called *functions* in accordance with this definition:

Function, its domain and range **Definition 10-3** A *function* is a relation such that to each first element there corresponds one and only one value of the second element. The set of all possible first elements is called the *domain* of the function, and the set of all possible second elements the *range*.

To put it another way and further clarify this definition, we can describe a function as a set of ordered pairs for which the formula or rule defining the relation is such that, when a value is assigned to the first element of the ordered pair, there is absolutely no doubt as to what the value is that should be assigned to the second element.

Example Consider these four sets of ordered pairs. Which represent functions?

1 {(1,2), (1,3), (1,4), (2,5), (2,6), (3,7)}
2 {(3,1), (4,1), (5,2), (6,2)}
3 {(1,3), (2,3), (3,3), (4,3), (5,3)}
4 {(1,2), (2,5), (3,4), (4,6), (5,7)}

Set 1 of ordered pairs represents a relation which is not a function, since to the first element 1 there correspond three values of the second element, namely, 2, 3, and 4. To the first element 2 there correspond the two values 5 and 6.

Each of the three remaining sets of ordered pairs is a function since to each first element in each set there corresponds one and only one second element. This means that once a value is selected for a first element, there is absolutely no ambiguity as to what value of the second element is to be associated with it. It is immaterial that in illustrations 2 and 3, the same value of the second element is associated with more than one value of the first element. This is not involved in the criterion which singles out and identifies a relation as being a function. It bears repeating that the criterion of distinction is that to each value of the first element selected from the domain of the relation, there corresponds one and only one value for the second element, selected from the range.

The complete graph of each of these relations, as shown in Fig. 10-6, supports the statement made about each particular relation in the preceding paragraph.

Because of its tremendous importance in mathematical thought, mathematicians have devised special symbols to represent a function. The fact that there are different symbols is not disturbing, since they connote the same basic idea. The differences are merely due to a desire to underscore clarification for a particular context being studied or investigated. The

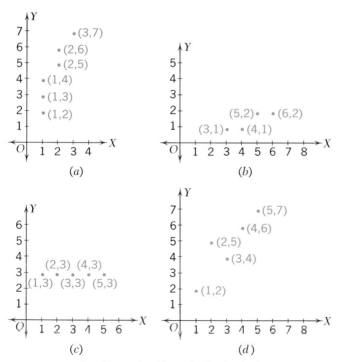

FIGURE 10-6 Complete graphs of four sets of ordered pairs

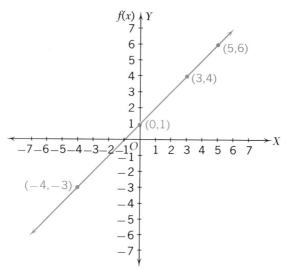

FIGURE 10-7 Incomplete graph of $f(x) = x + 1$

simplest and by far the most frequently used symbol in elementary mathematics is $f(x)$, which is read "function of x" or "function at x." The second of these two translations is possibly the more accurate of the two.

Example The relation "increased by one" is a function which may be represented by the symbol $\{[x, f(x)] \mid f(x) = x + 1\}$. If the domain is the set of all real numbers, the range is also the set of all real numbers. The incomplete graph of the function is that shown in Fig. 10-7. The symbol for the value of $f(x)$ at 0 is $f(0)$, and it is found by replacing x by 0 in the formula $f(x) = x + 1$, whence $f(0) = 1$. Similarly, the value of the function at $x = 3$ is 4, $f(3) = 3 + 1 = 4$; $f(5) = 5 + 1 = 6$; $f(-4) = -4 + 1 = -3$; continuing in this fashion, the value of the function at $x = a$ is $f(a) = a + 1$.

The graph of $f(x) = x + 1$ is known to be a straight line. It is continuous since there are no breaks in it. It can be obtained by constructing a table of values such as those of Figs. 10-3 and 10-4, plotting the points determined by the associated values, and then drawing a smooth curve through these points. However, Postulate 8-1 provides us with a much more refined technique for getting an accurate picture of the graph. Since "for any two distinct points in space there exists one and only one straight line which contains these two points," it follows that if we plot any two points, such as (0,1) and (3,4), which are determined by the function, the straight line which is the graph of the function may be drawn by fitting a straightedge to these two points and then drawing the line along its edge.

EXERCISES

1 Verify the statement that equality, congruency, and similarity are equivalence relations.

2 Why are concurrency, collinearity, and coplanarity not equivalence relations?

3 Distinguish between the concepts of constant, variable, and parameter.

4 Distinguish between the concepts of relation and function.

5 What is meant by the domain and range of a relation or function?

6 Which of the following are equivalence relations? Give reasons to support your answer in each case.

    (*a*) "Is less than or equal to" applied to numbers

    (*b*) "Is as tall as" applied to people

    (*c*) "Is shorter than" applied to distances

    (*d*) "Is an ancestor of" applied to people

    (*e*) "Is the square of" applied to numbers

    (*f*) "Has the same area as" applied to geometric figures

    (*g*) "Was born in the same town as" applied to people

    (*h*) "Is a factor of" applied to positive integers

    (*i*) "Is the supplement of" applied to plane angles

    (*j*) "Has the same shape as" applied to geometric figures

    (*k*) "Is perpendicular to" applied to straight lines in a plane

    (*l*) "Is not equal to" applied to natural numbers

    (*m*) "Has the same length as" applied to straight line segments

7 Which of these sets are functions?

    (*a*) $\{(x,2x) \mid x$ is a positive integer$\}$

    (*b*) $\{(x, x + 3) \mid x$ is a positive integer$\}$

    (*c*) $\{(x, \sqrt{2x^2}) \mid x$ is a real number$\}$

    (*d*) $\{(x, \pm \sqrt{2x}) \mid x$ is a positive real number$\}$

    (*e*) $\{(x, 2x + 5) \mid x$ is an integer$\}$

    (*f*) $\{(x, 2x \pm 3) \mid x$ is a real number$\}$

    (*g*) $\left\{ \left(x, \dfrac{1}{x}\right) \middle| x$ is a negative real number$\right\}$

8 What are the domain and range of each relation of Exercise 7?

9 Complete the expression $f(x) = $ ＿＿ for each function of Exercise 7.

10 For each relation of Exercise 7, list 10 ordered pairs which are members of the set describing the relation.

11 For each relation of Exercise 7, construct a table of values associated with it.

12 Draw an incomplete graph to represent each of the relations of Exercise 7*a* to *c*.

13 Explain how the symbol k in each of these formulas may be considered to be a parameter:

    (*a*) $y = x + k$ (*b*) $y = kx$ (*c*) $y = \dfrac{k}{x}$

*14 Write a formula representing a function and illustrating the concepts of constant, parameter, dependent variable, and independent variable.

*15 Given the set $S = \{0,1,2,3,4,5,6,7,8,9\}$. In each of the exercises select a dependent variable and an independent variable, then write the formula which expresses the relation involving the stated operation on each element of the set S as the domain of the relation:

(a) Multiplied by 3
(b) Divided by 5
(c) Increased by 4
(d) Decreased by 1
(e) Multiplied by $\frac{1}{2}$
(f) Multiplied by 2 and this product decreased by 1
(g) Decreased by 1 and this difference multiplied by 2
(h) Divided by 3 and this quotient increased by 4
(i) Increased by 4 and this sum divided by 3

*16 Write the set of ordered pairs for each relation of Exercise 15.

*17 Are any of the relations of Exercise 15 also functions? Why?

*18 Construct tables of paired values which represent each relation of Exercise 15.

*19 How can the sets of ordered pairs or the table of values help in deciding whether a given relation is also a function?

*20 Use the tables of values of Exercise 18 to draw the graph of each relation of Exercise 15.

*21 Are the graphs of Exercise 20 complete graphs or incomplete graphs?

22 How can the graph of a relation help in deciding whether a given relation is also a function?

23 Which of the graphs on page 405 are graphs of functions? Explain your decision for each relation.

24 Select a set of ordered pairs which belong to the graph of each relation of Exercise 23.

25 If each graph of Exercise 23 is considered as a complete graph, what are the domain and range of each relation? The presence of dots at the ends of line segments or curves indicates whether or not the point is on the graph.

26 Construct a table of values associated with the graph of each relation of Exercise 23.

27 Write a formula which expresses each of these relations:

(a) The distance an automobile will travel in a given time if it travels at an average speed of 40 miles per hour
(b) The simple interest at 6% for one year on a given principal
(c) A salesman's monthly salary if he is paid $250 a month plus a 5% commission on all sales for the month.
(d) The cost of gasoline at 30 cents per gallon
(e) The circumference of a circle is 2π times the radius
(f) The area of a square in terms of its side
(g) The perimeter of a regular hexagon in terms of its side

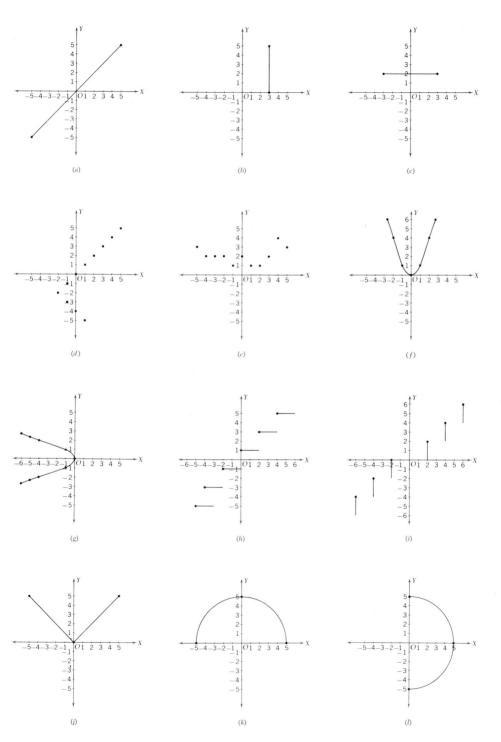

(a)

(b)

(c)

(d)

(e)

(f)

(g)

(h)

(i)

(j)

(k)

(l)

EXERCISE 23

28   Which, if any, of the relations of Exercise 27 are also functions?

29   Select a set of ordered pairs related to each relation of Exercise 27.

30   Construct a table of values for each relation of Exercise 27.

31   Indicate the dependent and independent variable for each relation of Exercise 27.

32   Use the table of values of Exercise 30 as an aid in drawing an incomplete graph of each relation of Exercise 27.

10-3   THE LINEAR FUNCTION

The function
$y = mx + k$

An important function in elementary mathematics is the *linear function*, which is defined by a formula of the form

$$y = mx + k \qquad (m,k \text{ real numbers, } m \neq 0) \qquad (10\text{-}1)$$

or $$f(x) = mx + k \qquad (m,k \text{ real numbers, } m \neq 0) \qquad (10\text{-}2)$$

In either of these two forms x is the independent variable and y, from Eq. (10-1), or $f(x)$, from Eq. (10-2), is the dependent variable. The symbols m and k are parameters, or arbitrary constants. It can be proved that the graph of such a function is always a straight line. We shall accept this as a true statement and proceed to investigate the role that x and y, as well as m and k, play in shaping this graph.

First, let us examine the part which x and y play by observing that for any given linear function the symbol $\{(x,y)\,|\,x \text{ and } y \text{ are real numbers}\}$ will represent the set of ordered pairs defining the function. These ordered pairs are coordinates of points of the line which is the graph of the function. In Fig. 10-8 the ordered pairs $(-3,-6)$, $(-1,-2)$, $(0,0)$, $(2,4)$, and $(4,8)$, selected by the linear function $y = 2x$, are seen to be coordinates of points all of which lie on the line determined by any two of them. If each of the several different forms of the linear functions shown as labels of lines in Figs. 10-8, 10-10, and 10-11 are tested in the same manner, it will be found that each function will select ordered pairs (x,y) as coordinates of points of the line associated with it. This statement should be verified by the reader. Thus we say that in any linear function, the x and y represent the x coordinate and y coordinate, respectively, of any point of the straight line which is the graph of the particular function. A particular point of interest on the graph of a linear function is the point where it crosses the x axis.

Zero of the function

Since the y coordinate of this point is 0, the x coordinate is called the *zero of the function*, or the x intercept of the graph.

Now, let us take a look at the parts m and k play in the formula. In Fig. 10-8 there are three lines, one representing the line $y = 2x$, another the line $y = 2x + 6$, and the third the line $y = 2x - 4$. Notice that in each of these functions, the value of m is 2, while there is a different value for k in each equation. These lines are parallel lines. For the line $\overleftrightarrow{ON}$, which

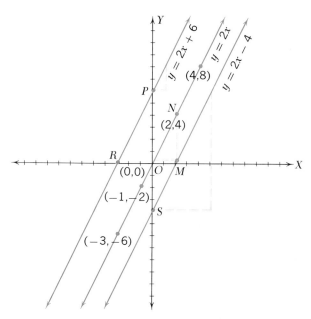

FIGURE 10-8

is the graph of the function $y = 2x$, count the number of units of length in the segment $\overline{OM}$ ($OM = 2$); now count the number of units of length in $\overline{MN}$ ($MN = 4$). The ratio $\dfrac{MN}{OM} = \dfrac{4}{2} = 2$. Repeat this experiment by starting at the points P and R on the line $\overleftrightarrow{RP}$, the graph of $y = 2x + 6$; then use the points S and M on the line $\overleftrightarrow{SM}$, the graph of $y = 2x - 4$. In each case the ratio which corresponds to MN/OM should be 2. This ratio is a very important characteristic used in describing straight lines in *Slope* the plane. It is called the *slope* of the line. In order to get a clearer picture of just what is meant by the slope of a straight line, we shall examine the line $\overleftrightarrow{ON}$ a bit more closely. This line can be described as the set of all ordered pairs (x, y) such that $y = 2x$ and x is a real number. This states that the domain of the given function is the set of all real numbers of the x axis and the range is the set of all real numbers of the y axis. From this fact it should be clear that, as the formula selects pairs of real numbers as coordinates of points of the line, for each change of one unit in the horizontal or x coordinate there are two units of change in the vertical or y coordinate. In other words, the ratio of the change in the y coordinate to the change in the x coordinate is constant and equal to 2. This describes what takes place in selecting coordinates of the points of any one of the three lines of Fig. 10-8 or any line parallel to them, as was verified by the experiment described previously in this paragraph.

It is well to raise at this point the question of how one can proceed to

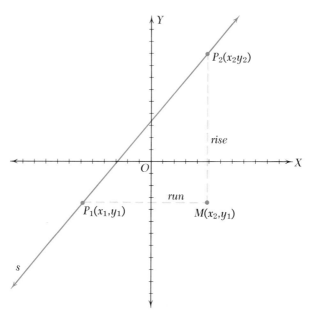

FIGURE 10-9 Slope of a line

find the slope of an arbitrary line. Consider the line s of Fig. 10-9, and let the points P_1 (read "P-one") and P_2 be any two arbitrarily selected points of the line. It is convenient to associate with P_1 and P_2, respectively, ordered pairs of coordinates (x_1, y_1) and (x_2, y_2). The line segment $\overline{P_1 M}$ is drawn parallel to the x axis, and the line segment $\overline{MP_2}$ is parallel to the y axis. It thus follows that the coordinates of M are (x_2, y_1). The change in the x coordinate in going from P_1 to P_2 is represented in the figure by the horizontal line segment $\overline{P_1 M}$, which is the *run* of the line segment $\overline{P_1 P_2}$ and may be expressed as $x_2 - x_1$. The change in the y coordinate is represented by the vertical line segment $\overline{MP_2}$, called the *rise* of the line segment $\overline{P_1 P_2}$. This change in the y coordinate, or the rise, may be expressed as $y_2 - y_1$. The slope of the line is then given by

$$m = \frac{\text{change in } y \text{ coordinate}}{\text{change in } x \text{ coordinate}} = \frac{\text{rise}}{\text{run}}$$

or $$m = \frac{y_2 - y_1}{x_2 - x_1} \qquad (10\text{-}3)$$

Since P_1 and P_2 were any two points of the line s, this formula gives the value of the slope of the line. In selecting two arbitrary points of a line in order to use Eq. (10-3), it is immaterial which is called P_1 and which is called P_2; the value of m will remain the same. The only thing to remember is this: after a y coordinate is selected for y_2 in the numerator of the formula, the x coordinate associated with it must be the x_2 of the denominator.

Example Use Eq. (10-3) to determine that 2 is the slope of line $\overleftrightarrow{ON}$ of Fig. 10-8.

The points (4,8) and (2,4) are two points of the line:

$$m = \frac{8 - 4}{4 - 2} = \frac{4}{2} = 2 \qquad \text{or} \qquad m = \frac{4 - 8}{2 - 4} = \frac{-4}{-2} = 2$$

Similarly, the points (4,8) and (−3,−6) are two points of the line:

$$m = \frac{8 - (-6)}{4 - (-3)} = \frac{8 + 6}{4 + 3} = \frac{14}{7} = 2$$

or

$$m = \frac{(-6) - 8}{(-3) - 4} = \frac{(-6) + (-8)}{(-3) + (-4)} = \frac{-14}{-7} = 2$$

The words "run" and "rise" are convenient words to use in discussing the slopes of lines. If we always think of passing from one point to another along a line segment so that the run is positive, then, if the segment rises as in Figs. 10-8 and 10-9, the rise is also positive. If the segment falls, as in Fig. 10-10, then the rise is negative.

Example Use Eq. (10-3) to find the slope of each line of Fig. 10-10.

1 The points (−2,8) and (4,−4) are two points of the line $y = -2x + 4$.

$$m = \frac{8 - (-4)}{-2 - 4} = \frac{12}{-6} = -2$$

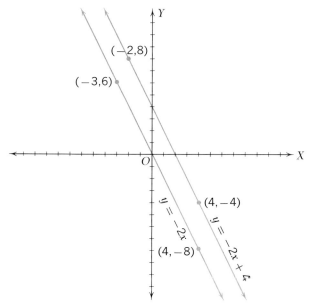

FIGURE 10-10

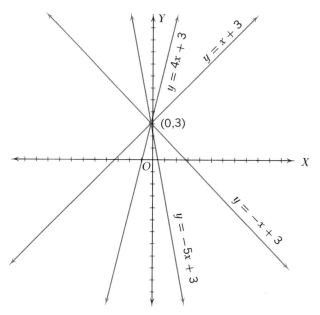

FIGURE 10-11

2   The points $(-3,6)$ and $(4,-8)$ are two points of the line $y = -2x$.

$$m = \frac{-8 - 6}{4 - (-3)} = \frac{-14}{7} = -2$$

If, in the linear function

$$y = mx + 3 \tag{10-4}$$

we assign x the value 0, then $y = 3$. This means that regardless of what value m has in Equation (10-4), the graph of the function will be a line passing through the point (0,3). This is indicated in Fig. 10-11, where four lines are shown passing through the point (0,3): one with $m = 4$, one with $m = 1$, one with $m = -1$, and one with $m = -5$. The point (0,3) where these lines all cross the y axis is at a distance of 3 units along the y axis from the origin. The 3 of the equation thus indicates the length of that portion of the y axis *intercepted* between the point (0,3) and the origin. *y intercept* For this reason it is said to be the y *intercept* of the line which is the graph of the function. For example, the line of Fig. 10-11, which is the graph of the function $y = 4x + 3$, is a line whose slope is 4 and whose y intercept is 3 units since it passes through the point (0,3). The parameters m and k of the linear function represent the slope and y intercept, respectively, of the line which is the graph of the function. Indeed, the formula for the *Slope-intercept* linear function is at times called *the slope-intercept form of the equation* *equation* *of a straight line,* since it is an equation whose graph is a straight line with slope m and whose y intercept is k units.

The lines of Fig. 10-11 also serve to illustrate a very important characteristic of the slope m of a line. When two lines with unequal slopes are compared, the steeper line is the one for which m has the larger numerical value, or absolute value (see Definition 7-1).

Example In Fig. 10-11 the lines which are the respective graphs of the functions $y = x + 3$, $y = 4x + 3$, and $y = -5x + 3$ are lines which have unequal slopes. The slope of $y = 4x + 3$ is 4 and that of $y = x + 3$ is 1. Since $|4| > |1|$, the first of these two lines is the steeper line. Also since $|-5| > |4|$ and $|-5| > |1|$, it follows that the line which is the graph of $y = -5x + 3$ is the steepest line of the three.

On the other hand, the lines $y = x + 3$ and $y = -x + 3$ have unequal slopes—one being the negative of the other—but are of the same degree of steepness. This is because their slopes have the same numerical value since $|-1| = |1| = 1$. As a result of the fact that their slopes are opposite in sign, the two lines are sloped in different directions.

While the graph of a linear function is always a straight line, it must not be inferred that the equation associated with a straight line is necessarily in the form of a linear function. For example, in Fig. 10-12 the line l through the point (0,4) and parallel to the x axis is such that 4 is the unique value of the second element y to be associated with any arbitrarily chosen value of the first element x in the ordered pair (x,y) which represents any one of the ordered pairs serving as coordinates of points of the line. In fact, the symbol

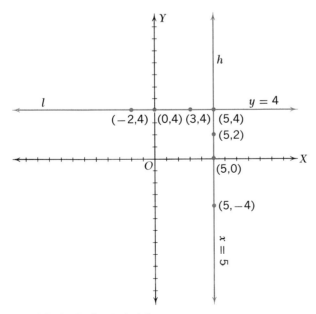

FIGURE 10-12 Constant relations

for the set of ordered pairs of this line can be written in better form as $\{(x,4)\,|\,x$ is a real number$\}$. As indicated in the figure, the ordered pairs $(-2,4)$, $(0,4)$, $(3,4)$, and $(5,4)$ are coordinates of a few selected points of this line. It is obvious that in changing from one point to another along this line, the vertical rise is always zero. To change from the point $(0,4)$ to the point $(3,4)$, for example, the rise is 0 and the run is 3. The slope of the line is thus given by $m = \frac{0}{3} = 0$. Since y is 4 for any x, we say that the equation of this line is $y = 4$. Also, since for each value of x there is one and only one value of y, we say that the line l of Fig. 10-12 is the graph of the constant function $y = 4$. Similarly, the formula $y = k$, where k is

The constant function   a constant, represents the *constant function* $\{(x,k)\,|\,k$ is a constant, the domain of x is the set of all real numbers$\}$. Although the graph of any constant function will be a straight line parallel to the x axis, the function cannot be considered as a special case of a linear function since its slope $m = 0$, while for a linear function $m \neq 0$.

The straight line h of Fig. 10-12 also demands particular attention. Since this line is parallel to the y axis, the values of x are the same for all points on the line. The formula $x = 5$ and the set of ordered pairs $\{(5,y)\,|\,y$ is a real number$\}$ may be used to associate real values of y with 5 in such a way as to get points on the line h. It follows then that in this case a relation, and *not* a function, is defined. Furthermore, an examination of the figure should reveal the fact that in changing from one point to another along this line, the run will always be zero. Since division by zero is undefined, we say that the slope of the graph of the relation $x = 5$ is undefined. Similarly,

The relation   the formula $x = c$, or the set of ordered pairs $\{(c,y)\,|\,c$ is a constant, the
$x = c$ domain of y is the set of all real numbers$\}$ defines a relation whose graph is a line parallel to the y axis at a distance of c units from it and whose slope is undefined.

10-4   THE LINEAR EQUATION

The equation

$$ax + by + c = 0 \tag{10-5}$$

Equation of a straight line   is the equation of a straight line if both a and b are not zero and the set of all real numbers is the domain of each variable x and y. If both a and b were zero, the equation would take the form

$$0 \cdot x + 0 \cdot y + c = 0$$

from which it would follow that $c = 0$. This would mean that the coordinates of any point in the plane would satisfy the equation, and it would lose its significance. There are three cases to consider:

Case I $a \neq 0$ and $b = 0$ Under this hypothesis Eq. (10-5) becomes

$ax + c = 0$, or $x = -(c/a)$. This is the equation of a line parallel to the y axis at a distance of $-(c/a)$ units from it. A particular example of this case is represented in Fig. 10-12 by the line h whose equation is $x = 5$. If, in Eq. (10-5), we place $a = 1$, $b = 0$, and $c = -5$, it becomes the equation $x - 5 = 0$, or $x = 5$, in which $-(c/a) = 5$.

Case II $a = 0$ and $b \neq 0$ Under this hypothesis Eq. (10-5) becomes $by + c = 0$, or $y = -(c/b)$. This is the equation of a line parallel to the x axis at a distance of $-(c/b)$ units from it. A particular example of this case is represented in Fig. 10-12 by the line l whose equation is $y = 4$. If, in Eq. (10-5), we place $a = 0$, $b = 1$, and $c = -4$, it becomes the equation $y - 4 = 0$, or $y = 4$, in which $-(c/b) = 4$.

Case III $a \neq 0$ and $b \neq 0$ Under this hypothesis, as in Case II, we may solve the equation for y, this time obtaining

$$y = -\frac{a}{b}x - \frac{c}{b}$$

This is in the form of the linear function where the slope $m = -(a/b)$ and the y intercept $k = -(c/b)$. Particular examples of this case are the lines of Figs. 10-8 to 10-11. For example, for $a = 2$, $b = 1$, and $c = -4$, Eq. (10-5) becomes $2x + y - 4 = 0$ or $y = -2x + 4$, whose graph is shown in Fig. 10-10; for $a = 2$, $b = -1$, and $c = 0$, Eq. (10-5) becomes $2x - y = 0$ or $y = 2x$. Its graph is shown in Fig. 10-8.

If in Eq. (10-3) we consider the ordered pair (x_1, y_1) as the coordinates of a fixed point of a line of slope m and use the ordered pair (x, y) to represent any other point of the line, we obtain this form of the equation:

$$m = \frac{y - y_1}{x - x_1}$$

or, in better form,

$$y - y_1 = m(x - x_1) \tag{10-6}$$

This represents the equation of a straight line with given slope m and passing *Point-slope* through the point (x_1, y_1). It is called *the point-slope form of the equation* *equation* *of a straight line.*

From Postulate 8-1 we know that for any two points in space, there exists one and only one line containing both of them. The equation of the line through the two points can be obtained by using Eq. (10-3) to determine the slope of the line and then using Eq. (10-6) with either point as the fixed point to determine the equation.

Example   Find the equation of the line. (1) with slope of $-\frac{1}{2}$ and passing through the point $(2, -3)$; (2) passing through the two points $(5,6)$ and $(8, -1)$.

1   Since $m = -\frac{1}{2}$ and the given point is $(2, -3)$, we have upon substitution in Eq. (10-6)

$$y - (-3) = -\tfrac{1}{2}(x - 2)$$

or
$$y + 3 = -\tfrac{1}{2}x + 1$$

whence
$$x + 2y + 4 = 0$$

2   From Eq. (10-3),

$$m = \frac{6 - (-1)}{5 - 8} = \frac{6 + 1}{-3} = -\frac{7}{3}$$

From Eq. (10-6), using $(5,6)$ as the fixed point and $-\frac{7}{3}$ as the slope, we get

$$y - 6 = -\tfrac{7}{3}(x - 5)$$

or
$$7x + 3y - 53 = 0$$

From Eq. (10-6), using $(8, -1)$ as the fixed point, we get

$$y - (-1) = -\tfrac{7}{3}(x - 8)$$
$$y + 1 = -\tfrac{7}{3}(x - 8)$$
$$7x + 3y - 53 = 0$$

Example   Given the equation $2x + 3y - 12 = 0$. (1) What are the slope and the y intercept of the line which is the graph of this equation? (2) What is the x intercept of the line?

1   To find the slope and y intercept, change the equation into the form of the linear function. This can be done by solving for y to get

$$y = -\tfrac{2}{3}x + 4$$

This line has a y intercept of 4 units and a slope of $-\frac{2}{3}$. It goes through the point $(0,4)$.

2   To find the x intercept of a line, it is necessary to find where it crosses the x axis. This can be accomplished by substituting $y = 0$ in the equation and then solving for x. In this example, $x = 6$. This line passes through the point $(6,0)$, and its x intercept is 6 units.

Example   The relation that exists between the set of odd integers and the set of all integers is a linear function. What is the formula for this function if the odd integer 1 corresponds to the integer 0 and the odd integer -9 corresponds to the integer -5?

If we use the symbol o to represent an odd integer and i to represent any integer, we may write the linear function which relates o to i in the form

$$o = mi + k$$

We are given that o is 1 when i is 0. From the equation we then have

$$1 = m0 + k$$

from which we get $k = 1$.

The relation between o and i may be written as

$$o = mi + 1$$

Since $o = -9$ when $i = -5$, we have

$$-9 = m(-5) + 1$$
$$-9 = -5m + 1$$
$$5m = 10$$
$$m = 2$$

Therefore the linear function which relates the odd integers to all integers such that 1 corresponds to 0, and -9 to -5, is

$$o = 2i + 1$$

The set of ordered pairs of real numbers (x, y) which represent the coordinates of points of the graph of a linear equation in the two variables *Solution set of* x and y is called the *solution set* of the equation.
an equation

The technique of finding ordered pairs which belong to the solution set *Equivalent* of a given equation makes use of *equivalent equations*, which are equations *equations* that have the same solution set. For any given equation the coefficients will be real numbers, and for each variable the domain will be the set of all real numbers. Since this is true, we may use all the properties of real numbers in manipulating equations. This means that, in order to derive an equation equivalent to a given equation, we may:

1   Add the same number to both members of an equation or subtract the same number from both members.

2   Multiply both members of an equation by the same nonzero number or divide both members by the same nonzero number.

These procedures have been used previously in this section and will be used throughout the chapter to derive equivalent equations.

Example

1   The equation $5x - 15 = 0$ is equivalent to the equation $x = 3$, since their common solution set is $\{3\}$. The derivation of the second equation from the first follows these steps:

$$5x - 15 = 0$$
$$(5x - 15) + 15 = 0 + 15$$
$$5x + [(-15) + 15] = 15$$
$$5x = 15$$
$$\tfrac{1}{5}(5x) = \tfrac{1}{5} \cdot 15$$
$$(\tfrac{1}{5} \cdot 5)x = 3$$
$$1 \cdot x = 3$$
$$x = 3$$

The reader should identify which properties from the field of real numbers are used.

2 In a previous example the equation $y = -\frac{2}{3}x + 4$ was used as an equation equivalent to $2x + 3y - 12 = 0$.

$$2x + 3y - 12 = 0$$
$$(2x + 3y - 12) + (-2x + 12) = 0 + (-2x + 12)$$
$$3y + [(2x - 12) + (-2x + 12)] = -2x + 12$$
$$3y = -2x + 12$$
$$\tfrac{1}{3}(3y) = \tfrac{1}{3}(-2x + 12)$$
$$y = -\tfrac{2}{3}x + 4$$

Again the reader should identify which properties from the field of real numbers are used.

EXERCISES

1 Write the formula for the linear function whose graph is described by each set of data.
   (a) Has slope 3 and y intercept of 2
   (b) Has slope $-\frac{1}{2}$ and passes through the point (0,6)
   (c) Has slope $\frac{3}{2}$ and passes through the origin
   (d) Has slope -5 and y intercept -3
   (e) Has slope 1 and passes through the point $(-1,2)$
   (f) Has slope -3 and x intercept of 2

2 Describe a technique other than that used in the example on page 414 for finding the y intercept of a given linear function.

3 Find the zero of each linear function of Exercise 1.

* 4 Show that the slope-intercept form of the equation of a straight line can be obtained as a special case of the point-slope form of the equation.

5 Find the equation of the straight line whose graph is described by each set of data.
   (a) Has slope $-\frac{2}{3}$ and passes through the point (0,1)
   (b) Has slope 4 and passes through the point (2,3)
   (c) Has slope $-\frac{5}{4}$ and passes through the point $(-4,-2)$
   (d) Passes through the two points (1,2) and (4,5)
   (e) Passes through the two points $(-2,4)$ and $(3,-1)$
   (f) Has x intercept 5 and y intercept 2
   (g) Has slope 4 and x intercept 4
   (h) Has zero slope and passes through the point $(1,-3)$
   (i) Has slope $-\frac{5}{3}$ and passes through the origin

* 6 Show that the equation of a straight line whose x intercept is h and y intercept is k, with $hk \neq 0$, may be written in the form $\dfrac{x}{h} + \dfrac{y}{k} = 1$.

7 Write the formula for the linear function which relates the odd integers to all the integers under the conditions specified.

(a) -1 corresponds to 0 and 5 to 3.

(b) 3 corresponds to 0 and 9 to 3.

8   Write the formula for the linear function which relates the even integers to all the integers under the conditions specified.

(a) 0 corresponds to 0 and 10 to 5.

(b) 2 corresponds to 0 and -12 to -7.

(c) 4 corresponds to 0 and 12 to 4.

* 9   The relation between Fahrenheit temperature F and centigrade temperature C is a linear function. What is the formula for this function if $F = 32$ when $C = 0$ and $F = 212$ when $C = 100$?

*10   Draw the graph of the function of Exercise 9.

11   The relation between the circumference C of a circle and its radius r is a linear function. What is the formula for this function if $C = 0$ when $r = 0$, and $C = 6\pi$ when $r = 3$?

12   The relation between the total distance d traveled and the time t spent in traveling at a uniform speed is a linear function. What is the formula for this function if $d = 0$ when $t = 0$ and $d = 120$ when $t = 2$?

13   What is the equation of the x axis? Of the y axis?

10-5   SYSTEMS OF LINEAR EQUATIONS

A system of linear equations in two variables consists of two or more equations of the form

$$a_1 x + b_1 y = c_1$$
$$a_2 x + b_2 y = c_2$$

The common solution of such a system will be the intersection of the solution sets of the respective equations. The system is *consistent* or *inconsistent* depending upon whether this intersection is nonempty or empty.

Consistency of equations   **Definition 10-4**   A system of equations is *consistent* if the equations have at least one solution in common. It is *inconsistent* if they have no common solution.

The discussion of Secs. 10-3 and 10-4 has established the fact that the graph of a linear equation in two variables is a straight line. The graph of a consistent system of such equations will consist of coincident lines (as illustrated in Fig. 10-13) or intersecting lines (as illustrated in Figs. 10-11, 10-12, and 10-14).

Example   Find the solution set of the system

$$3x - 4y = 12$$
$$6x - 8y = 24$$

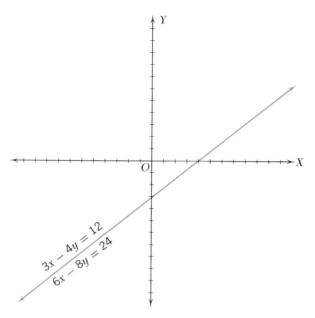

FIGURE 10-13   Coincident lines

An examination of these two equations reveals the fact that a constant ratio exists between corresponding coefficients.

$$\frac{3}{6} = \frac{-4}{-8} = \frac{12}{24} = \frac{1}{2}$$

Each equation may be converted to the same equivalent equation which may be expressed in the linear function form

$$y = \tfrac{3}{4}x - 3$$

The graph of each line has a slope of $\tfrac{3}{4}$ and goes through the point $(0, -3)$. The lines may be represented by the same line (Fig. 10-13). They are said *Coincident* to be *coincident lines*. The solution set of either equation is also the solution *lines* set of the other and, consequently, is the solution set of the system.

Example   Find the solution set of the system

Intersecting
lines

$$2x + 3y = 1 \tag{1}$$
$$7x + 5y = 9 \tag{2}$$

Multipliers may be chosen, first to obtain two equations in which the coefficients of y are equal, and then two equations in which the coefficients of x are equal. This process produces these two equivalent systems.

$5 \times$ Eq. (1): $10x + 15y = 5$ $7 \times$ Eq. (1): $14x + 21y = 7$
$3 \times$ Eq. (2): $21x + 15y = 27$ $2 \times$ Eq. (2): $14x + 10y = 18$

In each system subtract the second equation from the first to obtain the equivalent system.

$$-11x = -22$$
$$11y = -11$$

The coefficients of x and y are different from zero, so that we may divide to get

$$x = 2 \quad \text{and} \quad y = -1$$

The substitution of these values in the original system of equations will verify that the ordered pair $(2, -1)$ is the unique solution of the original system. Figure 10-14 shows that the lines which are the graphs of the two equations intersect in the point $(2, -1)$.

The graph of an inconsistent system of linear equations will be a set of *parallel lines* (as illustrated in Figs. 10-8, 10-10, and 10-15). The lines will all have the same slope but different y intercepts.

Parallel lines

Example   Find the solution set of the system

$$3x - 4y = 12$$
$$6x - 8y = 48$$

Following the same process as that used in the previous example, we obtain

$$1 \times \text{Eq. (1): } 3x - 4y = 12$$
$$\tfrac{1}{2} \times \text{Eq. (2): } 3x - 4y = 24$$

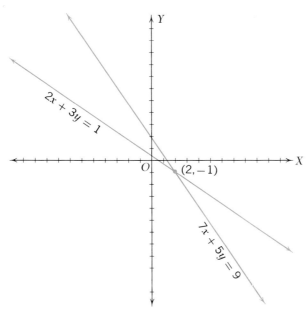

FIGURE 10-14   Intersecting lines

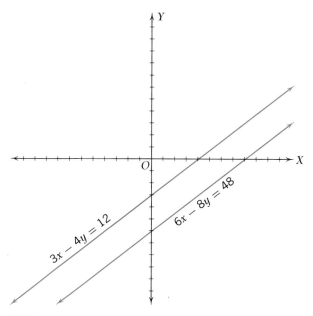

FIGURE 10-15 Parallel lines

It should be evident that no values of x and y can satisfy these two equations. The system is inconsistent and there exists no solution set.

If each of the two equations is expressed in the form of a linear function, we have the equivalent system

$$y = \tfrac{3}{4}x - 3$$
$$y = \tfrac{3}{4}x - 6$$

The graphs of these two functions are parallel lines, since they have the same slope but different y intercepts. The graphs are shown in Fig. 10-15.

EXERCISES

Determine whether each system of equations is or is not consistent. For each consistent system, determine the solution set. When possible, draw the graph of each system.

1 $3x + 4y = -1$
   $2x - y = 3$

2 $4x - y = 8$
   $8x - 2y = 16$

3 $4x - y = 8$
   $12x - 3y = 16$

4 $\tfrac{1}{2}x + \tfrac{1}{3}y = 4$
   $x - y = -2$

5 $\tfrac{2}{3}x - \tfrac{3}{2}y = -5$
   $\tfrac{1}{2}x + \tfrac{1}{3}y = 5$

6 $x - y = -3$
   $-\tfrac{1}{3}x + \tfrac{1}{3}y = 1$

7   $0.05x + 0.2y = 0.25$
    $0.1x + 0.4y = 0.5$

* 8 $ax + by = c$
      $3ax + 3by = 4c$

* 9 $lx + my = h$
      $klx + kmy = hk$

10   $1.2x - 4.8y = 9.6$
     $3.6x - 14.4y = 28.8$

11   $0.3x + 0.4y = 11$
     $x - y = -10$

12   $4x - 8y = 14$
     $x + 5y = -7$

13   $8x - y = 16$
     $4x - \frac{1}{2}y = 7$

14   $7x - 3y = -12$
     $-\frac{7}{3}x + y = 4$

15   $3x + 5y = 7$
     $2x + 3y = 8$

*16 $\frac{2}{3}x - \frac{4}{5}y = \frac{5}{6}$
      $\frac{1}{2}x - \frac{3}{5}y = \frac{5}{3}$

*17 $\frac{2}{3}x - \frac{2}{9}y = \frac{2}{3}$
      $\frac{4}{5}x + \frac{4}{15}y = 0$

*18 $0.5x - 0.4y = 1$
      $2.5x - 2y = 5$

*19 $0.25x + 0.20y = 0.90$
      $0.4x + 0.3y = 1.39$

*20 $0.25x + 0.20y = 0.90$
      $x + 0.80y = 4$

10-6   THE RELATION OF INEQUALITY

Law of trichotomy

Attention has been called previously to the fact that for any two real numbers a and b, the law of trichotomy always holds. In other words, whether a and b are integral, rational, or irrational, it is always true that exactly one of these three relations holds: $a = b$, a is greater than b ($a > b$), or a is less than b ($a < b$). Furthermore, if $a > b$, then it is true that $a = b + c$ for some positive real number c; and if $a < b$, then $a + d = b$ for some positive real number d.

The relations $<$ and $>$ are neither reflexive nor symmetric, but they are transitive. (See Theorem 3-5. Although this theorem is stated for natural numbers, it holds true for any real numbers.) These relations are called

Inequalities

strict inequalities in contrast to $\leq$ (*is less than or equal to*) and $\geq$ (*is greater than or equal to*), which are called *mixed inequalities*.

In Chap. 3 certain other theorems concerning inequalities were either proved or stated as exercises. The same theorems can be proved for real numbers. They will be stated here, without proof, for real numbers.

Theorem 10-1   If a, b, c are real numbers such that $a < b$ and $b < c$, then $a < c$.

Theorem 10-2   If a, b, c are real numbers such that $a < b$, then $a + c < b + c$.

Theorem 10-3 If a, b, c, d are real numbers such that $a < b$ and $c < d$, then $a + c < b + d$.

Theorem 10-4 If a and b are real numbers such that $a < b$ and c is a positive real number, then $ac < bc$.

Theorem 10-5 If a and b are real numbers such that $a < b$ and c is a negative real number, then $ac > bc$.

Another true theorem can be derived from each of these theorems by replacing $<$ by $>$ and $>$ by $<$. These proofs are left as exercises for the reader. The theorems also hold if each strict inequality is replaced by its corresponding mixed inequality.

Example Replace these inequalities with equivalent inequalities of the form $-a < x < a$: (1) $|x| < 3$; (2) $|x - 5| < 4$.

1 $|x| < 3$ means that x must be a number which is larger than -3 but smaller than $+3$. Any such number has an absolute value (numerical value) which is less than 3. The domain of values for x is thus the portion of the number line between -3 and $+3$, as shown by the colored portion of line l of Fig. 10-16. In other words, $-3 < x < 3$.
2 $|x - 5| < 4$ means that x must be such that $-4 < x - 5 < 4$. By Theorem 10-2, $-4 + 5 < (x - 5) + 5 < 4 + 5$ or $1 < x < 9$.

The interval of the number line which is the domain of x is the colored portion of line m of Fig. 10-16.

In each case of this example the domain of x does not include the end
Intervals   points of the interval. For this reason it is called an *open interval*. Strict inequalities always imply open intervals. If the inequalities had been mixed, as in $|x| \leq 3$ and $|x - 5| \leq 4$, then the domain of x in each case would have included the end points of the interval, in which case the interval would have been called a *closed interval*.

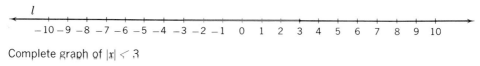

l

Complete graph of $|x| < 3$

m

Complete graph of $|x - 5| < 4$
FIGURE 10-16

Example Express each of these closed intervals in an equivalent form using the absolute-value sign: (1) $-1 \leq x \leq 1$; (2) $-2 \leq x \leq 6$.

1 $-1 \leq x \leq 1$ states that x must be numerically less than or equal to 1, and this condition may be written $|x| \leq 1$.

2 $-2 \leq x \leq 6$. In order to use the absolute-value sign it is necessary to have the two limits equal but opposite in sign. This may be accomplished here in this manner:

(a) Take one-half the sum of the interval end points:

$$\frac{(-2) + 6}{2} = 2$$

(b) Add the additive inverse of 2 to each member of the inequality. Theorem 10-2 states that the inequality is not disturbed. Thus, $(-2) + (-2) \leq x + (-2) \leq 6 + (-2)$ or $-4 \leq x - 2 \leq 4$, which may be written as $|x - 2| \leq 4$.

Inequality in two variables   An inequality in two variables is of the form

$$ax + by > c \quad \text{or} \quad ax + by < c$$

where a, b, and c are real numbers. If $b > 0$, we may use Theorems 10-2 and 10-4 to get

$$y > -\frac{a}{b}x + \frac{c}{b} \quad \text{or} \quad y < -\frac{a}{b}x + \frac{c}{b}$$

If $b < 0$, we use Theorems 10-2 and 10-5 to get

$$y < -\frac{a}{b}x + \frac{c}{b} \quad \text{or} \quad y > -\frac{a}{b}x + \frac{c}{b}$$

Thus in either case we have a formula which gives many values of y for each value of x. We have in each case, therefore, a formula that expresses *Graph of an inequality* a relation which is not a function. The graph of such a relation can be obtained by first plotting the line which is the graph of $ax + by = c$. For the strict inequality $y > -\frac{a}{b}x + \frac{c}{b}$, the graph will be that half plane which contains those points such that, for a given x, the y coordinate is greater than the corresponding y coordinate of the point of the line. For the strict inequality $y < -\frac{a}{b}x + \frac{c}{b}$, the y coordinate of each point of the half plane will be less than the corresponding y coordinate of the point on the line. For any mixed inequality the graph will also include the line.

Example Plot the graph of the strict inequality $2x + 3y > 6$.

Using Theorems 10-2 and 10-4, we may write this inequality as $y > -\frac{2}{3}x + 2$.

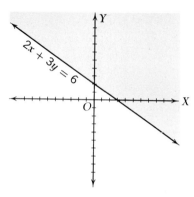

FIGURE 10-17 Incomplete graph of $2x + 3y > 6$

To plot the graph of the equation $2x + 3y = 6$ we write it in the form $y = -\frac{2}{3}x + 2$. The graph is therefore a line with slope $-\frac{2}{3}$ and passing through the point (0,2). The colored portion of Fig. 10-17 indicates the half plane which would be the graph of $2x + 3y > 6$. Note that the line which is the graph of $2x + 3y = 6$ is not colored. Since this is a strict inequality, the points of the line are not part of the graph. To check the graph, select points in the colored half plane and test their coordinates in the inequality. For example, the point (5,2) gives $2(5) + 3(2) = 16$, and $16 > 6$.

Example Plot the graph of the mixed inequality $10x - 2y \geq 7$.

Using Theorems 10-2 and 10-5, we may write this inequality in the form $y \leq 5x - \frac{7}{2}$. Notice that when both sides of the inequality are multiplied by $-\frac{1}{2}$, the inequality changes its sense. It changes from $\geq$ to $\leq$, which is in accordance with Theorem 10-5.

To plot the graph of the equation $10x - 2y = 7$ we may write it in the form $y = 5x - \frac{7}{2}$. The graph is a line with a slope of 5 and a y intercept of $-\frac{7}{2}$. An incomplete graph of the inequality is the indicated colored half plane of Fig. 10-18, *including* the line which is the graph of the equation $10x - 2y = 7$. To check the graph, use points (2,2) and (3, −5). We find

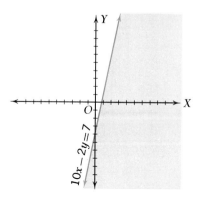

FIGURE 10-18 Incomplete graph of $10x - 2y \geq 7$

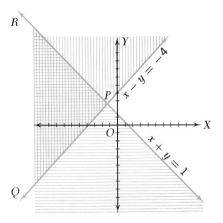

FIGURE 10-19   Incomplete graphs of $x + y \leq 1$ and $x - y \leq -4$

for (2,2): $10(2) - 2(2) = 20 - 4 = 16$, and $16 > 7$; for $(3, -5)$: $10(3) - 2(-5) = 30 + 10 = 40$, and $40 > 7$.

System of inequalities The solution set for two inequalities in two unknowns can be found by graphing the solution set of each inequality and then taking the intersection of the two sets.

Example Find the solution set of the system of inequalities

$$x + y \leq 1$$
$$x - y \leq -4$$

The portion of Fig. 10-19 which is shaded with horizontal lines is an incomplete graph of $x + y \leq 1$. The portion shaded with vertical lines is an incomplete graph of $x - y \leq -4$. The cross-hatched portion of the graph then represents the intersection of the two sets. The coordinates of any point lying within this area will satisfy both inequalities. It should be noted that, since both inequalities are mixed, the line is to be included as a portion of the graph of each inequality. The rays $\overrightarrow{PQ}$ and $\overrightarrow{PR}$ are included in the graph.

EXERCISES

Express each of the following in the equivalent form using absolute values.

1	$-2 < x < 2$	2	$-5 \leq x \leq 5$
3	$-1 \leq x \leq 5$	4	$-1 < x - 1 < 2$
5	$0 \leq x + 2 \leq 4$	6	$2 \leq x - 3 \leq 8$
7	$4 < x - 5 < 8$	8	$-3 \leq x + 1 \leq 2$
9	$-5 \leq x + 2 \leq 6$		

Express each of the following in the equivalent interval form ($a \leq x \leq b$ or $a < x < b$).

10   $|x| \leq 1$ 11 $|x| \leq 8$ 12 $|x - 1| \leq 2$
13   $|x + 3| < 4$ 14 $|x + 5| < 2$ 15 $|x - 2| < 2$

For what values of x are these inequalities true?

16   $2x + 1 < 0$ 17 $3x - 5 < 0$
18   $2x + 1 < x + 2$ 19 $-4x + 2 \geq x - 8$

20   $2x - \dfrac{1}{3} \leq x + \dfrac{2}{3}$ 21 $x + 2 \geq 3x + 4$

Draw the graph of each inequality.

22   $x \geq 2$ 23 $x \leq -3$ 24 $y > 1$
25   $-2x > 1$ 26 $2x + y > 4$ 27 $x - y < 6$
28   $2x - y \leq 6$ 29 $3x + 2y \geq -12$ 30 $x - 4y \leq -12$

Draw the graph of the solution of each system of inequalities.

*31 $x + 2y > 2$ 32 $y \leq x$ 33 $y \leq x - 3$
        $2x - y > 2$ $x + y \leq 1$ $x + y \leq 2$

34   $x + y \geq -6$ 35 $x - y > -3$ *36 $3y - x < 6$
        $y \leq 0$ $x + y < 3$ $x + 2y > 8$

*37   Prove Theorem 10-1.
*38   Prove Theorem 10-2.
*39   Prove Theorem 10-3.
*40   Prove Theorem 10-4.
*41   Prove Theorem 10-5.

10-7   VARIATION

Whether in business or in ordinary experience, one of the most common ways of describing relationships that exist is in terms of *variation*. We shall be concerned with three distinct types of variation: *direct variation, inverse variation*, and *joint variation*.

Direct variation

In *direct variation* two variables are so related that the ratio between them remains constant. As one variable increases the other increases, and as one decreases the other decreases. In other words, if y varies directly as x, the relationship can be expressed in either of two ways: $y/x = k$ or $y = kx$, where k is a constant. The constant k is called the *constant of proportionality* or *constant of variation*.

Constant of proportionality (variation)

Example   The circumference C of a circle varies directly as its diameter d. Find the value of C when $d = 6$, if $C = 12.56$ when $d = 4$.

The formula is $C = kd$.

Since $C = 12.56$ when $d = 4$, we have

$$12.56 = k(4) \qquad \text{or} \qquad k = 3.14$$

The equation may now be written

$$C = 3.14d$$

so that when $d = 6$, we have

$$C = 3.14(6) = 18.84$$

Inverse variation   In *inverse variation* two variables are so related that their product remains constant. If y varies inversely as x, then $xy = k$, where k is the constant of proportionality. The same relation may be written in the form $y = k/x$, which states that y varies directly as the reciprocal of x. As one variable increases the other decreases, or as one decreases the other increases.

Example The time it takes a person to travel a fixed distance varies inversely with the average speed at which he travels. How long will it take a person to drive a given distance at an average speed of 45 miles per hour, if it would take him 9 hours traveling at an average speed of 40 miles per hour?

If s represents average speed and t the time traveled, the formula is

$$k = st \quad \text{or} \quad t = \frac{k}{s}$$

When $t = 9$, $s = 40$, and so $k = 360$ and

$$360 = 45t \quad \text{or} \quad t = \frac{360}{45}$$

$$t = 8$$

Joint variation   In *joint variation* three variables are so related that one varies directly as the product of the other two. If z varies jointly as x and y, then $z = kxy$, where k is the constant of proportionality.

Example The area A of a triangle varies jointly as its base b and its altitude h. What is the area of a triangle whose base is 15 feet and whose altitude is 8 feet if 84 square feet is the area of a triangle whose base is 24 feet and whose altitude is 7 feet?

The formula is $A = kbh$.

$$84 = k(24)(7)$$
$$84 = 168k$$
$$k = \frac{1}{2}$$

Therefore the equation is $A = \frac{1}{2}bh$. When $b = 15$ and $h = 8$,

$$A = \frac{1}{2}(15)(8)$$

$$A = 60 \text{ sq ft}$$

EXERCISES

1 The perimeter of a certain type of regular polygon varies directly as the length of one of its equal sides. What is the perimeter of such a polygon whose sides are 8 inches if the perimeter of a similar polygon is 72 inches when its side is 12 inches?

2 The volume V of a sphere varies directly as the cube of its radius r. When $r = \frac{5}{2}$, $V = 125\pi/6$. What is the value of V for $r = 6$?

3 The pressure of a gas of constant volume varies directly as its absolute temperature. If the pressure is 30 pounds per square inch when the absolute temperature is 600°F, what will be the pressure if the temperature falls to 500°F?

4 The current I in amperes in an electric circuit varies inversely as the resistance R in ohms when the electromotive force is constant. If, in a given circuit, I is 24 amperes when R is 1.5 ohms, what is I when R is 0.5 ohm? What is R when I is 75 amperes?

5 The weight of an object above the surface of the earth varies inversely as the square of its distance from the center of the earth. If a person weighs 147 pounds on the earth's surface, what will be his weight at a distance of 160 miles above the earth? Take the earth's radius to be 3,960 miles and compute the answer correct to the nearest pound.

6 At a constant rate of simple interest the return on an investment varies jointly as the principal invested and the time. If the return on $1,200 for 1 year is $54, what would be the return on $2,500 for 3 years. On $850 for 9 months?

7 The force f of the wind, blowing against a flat surface at right angles to the direction of the wind, varies jointly as the area A of the surface and the square of the speed v of the wind. The force with which a 15 mile per hour wind strikes a surface of 20 square feet is 20 pounds. What is the force against an area of 15 square feet if the wind's speed is 25 miles per hour?

8 The variable y varies directly as x and inversely as z. When $x = 2$ and $z = 10$, $y = 10$. What is the value of y when $x = 3$ and $z = 15$?

9 The intensity I of light received from a source varies directly as the candlepower c and inversely as the square of the distance d from the source. How far from a 200-candlepower light would a screen have to be to receive the same amount of light as a screen placed 25 feet from a 50-candlepower light?

10-8 STATISTICAL GRAPHS

In Sec. 10-2 five methods were presented for describing a relation that exists between variables. These methods are verbal description, formula, set of ordered pairs, table of values, and graph. In contradistinction to the types

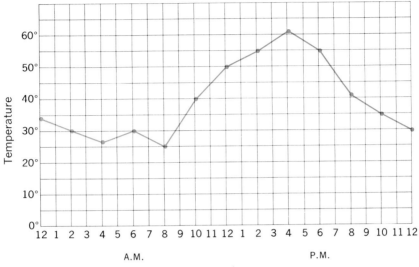

FIGURE 10-20   Temperatures at 2-hour intervals

of relations discussed so far, there are those sets of data related to each other in patterns not subject to formula representation. The graphs used to depict such relationships are often referred to as *statistical graphs* and are usually grouped in four distinct classifications: the *broken-line graph*, the *bar graph*, the *circle graph*, and the *pictograph*.

Broken-line graph

The *broken-line graph* is used primarily to show trends. Comparisons also can be read from it rather easily. Figure 10-20 shows temperature readings at 2-hour intervals for a 24-hour period. A casual glance will tell at what times the temperature was rising or falling. More careful inspection will answer such questions as: At what hours was the temperature the same? At what hour was the temperature the lowest? The highest? Between what hours did the most rapid rise occur? The most rapid fall? There are many other questions that can be answered. There are also some questions which cannot be answered. For example, although the graph tells what the temperature was at every even hour during the 24-hour period, it does not tell what it was at any odd-hour period. The line segment which joins the dots that indicate the temperatures at 8 A.M. and 10 A.M. simply indicates that, in general, the temperature was rising during this 2-hour interval. It does not say anything about what the temperature was at any specific time within this interval.

Another precaution is necessary in interpreting broken-line graphs. The graph of Fig. 10-21 tells the story of the growth in population in the United States during the first seven decades of the twentieth century. It is a very simple matter to observe from the graph that the general characteristic has been an increasing population. Furthermore, it is evident that the increase has been rather uniform over each 10-year period. But when the compari-

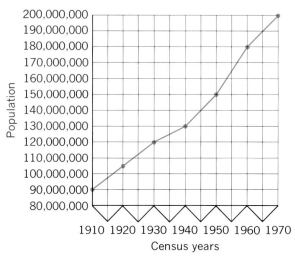

FIGURE 10-21 Population growth in the United States

son between the population figure for one year with that for another is read just from the graph, there can be serious distortion in interpretations. For example, the population for 1960 might seem to be more than twice that for 1930 and more than three times that for 1920. A look at the actual population figures for these years reveals the fact that these ratios are far from correct. This false impression is due to the fact that the population figures are large, all in excess of 75,000,000, and yet, in order to get any significant picture at all, a scale with a unit no larger than 10,000,000 had to be used. As a result, in order to get the picture in the space available

False zero line it became necessary to use 80,000,000 as a *false zero line*. This is indicated by the wavy line rather than the usual straight line and by the legend on the vertical scale. Careless attention to such details of graph structure frequently leads to grossly erroneous interpretations of the message of the graph.

Bar graph   The *bar graph* is used essentially for making comparisons. There are three types to be considered: the *vertical bar*, the *horizontal bar*, and the *composite bar* or *100% bar*. The vertical bar graph derives its name from the fact that the bars are drawn in a vertical position. From such a graph, when appropriate, trends can also be read by noting the characteristic behavior of the lengths of the bars. For example, a vertical bar graph might be used to portray the facts about wheat production in the United States for a given period of time. In Fig. 10-22 a comparison of the heights of the bars will answer pertinent questions concerning relative crop production for the years given. Also by noting the tendency in the heights of the bars one can get an answer to the question as to the trend in wheat production over the period of time pictured. On the other hand, a similar graph might be used to compare wheat production in each of a selected number of different states during the period of 1 year. In this case there would be no pertinent interpretation related to trends.

If the bars of a graph are in the horizontal position, then the graph is called a *horizontal bar graph*. Such a graph can be used effectively for comparisons but not very well for picturing trends. In Fig. 10-23 a horizontal bar graph is used to present the rainfall picture for a certain city over a period of 30 years. The average number of inches in each 3-year period is indicated by the appropriate length of a bar.

In most cases the decision as to whether a vertical or a horizontal bar graph is to be used is quite arbitrary. There are times when the nature of the data might make one type quite qppropriate and the other just as decidedly inappropriate. Just as in the case of the broken-line graph, in order to protect against distorted and unwarranted interpretations from either type of bar graph, it is very important to pay careful attention to the position of the zero line.

Circle graph   The *composite bar graph* and the *circle graph* are used for the same basic purpose, namely, to compare parts with a whole or parts with parts. The only distinction between the two is illustrated in Fig. 10-24, in which the two types of graph are used to picture an estimated national expenditure budget. In the bar graph a bar of arbitrary size is used to represent 100%, while in the circle graph a circle of arbitrary radius is used.

Pictograph   The *pictograph* is merely a bar graph in which small pictures instead of solid bars are used to emphasize the comparisons. Each small figure represents a selected number of units of data. The pictograph of Fig. 10-25

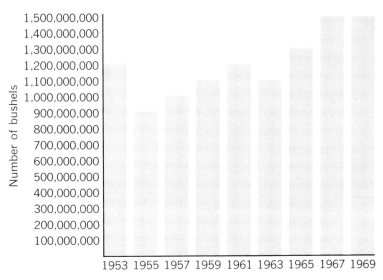

FIGURE 10-22   Wheat production in the United States (correct to the nearest 100,000,000 bushels). (*Source:* 1971 World Almanac)

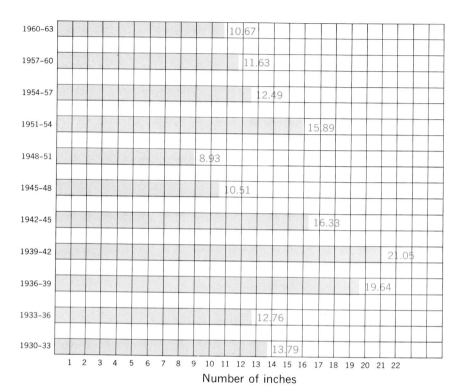

FIGURE 10-23 Average rainfall for a certain city for 3-year periods

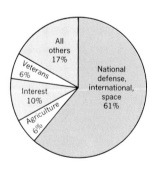

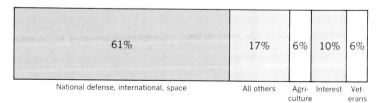

FIGURE 10-24 Estimated national budget

YEAR	NUMBER OF HOMES	
1930	12,000,000	
1935	23,000,000	
1940	29,000,000	
1945	34,000,000	
1950	45,000,000	
1955	52,000,000	
1960	55,000,000	
1965	57,000,000	

Scale: 1 picture for each 5,000,000 homes
Source: 1962 World Almanac

FIGURE 10-25 American homes with radios

shows in 5-year intervals over a 35-year period the total number of houses with radios in the United States. Each figure represents 5,000,000 homes.

EXERCISES

1 What are the relative advantages and disadvantages of the table of values, the formula, and the graph as a means for the presentation of quantitative data?

2 What are the distinguishing characteristics of the bar graph, the broken-line graph, and the circle graph?

3 What are some of the basic similarities and dissimilarities between the circle graph and the composite, or 100%, bar graph?

4 Draw a broken-line graph of the data of the table showing the normal monthly temperatures of a certain southern city.

MONTH	TEMPERATURE	MONTH	TEMPERATURE	MONTH	TEMPERATURE
Jan.	40	May	72	Sept.	73
Feb.	42	June	77	Oct.	62
March	50	July	80	Nov.	49
April	60	August	79	Dec.	42

EXERCISE 4

5 Answer these questions about the graph of Exercise 4.
   (a) Between what months was the temperature rising?
   (b) Between what months was the temperature falling?
   (c) When was the most abrupt rise in temperature?
   (d) When was the most abrupt fall in temperature?
   (e) For what two months was the normal temperature the same?

6 Draw a bar graph of the data of the table showing the number of tons of walnuts produced in the United States at 3-year intervals over a period of 18 years.

YEAR	TONS	YEAR	TONS
1951	77,400	1963	83,100
1954	75,400	1966	96,100
1957	66,600	1969	102,500
1960	72,800		

EXERCISE 6

7 A recommended state expenditure budget of $1,670,000,000 was allotted according to the pattern of the circle graph. Correct to the nearest million dollars, how much money was recommended for each type of expenditure? What can be used as an appropriate check on your computations?

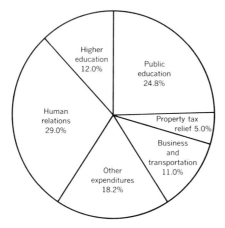

EXERCISE 7

8 Draw both a circle graph and a composite bar graph of the family budget plan shown here. (*Hint:* 10% of 360° = 36°.)

Savings	10%	Health	9%
Food	26%	Church	10%
Rent	18%	Household expenses	12%
Clothing	15%		

EXERCISE 8

∗ 9 Select data from reliable sources suitable for presentation by (*a*) horizontal bar graph; (*b*) vertical bar graph; (*c*) broken-line graph; (*d*) circle graph.

∗10 Draw the graph to represent each set of data selected in Exercise 9.

∗11 Draw a composite bar graph to represent the data used for the circle graph of Exercise 9.

10-9 THE NATURE OF PROBLEM SOLVING

Whether a problem is simple or complex, it is distinctly characterized by three significant components, namely, the data, the unknown elements, and the particular relations which exist between the data and the unknown. Consideration of any specific problem thus gives rise to critical analysis of words and sentences for accurate interpretation, recognition of possible relations between known elements as well as between known and unknown elements, recall of the basic functions of different operational procedures, and the search for implications sufficient to justify the result desired.

Whatever procedure is followed, problem solving is always characterized by difficulties of at least four distinct types: *comprehension, structure, operation,* and *judgment.*

The first major comprehension difficulty frequently is that of understanding the language in which the problem is stated. At times this can be overcome by the mere use of a dictionary, but there can be situations which call for reference to source materials to clarify technical terminology. There also can be difficulty in analyzing the problem situation to determine what data are essential, whether additional data are needed, what relations are stated or implied, and what missing information is needed before the desired solution can be obtained.

Example   After traveling for $1\frac{1}{4}$ hours Mr. Jones noticed that the number 70 showed on the odometer of his new car. If he continued to travel at this average rate of speed, how far would he travel in the next 2 hours?

The correct solution of this problem depends, in part, on knowing what the word "odometer" means. A careless reader might misread or misinterpret the word as "speedometer"; overlook, discard, or recognize the $1\frac{1}{4}$ hours as nonessential data; recognize the familiar and pertinent distance-rate-time relation; and proceed to the incorrect solution of $2 \times 70 = 140$ miles as the distance to be traveled "in the next 2 hours." There are also those who, in the same situation, would find some way to use the "$1\frac{1}{4}$ hours" simply due to the presence of this quantity in the statement of the problem. The simple use of a dictionary and a critical attitude toward data evaluation would help to avoid such difficulties.

A careful reader would look up the meaning of the word, if he were not sure of its meaning. Further analysis of the problem would reveal that additional data are needed before a solution can be obtained. There is a hidden question to which an answer is needed. Since an odometer is an instrument which measures distance traversed, the pertinent question is: What mileage did the odometer show at the start of the $1\frac{1}{4}$-hour period of travel? If the trip happened to have started at 0 miles, the unknown average speed would be given by the formula $r = d/t$ or $r = 70 \div 1\frac{1}{4} = 56$ miles per hour. At this average speed Mr. Jones would travel $2 \times 56 = 112$ miles. Thus two forms of the basic distance-rate-time formula would be required. Any other initial mileage indication would, of course, change the solution. If this information were not available, the problem, as stated, would be unsolvable.

Structure difficulty

After the statement of a problem is clearly understood, one is then confronted with the difficulty of determining the operational structure needed to provide the solution desired. Answers are needed to such questions as: What are the functions of each basic operation (addition, subtraction, multiplication, division)? What facts and relationships are implied by the data? What formulas are suggested? How are the data related? Which data are essential and which are nonessential? What data, if any, are missing?

Intuitive reasoning

In seeking answers to such questions intuition and induction can play a very important role. Following one's intuition is largely a process of attempting to capitalize on significant hunches. Such guesses, however, should have some form of authoritative justification: they should not be mere chance occurrences. While intuitive reasoning, through its appeal to analogy and specialization, can be very helpful in the discovery of effective leads to be followed in the search for the solution of a given problem, it is by no means a conclusive process. It can lead through analogy to the recognition of special cases which can vary from those with the same problem structure but simpler data to analogous problem situations with simpler structure. Such analogous situations may take the form of diagrams illustrating the basic relations between the data and the unknown, concrete examples similar to the more abstract problem situation, or simpler special cases which are significant components of the more difficult general problem.

Induction

Induction is fundamentally a process of reasoning that carries one from the particular to the general. Such a pattern of thinking can lead one to examine the intuitively recognized special cases in the quest for a characteristic thread of similarity, regularity, or coherence which might suggest techniques for dealing with the more general situation. Just as is the case with intuition, the most significant techniques of induction are analogy, specialization, and generalization. Through the analogies of special cases, generalizations are sought in order that they might be checked against other specializations. Such generalizations lead to the formalization of formulas or rules which summarize and characterize recognized basic relations that exist between data, and between data and unknowns. Great care must be exercised, however, in drawing conclusions from any inductive process. Such conclusions should never be stated as actualities, but only as possibilities, probably worthy of further investigation.

Operational difficulty

After a problem has been carefully structured there still remains the threat of difficulty in performing the operations required to gain a solution. There may be an inadequate knowledge of the basic operational algorithms; a lack of understanding of the fundamental properties which control and provide flexibility in operational mechanics; inadequacy in dealing with formulas and equations; an inclination toward careless and shoddy performance. Review and practice can help one become more capable in operational performance. Pride in one's work and careful checking can help to remove the threat of inexcusable error and inadequate performance due to work done in a sloppy manner.

After a solution has been obtained there always remain the questions:
Judgment Does it conform to the requirements of the problem situation? What inter
difficulty pretation is to be given to the solution? What are the most effective methods for checking?

Example During a recent year the total amount of money spent in the United States on new construction was $84,690,000,000. Of this amount $27,694,000,000 was spent on public construction, $28,823,000,000 on private residential construction, and $28,173,000,000 on private nonresidential construction. Correct to the nearest tenth what percent of the total construction expenditure was for private residential construction? The data are:

Total expenditures for new construction	$84,690,000,000
Public construction	27,694,000,000
Private residential construction	28,823,000,000
Private nonresidential construction	28,173,000,000

The requested information is the following: What percent of the total expenditure is the amount for private residential construction?

When this question is analyzed two pertinent bits of information are recognized:

1 The amounts spent for private nonresidential and public construction are nonessential data.
2 The basic formula needed is the percentage formula in the form $r = p/b$.

Further critical study of the problem reveals that the question calls for no more than three significant digits in the answer. Since the data involves quite a few nonsignificant digits, scientific notation can be used to great advantage. With these facts in mind the solution may be expressed in the form

$$r = \frac{2.882 \times 10^{10}}{8.469 \times 10^{10}} = \frac{2.882}{8.469} \times \frac{10^{10}}{10^{10}}$$

$$= \frac{2.882}{8.469} = 0.340$$

$$r = 34.0\%$$

(Why is $r = 34\%$ an *incorrect* answer?)

There are several ways to check this solution. Only two methods will be mentioned here. (Can you suggest at least one other check?)

1   Since division was used in the solution, multiplication provides a check.

$$34.0\% \text{ of } \$84,690,000,000 = (3.4 \times 10^{-1}) \times (8.469 \times 10^{10})$$
$$= (3.4 \times 8.469) \times 10^{9}$$
$$= \$28,795,000,000$$

Correct to three significant digits this is $28,800,000,000.

   This is a very satisfactory check since the error (due to rounding) is less than 0.04% (four hundredths of 1%).

2   The data not essential to the solution of the problem can be used here as an aid in checking. $100\% - 34.0\% = 66.0\%$. The suggested check is that 66.0% of $84,690,000,000 should closely approximate $55,867,000,000, the sum of the amounts for private nonresidential and public construction. Again the error is less than 0.04%.

EXERCISES

Find the solution set of each exercise.

1   At simple interest what is the interest on $1,200 at 4.8% for 3 months?
2   Two trains depart from the same station at the same time. One train travels east at 60 miles per hour and the other travels west at 50 miles per hour. How long will it be before they are 495 miles apart?
3   The sum of a number, its half, and its third is 77. What is the number?
4   An air conditioner was installed at a total cost of $395.20. This included a sales tax of 4%. What was the amount of the tax?
5   The sum of two consecutive integers is 83. What are the two integers?
6   The sum of two consecutive even integers is 758. What are the two integers?
7   The sum of two consecutive odd integers is 1,000. What are the two integers?
8   The sum of two integers is 24. Twice the smaller number less the larger number is 3. What are the two integers?
9   The difference between two numbers is 75. If the larger number is divided by the smaller number there is obtained a quotient of 6 and a remainder of 5. What are the two numbers?
10   A refrigerator was purchased on a 3-month term basis for $339.95. The cash price was $325. What rate of interest was paid?
11   A department store bought 25 television sets. Each set cost $185.25. At what price must these sets sell if the cost is to be 65% of the selling price?
12   Use the formula for the linear function to find the equation of the line through the points $(2, -3)$ and $(-5, 4)$.
13   Find the point of intersection of the line of Exercise 12 and the line which has a y intercept of 2 and an x intercept of 4.
14   The length of a rectangle is 30 yards longer than its width. The perimeter is 200 yards. What are the dimensions of the rectangle?

15 The tens digit of a given number is 5 more than the ones digit. If the digits are reversed, the new number is 9 less than one-half the given number. What is the given number?

16 In the graduating class of City High School there are 180 students. The ratio of boys to girls is 5 to 4. How many of each sex are in the graduating class?

17 A square and an equilateral triangle are to be constructed so that they have the same perimeter. The side of the triangle is to be 2 feet longer than the side of the square. What will be the length of a side of each figure?

18 An investment of $12,000 brings a total yearly return of $625. Part of the investment bears 5% interest and the rest $5\frac{1}{2}$%. How much is invested at each rate?

19 Sidewalks are being constructed along the streets of a residential subdivision. The dimensions of the two street sides of an irregular corner lot are 56 feet and 126 feet. The lengths of the other two sides are 70 feet and 84 feet. How many square feet of cement will be required to build a concrete sidewalk 3 feet wide along the street border of the lot?

20 A firm mailed 300 letters, some of which went by airmail and required 11 cents' postage. The balance of the letters required 8 cents' postage. The total amount of postage was $26.16. How many letters of each kind were mailed?

21 The total of the basic charge and tax on two long-distance telephone calls was $12.43. The tax rate was 10% of the basic charge. The basic charge on one call was $2.90 less than that of the other. What were the basic charge and the amount of the tax on each charge?

22 The average of the four highest speeds attained in the qualifying trials for a recent Indianapolis Speedway Race was slightly more than 150 miles per hour, the speed necessary to make one lap of the track in 1 minute. The average of the four highest speeds of the race was slightly more than 142 miles per hour. Correct to the nearest second, how long did it take a car to make one lap of the track when traveling at this speed?

∗23 The total receipts for a certain baseball World Series were $1,995,189.09. Fifty-one percent of this amount was the players' pool, of which 30% was divided among the second, third, and fourth teams of the two leagues. The balance was divided so that the winning team received 60%. After assigning smaller portions totaling $11,565, the team voted to divide the balance of their pool into $32\frac{1}{2}$ full shares. The losing team, after assigning $3,406 in smaller portions, voted to divide the balance of their pool into $35\frac{3}{4}$ full shares. To the nearest dollar, what was the value of one full share to the recipient on each team?

∗24 Use the techniques of modular arithmetic to prove this theorem for a number system in base **twelve**: The excess of **e**'s in a product is equal to the excess in the product of the excesses of the two factors.

*25 A salesman, having made a call in city A, needs to drive a distance of 108 miles in 2 hours' time to meet an appointment in city B. There are three areas of traffic congestion through which he must pass: the first is 1 mile in length, the second is 2 miles in length, and the third is 5 miles in length. He estimates that the average speeds he will be able to maintain in each of the congested areas are: 1-mile area, 15 miles per hour; 2-mile area, 30 miles per hour; and 5-mile area, 25 miles per hour. What average speed must he maintain on the open road (outside the congested areas) in order to be on time for his appointment?

10-10 REVIEW EXERCISES

1–28 Write a carefully constructed answer to each of the guideline questions of this chapter.

29 Which of these relations are functions?
   (*a*) $\{(2,1),\ (3,1),\ (4,1),\ (5,1)\}$
   (*b*) $\{(2,y)\,|\,y$ is a real number$\}$
   (*c*) $\{f(x) = 2x - 4\,|\,|x| \leq 3\}$
   (*d*) $\{y \geq 4x + 1\,|\,x,y$ are real numbers$\}$
30 What are the domain and range of each relation of Exercise 29?
31 Draw a graph of each relation of Exercise 29. In each case state whether it is a complete or an incomplete graph.
32 For the set $\{(x, f(x)\,|\,x$ is an integer and $-4 < x < 3\}$, first write a formula which expresses the relation between the dependent and independent variables as described in (*a*) and (*b*); then draw a graph of each relation.
   (*a*) Multiplied by 2 and this product increased by 4
   (*b*) Decreased by 1 and the difference divided by 3
33 Are the graphs of Exercise 32 complete or incomplete?
34 Write the sets of ordered pairs for each relation of Exercise 32.
35 Is either relation of Exercise 32 a function?
36 What is the basic distinction between the formula for the constant function and that for the linear function?
37 Write the formula for each linear function whose graph is described.
   (*a*) Has slope $\frac{2}{3}$ and y intercept -1
   (*b*) Has slope 3 and x intercept 2
   (*c*) Has x intercept -2 and y intercept 2
   (*d*) Has slope -4 and passes through the point (3, 2)
   (*e*) Passes through the points $(-1,5)$ and $(6,-4)$
38 Write these equations in the linear function form.
   (*a*) $2x + 5y = 10$
   (*b*) $x - 2y = 4$
   (*c*) $12x - 6y = -24$

39 For each system of equations: Determine whether it is consistent or inconsistent; if consistent, find the solution set; draw the graph of each equation.

(a) $2x + 3y = 4$
$x - 4y = -9$

(b) $3x - 4y = 12$
$9x - 12y = 24$

(c) $3x - 4y = 12$
$6x - 5y = 24$

(d) $4x + y = 9$
$8x + 2y = 18$

40 Express each of the following in absolute-value form:

(a) $-5 \leq x \leq 5$
(b) $-5 \leq x \leq 13$
(c) $-4 < x < 4$
(d) $-2 \leq x - 2 \leq 2$
(e) $1 \leq x - 3 \leq 5$
(f) $-8 < x + 2 < 0$

41 Express each of the following in the equivalent interval form ($a \leq x \leq b$ or $a < x < b$).

(a) $|x| \leq 7$
(b) $|x - 2| \leq 3$
(c) $|x + 5| < 4$
(d) $|x - 1| \leq 1$
(e) $|x + 3| < 3$
(f) $|x - 9| < 5$

42 Find the solution set of each system of inequalities

(a) $4x + y < 8$
$3x + 2y > 6$

(b) $6x - 3y \leq 18$
$x - y \geq 0$

43 The distance fallen by a body in a vacuum varies as the square of the time spent in falling. If a body falls 144 feet in 3 seconds, how far will it fall in 5 seconds?

44 The volume of a right circular cone varies jointly as its height and the square of the radius of its base. A cone whose height is 20 inches and which has a base with a radius of 10 in. has a volume of 2,094 cubic inches. Correct to the nearest cubic inch, what is the volume of a cone with a circular base with a 6-inch radius and with a height of 12 inches?

45 The number r of revolutions per minute of a ball governor necessary to keep the balls suspended h inches below the point of suspension varies inversely as $\sqrt{h}$. If a speed of 60 revolutions per minute will keep the balls suspended 4 inches below the point of suspension, what speed will be necessary to keep them suspended $2\frac{1}{4}$ inches below?

∗46 An estimated national budget recommended the expenditure of $77,930,000,000 for National Defense. This represented 34% of the total budget. Other recommended expenditures were the following: Human Resources, 42%; Physical Resources, 11%; Miscellaneous Expenditures, 13%. This budget exceeded by $11,600,000,000 the estimated income to be received from these sources: Individual Income Taxes, $89,220,000,000; Social Income Taxes, $54,400,000,000; Corporation Taxes, $34,820,000,000; Excise Taxes, $17,410,000,000; Miscellaneous Sources, $21,760,000,000.

Use scientific notation in seeking answers to these questions:

(a) Correct to the nearest $10 million, what was the total national budget?

(b) Correct to the nearest $10 million, what amount of money was allotted to each expenditure area?

(c) By what percent of the total budget does the expenditures for Human Resources exceed that for National Defense?

(d) Correct to the nearest $10,000,000, what was the total estimated income?

(e) Correct to the nearest whole percent, what percent of the total income was estimated to come from each source?

(f) The percent of the total income to be received from Individual Income Taxes is the same as the total percent estimated to be received from what other two sources?

(g) Draw a circle graph of the recommended expenditure budget.

(h) Draw a composite bar graph of the estimated income budget.

*INVITATIONS TO EXTENDED STUDY

1   Write the equation of the line through the two points $P_1(x_1, y_1)$ and $P_2(x_2, y_2)$.

2   A linear equation in two variables is said to represent a *family of lines* if it contains one parameter. Describe the family of lines represented by each equation.

(a) $4x + 2y + k = 0$ (b) $kx + 3y - 6 = 0$
(c) $5x - ky + 10 = 0$ (d) $x - ky + k = 0$
(e) $kx + y - k = 0$ (f) $kx + ky + 1 = 0$

3   Name at least two techniques for finding the solution set of three equations in three variables.

4   Discuss the geometric interpretations of systems of three linear equations in three variables.

5   Discuss matrix techniques for dealing with systems of three linear equations in three variables.

6   Prepare a paper on the elementary properties of determinants and matrices.

7   Discuss matrix techniques for dealing with systems of m linear equations in n variables. There are three cases: $m < n$; $m = n$; $m > n$.

8   Present to the class an analysis of the quadratic function and its graph.

9   Develop and illustrate the use of at least four techniques for finding the zeros of the quadratic function of one variable.

10   Prepare a paper on techniques for finding the solution set of a system of two equations in two variables involving (a) one quadratic and one linear equation; (b) two quadratic equations.

11   Solve these inequalities:

(a) $\dfrac{2x - 2}{x + 4} \le 1$ (b) $\dfrac{x + 4}{x - 1} \ge 0$

(c) $\dfrac{2x - 3}{x - 5} > 2$ (d) $\dfrac{3x + 8}{2x - 1} < 1$

12   Plot these inequalities:
   (a) $f(x) \ge |x|$ for $-5 \le x \le 5$
   (b) $f(x) \le |x - 2|$ for $-4 < x < 8$

13   Plot these systems of inequalities:

(a) $|x| \le 4$ (b) $|x - 3| \le 2$
   $|x| \ge 2$ $|x + 4| \ge 1$

(c) $x \ge 0$ (d) $x \ge 2$
   $y \ge 0$ $y \ge 1$
   $x + y \le 2$ $x + 2y \le 12$
                          $y \le x$

14   Prepare a discussion of inequalities in two variables involving quadratic expressions.

GLOSSARY OF SYMBOLS
BIBLIOGRAPHY
ANSWERS TO ODD-NUMBERED EXERCISES

GLOSSARY OF SYMBOLS

Symbol	Interpretation	Page
$a_1, a_2, a_3, \ldots, a_n$	The elements a-one, a-two, a-three, and on to include a-sub n.	33
$S = \{s_1, s_2, s_3, \ldots, s_n\}$	S is a finite set whose elements are s_1, s_2, s_3, and so on to s_n	33
$S = \{s \mid s$ is one of the elements $s_1, s_2, \ldots, s_n\}$	S is the set of all s such that s is one of the elements $s_1, s_2, \ldots, s_n$	33
$S = \{s_1, s_2, s_3, \ldots\}$	S is an infinite set whose elements are $s_1, s_2, s_3,$ and so on	33
$x \in S$	x is an element of the set S	33
$x \notin S$	x is not an element of the set S	33
U	The universal set	35
$\sim S$	The complement of S	41
$P - Q$	The relative complement of set Q in set P	42
$P \subseteq Q$	P is a subset of Q	35
$P \subset Q$	P is a proper subset of Q	35
$P \nsubseteq Q$	P is not a subset of Q	38
$P \not\subset Q$	P is not a proper subset of Q	38
$P \supseteq Q$ or $P \supset Q$	Set P contains set Q	36
$P = Q$	Sets P and Q have the same elements	36
$P \cup Q$	The union of sets P and Q (P cup Q)	42
$P \cap Q$	The intersection of sets P and Q (P cap Q)	43
$\varnothing$	The empty, or null, set	43
$P \cap Q = \varnothing$	P and Q are disjoint sets	43
$n(P)$	The cardinal number of set P	59
$a \mathrel{R} b$	a is R-related to b	37
$a \mathrel{\not R} b$	a is not R-related to b	37
-base	The numeral immediately preceding the dash is written in the base indicated (e.g., **234—five** is in base **five**)	21
(x, y)	The ordered pair of elements x and y	54
(x, y, z)	The ordered triple of elements x, y, and z	295
$P \times Q$	Cartesian product of sets P and Q	54
$a < b$	a is less than b	98
$a \leq b$	a is less than or equal to b	119
$a \leq b < c$	a is less than or equal to b, which is less than or equal to c	119, 256
$a > b$	a is greater than b	98

447

Symbol	Interpretation	Page
$a \geq b$	a is greater than or equal to b	119
$a \geq b \geq c$	a is greater than or equal to b, which is greater than or equal to c	256
$p \rightarrow q$	Proposition p implies proposition q	101
0	The additive identity; also the cardinal number of the empty set	59, 122
1	The multiplicative identity; also the cardinal number of any set whose elements can be placed in one-to-one correspondence with the elements of the set $\{a\}$	59, 95
^-a	The additive inverse of a	122
$-a$	A number less than zero	123
$a \neq b$	a is not equal to b	119
a^{-1} or $\frac{1}{a}$	The multiplicative inverse of a ($a \neq 0$)	200
$\pm a$	Plus or minus a	124
$a + b$	The sum of a and b	74, 75
$a - b$	The difference between a and b	118
$a \times b$ or $a \cdot b$ or ab	The product of a and b	89, 90
$\frac{a}{b}$ or $a \div b$	The quotient of a divided by b ($b \neq 0$)	119, 193
$\frac{a}{b}$	A rational number or a fraction if a and b are integers ($b \neq 0$)	193, 211
$\frac{a}{b}$ or $a:b$	The ratio of the two numbers a and b ($b \neq 0$)	212, 213
g.c.f. (g.c.d.)	Greatest common factor (divisor) of two or more integers	107, 156
l.c.m.	Least common multiple of two or more integers	107
a^n	a is used n times as a factor if and only if n is a natural number or a positive integer	15
$a^{-m} = \frac{1}{a^m}$ for $a \neq 0$	By definition	14
$a^0 = 1$ for $a \neq 0$	By definition	14
$a \approx b$	a is approximately equal to b	222
$\sqrt{a}$	The positive square root of a	246
$\sqrt[3]{a}$	A cube root of a	246
0.75	A finite (terminating) decimal	220
$0.\overline{142857}$	An infinite (nonterminating) repeating decimal	220
$3.14159 \cdots = \pi$	An infinite nonrepeating decimal	221

Symbol	Interpretation	Page
$r\%$	The ratio of r to 100, or the rate percent	232
$p = rb$	Percentage is the product of the rate percent and the base	234
$|a|$	The absolute, or numerical, value of a	255
i	The imaginary unit such that $i^2 = -1$	258
$a + bi$	A complex number if a and b are real numbers and $i^2 = -1$	258
$a \equiv b \pmod{m}$	a is congruent to b modulo m	177
$a \equiv b$	a is identically equal to b	378
$\overleftrightarrow{AB}$	The line of indefinite extent on which A and B are two distinct points	274
$\overline{AB}$	The line segment joining points A and B	275
AB	The length of the line segment $\overline{AB}$	277
$\overrightarrow{AB}$	The ray which extends indefinitely in one direction from its end point A	276
$\overparen{AB}$	The arc of a circle	303
$\overline{AB} \cong \overline{CD}$	Line segment $\overline{AB}$ is congruent to line segment $\overline{CD}$	278
$a \sim b$	a is similar to b	309
$\angle ABC$	The angle of which B is the vertex and $\overrightarrow{BA}$ and $\overrightarrow{BC}$ are the rays which form the angle	285
$m \angle ABC$	The measure of $\angle ABC$	286
$\angle ABC \cong \angle PQR$	$\angle ABC$ is congruent to $\angle PQR$	286
$1°$ (one degree)	Unit of angle measure	285
$\overleftrightarrow{X'X}$, or x axis	The horizontal axis, or axis of abscissas, in a two-dimensional cartesian frame of reference	294
$\overleftrightarrow{Y'Y}$, or y axis	The vertical axis, or axis of ordinates, in a two-dimensional cartesian frame of reference	294
$\overleftrightarrow{Z'Z}$, or z axis	The third axis, with $\overleftrightarrow{X'X}$ and $\overleftrightarrow{Y'Y}$, in a three-dimensional cartesian frame of reference	294
$\triangle ABC$	Triangle ABC	305
$\triangle ABC \cong \triangle DEF$	Triangle ABC is congruent to triangle DEF	314
$a \leftrightarrow b$	a corresponds to b	59, 311
$f(x)$	Function of x, or function at x	402
$f(x) = mx + k$	The linear function of x where $m \neq 0$	406

Symbol	Interpretation	Page
$P(x, y)$	The point P whose two-dimensional rectangular coordinates are x and y	296
$P(x, y, z)$	The point P whose three-dimensional coordinates are x, y, and z	295
$m = \dfrac{y_2 - y_1}{x_2 - x_1}$	The slope of the straight line passing through the two points $P_1(x_1, y_1)$ and $P_2(x_2, y_2)$	408
$ax + by + c = 0$	The linear equation in x and y	412
$y = mx + k$	The slope-intercept form of the equation of a straight line	410
$y - y_1 = m(x - x_1)$	The point-slope form of the equation of a straight line	413
σ^2	Variance of a distribution	384
σ	Standard deviation of a distribution	384
$\{(x, k)\}$	The constant function $y = k$	412
$\{(k, y)\}$	The constant relation $x = k$	412
$ax + by < c$, or $ax + by > c$	Inequality relation in the two variables x and y	423
$\dfrac{y}{x} = k,\ y = kx$	y varies directly as x	426
$y = \dfrac{k}{x}$, or $xy = k$	y varies inversely as x	427
$y = kxz$	y varies jointly as x and z	427

BIBLIOGRAPHY

Adler, Irving, and Ruth Adler: "Numbers, Old and New," The John Day Company, Inc., New York, 1960.

Allendoerfer, Carl B.: "Principles of Arithmetic and Geometry for Elementary School Teachers," The Macmillan Company, New York, 1971.

Backman, Carl A., and Robert G. Cromie: "Introduction to Concepts of Geometry," Prentice-Hall, Inc., Englewood Cliffs, N.J., 1971.

Banks, J. Houston: "Learning and Teaching Arithmetic," 2d ed., Allyn and Bacon, Inc., Boston, 1964.

Brumfiel, Charles, and Eugene Krause: "Elementary Mathematics for Teachers," Addison-Wesley Publishing Company, Inc., Reading, Mass., 1969.

—— and I. E. Vance: "Algebra and Geometry for Teachers," Addison-Wesley Publishing Company, Inc., Reading, Mass., 1970.

Chace, A. B., L. S. Bull, H. P. Manning, and R. C. Archibald: "The Rhind Mathematical Papyrus," vols. I and II, Mathematical Association of America, Inc., Buffalo, N.Y., 1927 and 1929.

Dantzig, Tobias: "Number, the Language of Science," 3d ed., The Macmillan Company, New York, 1945.

Dwight, Leslie A.: "Modern Mathematics for the Elementary Teacher," Holt, Rinehart and Winston, Inc., New York, 1966.

Eves, Howard: "An Introduction to the History of Mathematics," 3d ed., Holt, Rinehart and Winston, Inc., New York, 1969.

Heinke, Clarence H.: "Fundamental Concepts of Elementary Mathematics," Dickenson Publishing Company, Inc., Encino, Calif., 1970.

Johnson, David C., and Louis S. Cohen: Functions, *The Arithmetic Teacher*, **17**:305–315 (1970).

Johnson, Paul B., and Carol Herdina Kipps: "Geometry for Teachers," Brooks/Cole Publishing Company, Belmont, Calif., 1970.

Keedy, Mervin L.: "Number Systems: A Modern Introduction," Addison-Wesley Publishing Company, Inc., Reading, Mass., 1969.

Kelley, John L., and Donald B. Rickert: "Mathematics for Elementary Teachers," Holden-Day, Inc., Publisher, San Francisco, 1970.

Menninger, Karl: "Number Words and Number Symbols: A Cultural History of Numbers," The M.I.T. Press, Cambridge Mass., 1969.

Moise, Edwin E.: "The Number Systems of Elementary Mathematics," Addison-Wesley Publishing Company, Inc., Reading, Mass., 1966.

Moser, James M.: "Modern Elementary Geometry," Prentice-Hall, Inc., Englewood Cliffs, N.J., 1971

National Council of Teachers of Mathematics: The Metric System of Weights and Measures, *Twentieth Yearbook*, 1948; Emerging Practices in Mathematics Education, *Twenty-second Yearbook*, 1954; Growth of Mathematical Ideas, Grades K–12, *Twenty-fourth Yearbook*, 1959; Enrichment Mathematics for the Grades, *Twenty-seventh Yearbook*, 1963; Enrichment Mathematics for the High School, *Twenty-eighth Yearbook*, 1963; Topics in Mathematics for Elementary School Teachers, *Twenty-ninth Yearbook*, 1964; More Topics in Mathematics for Elementary Teachers, *Thirtieth Yearbook*, 1971; Washington.

Nelson, Jeanne: Percent: A Rational Number of a Ratio, *The Arithmetic Teacher*, **16:**105–109 (1969).

Ohmer, Merlin M.: "Elementary Geometry for Teachers," Addison-Wesley Publishing Company, Inc., Reading, Mass., 1969.

Peterson, John A., and Joseph Hashisaki: "Theory of Arithmetic," 3d ed., John Wiley & Sons, Inc., New York, 1971.

Piaget, Jean: "The Child's Conception of Number," Butler and Tanner, Ltd., London, 1952.

———, Barbel Inhelder, and Alina Szemknska: "The Child's Conception of Geometry," Basic Books, Inc., Publishers, New York, 1960.

Rudnick, Jesse A.: Numeration Systems and Their Classroom Roles, *The Arithmetic Teacher*, **15:**138–147 (1968).

Smart, James R.: "New Understanding in Arithmetic," Allyn

and Bacon, Inc., Boston, 1963.

———, "Introductory Geometry: An Informal Approach," 2d ed., Brooks/Cole Publishing Company, Belmont, Calif., 1972.

Smith, Frank: Divisibility: Rules for the First Fifteen Primes, *The Arithmetic Teacher*, **18:**85–87 (1971).

Smith, Lewis B.: Venn Diagrams Strengthen Children's Mathematical Understanding, *The Arithmetic Teacher*, **13:**92–99 (1966).

Stenger, Donald J.: Prime Numbers from the Multiplication Table, *The Arithmetic Teacher*, **16:**617–630 (1969).

Vaughan, Herbert E.: What Sets Are Not, *The Arithmetic Teacher*, **17:**55–60 (1970).

Wheeler, Ruric E.: "Modern Mathematics: An Elementary Approach," 2d ed., Brooks/Cole Publishing Company, Belmont, Calif., 1970.

Wilson, Patricia, Delbert Mundt, and Fred Porter: A Different Look at Decimals, *The Arithmetic Teacher*, **16:**95–98 (1969).

Wren, F. Lynwood: It's Not How New You Make It, but How You Make It New, *The Arithmetic Teacher*, **18:**7–9 (1971).

———: The "New Mathematics" in Historical Perspective, *The Mathematics Teacher*, **62:**579–585 (1969).

——— and John W. Lindsay: "Basic Algebraic Concepts," McGraw-Hill Book Company, New York, 1969.

Zink, Mary Hart: Greatest Common Divisor and Least Common Multiple, *The Arithmetic Teacher*, **13:**138–140 (1966).

ANSWERS TO ODD-NUMBERED EXERCISES

Sec. 1-3, page 8

1   Babylonian: a, b, c, d, e, f
Chinese-Japanese: a, b, c, d, g
Egyptian: a, b, c, f, g
Greek: a, b, c, g
Roman: a, b, c, e, f, g

Sec. 1-4, page 11

1   (a) 56; (b) 301; (c) 499; (d) 20,094; (e) 2,003,609; (f) 1,150,045
3   (a) 247; (b) 1,001; (c) 1,962; (d) 1,492; (e) 1,002,004; (f) 1,610,210
5   It provided a technique for using the concept of place value as an aid in computation.
7   (a) 35; (b) 607; (c) 981; (d) 7,426; (e) 8,982
9   (See table.)

EGYPTIAN	GREEK	ROMAN	MAYAN
(a)	$\nu\xi\zeta$	CDLXVII	
(b)	$\beta'\phi\eta$	MMDVIII	
(c)	$\mu'\psi\kappa$	$\overline{\text{X}}$LDCCXX	
(d)	$\alpha'\nu \varphi\beta$	MCDXCII	
(e)	$\gamma'\delta$	MMMIV	
(f)	$\nu'\phi$	$\overline{\text{CD}}$D	
(g)	$\rho\xi\alpha\text{M}\nu\lambda$	$\overline{\text{MDCXCD}}$XXX	
(h)	$\nu\kappa\text{M}\alpha'\omega \varphi\zeta$	$\overline{\text{MMMMCCI}}$DCCCXCVII	

Sec. 1-6, page 16

1 (a) 729; (b) 100,000,000; (c) 512; (d) 1,000,000; (e) 78,125; (f) 531,441
3 (a) 64; (b) 1,000,000,000,000; (c) 1,000,000; (d) 100,000,000
5 (a) 4; (b) 1,000; (c) 27; (d) 125; (e) $\frac{1}{10}$; (f) $\frac{1}{64}$; (g) $\frac{1}{100}$; (h) $\frac{1}{8}$
7 (a) 128; (b) 10,000,000,000; (c) 1,000; (d) 100,000,000; (e) 65,536; (f) 729
9 (a) $3(1,000) + 4(100) + 2(10) + 1$ or $3(10^3) + 4(10^2) + 2(10) + 1$
   (b) $2(100) + 0(10) + 2$ or $2(10^2) + 0(10) + 2$
   (c) $2(1,000,000) + 6(100,000) + 2(10,000) + 4(1,000) + 3(100) +$
     $2(10) + 1$ or $2(10^6) + 6(10^5) + 2(10^4) + 4(10^3) + 3(10^2) + 2(10) + 1$
   (d) $1(100,000) + 0(10,000) + 0(1,000) + 2(100) + 0(10) + 0$ or $1(10^5) +$
     $0(10^4) + 0(10^3) + 2(10^2) + 0(10) + 0$
   (e) $7(100) + 8(10) + 0$ or $7(10^2) + 8(10) + 0$
   (f) $5(1,000,000) + 0(100,000) + 0(10,000) + 6(1,000) + 7(100) +$
     $0(10) + 3$ or $5(10^6) + 0(10^5) + 0(10^4) + 6(10^3) + 7(10^2) + 0(10) + 3$

Sec. 1-8, page 20

1 History reveals that it first originated with the Hindus but was brought to Western Europe by the Arabs.
3 $1,056.013 = 1(10)^3 + 0(10)^2 + 5(10) + 6 + 0(10)^{-1} + 1(10)^{-2} + 3(10)^{-3}$
5 (a) *Mayan:* simpler notation; (b) *Babylonian:* place value and simpler notation; (c) *Egyptian:* place value and simpler notation; (d) *Roman:* place value and simpler notation
7 Its place value is divided by 10.
9 1 dime = 10 pennies; 1 dollar = 10 dimes = 100 pennies

Sec. 1-12, page 24

1 (a) **222**; (b) **42**; (c) **35**; (d) **32**; (e) **11010**
3 Base **five:** (a) **100**; (b) **140** (self-consistent is one word); (c) **111**; (d) **133**
   Base **twelve:** (a) **21**; (b) **39**; (c) **27**; (d) **37**
   Base **eight:** (a) **31**; (b) **55**; (c) **37**; (d) **53**
   Base **two:** (a) **11001**; (b) **101101**; (c) **11111**; (d) **101011**
5 (a) **23** in.; (b) **8t** in.; (c) **40** in.; (d) **ee** in.
7 (a) **20**; (b) **40**
9 (a) **3t** (where **t** is the symbol for ten); (b) $\overset{\cdots}{=\!=}$
11 **b** symbols will be needed (0 and **b** − 1 nonzero symbols).

Sec. 1-13, page 26

17 (a) 4,569; (b) 320,050; (c) 2,201,001; (d) 30,246
19 (a) 89; (b) 778; (c) 211; (d) 8,008; (e) 9,860; (f) 334; (g) 88,535; (h) 808,080
21 (a) 370; (b) 86; (c) 9,706; (d) 5,412; (e) 2,903
23 (a)   (b)   (c)   (d)   (e)   (f)

(g) 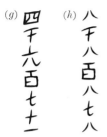 (h)

25 (a) 15,625; (b) 1,000,000,000; (c) 15,625; (d) $\frac{1}{10}$; (e) $\frac{64}{125}$; (f) 1; (g) 100; (h) $\frac{1}{36}$; (i) $\frac{1}{25}$

27 (a) **32**; (b) **11010**; (c) **20**; (d) **28**; (e) **1e**

Sec. 2-2, page 38

1 a, b, d, f, g, h, i, j

3 True: a, c, e False: b, d

5 *Roster form*
   (a) $V = \{e,u,a,i,o\}$
   (b) $W = \{6,7,8,9,10,11,12,13,14,15,16,17,18,19\}$
   (c) $O = \{1,3,5,7,9,11,13,15\}$
   (d) $L = \{c,o,l,e,g\}$
   (e) $M = \{$January, February, March, April, May, June, July, August, September, October, November, December$\}$
   (f) $P = \{a,i,u,o\}$
   (g) $F = \{$February$\}$
   (h) $E = \{2,4,6,8,10\}$
   Set-builder form
   (a) $V = \{*|*$ is a vowel in "equation"$\}$
   (b) $W = \{*|*$ is a whole number larger than 5 and less than 20$\}$
   (c) $O = \{*|*$ is an odd number between and including 1 and 15$\}$
   (d) $L = \{*|*$ is a letter in "college"$\}$
   (e) $M = \{*|*$ is the name of a month of a year$\}$
   (f) $P = \{*|*$ is a vowel in "ambiguous"$\}$
   (g) $F = \{*|*$ is the name of a month of less than 30 days$\}$
   (h) $E = \{*|* = 2k$ where $k \in \{1,2,3,4,5\}\}$

7 $C = \{a,b,c,r\}$

9 *Description form*
   (a) I is the set of whole numbers greater than 0 and less than 6.
   (b) N is the set of all counting numbers.
   (c) R is the set of all even numbers between 1 and 19.
   (d) S is the set of names of oceans of the world.
   (e) U is the set of names of all states of the United States which start with the letter U.
   (f) T is the set of names of months with exactly 30 days.
   Set-builder form
   (a) $I = \{\#|\#$ is a whole number greater than 0 and less than 6$\}$
   (b) $N = \{\#|\#$ is a counting number$\}$
   (c) $R = \{\#|\#$ is an even number between 0 and 19$\}$
   (d) $S = \{\#|\#$ is the name of an ocean of the world$\}$

(e) $U = \{\# \mid \#$ is the name of a state of the United States starting with $U\}$

(f) $T = \{\# \mid \#$ is the name of a month of exactly 30 days$\}$

11 $\subseteq$ is not symmetric; $\subset$ is neither reflexive nor symmetric.

13 $c, e, g, h, j, k, l, m, n, o, s$

15 (a) No; is neither reflexive nor transitive

(b) No; is neither reflexive nor symmetric

(c) Yes; is reflexive, symmetric, and transitive

(d) No; is not transitive

(e) Yes; is reflexive, symmetric, and transitive

(f) No; is neither reflexive nor symmetric

(g) Yes; is reflexive, symmetric, and transitive

(h) No; is neither reflexive nor symmetric

(i) Yes; is reflexive, symmetric, and transitive

(j) No; is not reflexive, not symmetric, and not transitive

(k) No; is not symmetric

(l) Yes; is reflexive, symmetric, and transitive

(m) Yes; is reflexive, symmetric, and transitive

Sec. 2-3, page 45

1 (a) $I = \{1,2,3,4, \ldots\}$

(b) $S = \{6,7,8,9, \ldots, 19\}$

(c) $G = \{$Florida, Alabama, Mississippi, Louisiana, Texas$\}$

(d) $L = \{$Superior, Michigan, Huron, Ontario, Erie$\}$

(e) $E = \{2,4,6,8,10, \ldots\}$

(f) $M = \{$Texas, Arizona, California, New Mexico$\}$

3 No. $M \cap N = \{0\}$, and so it is not the empty set

5 (a) False; (b) false; (c) false; (d) true; (e) false; (f) true; (g) false; (h) true; (i) false; (j) true; (k) true; (l) false; (m) false; (n) true; (o) true

7 (a) False; (b) false; (c) true; (d) false; (e) false; (f) false; (g) true

9 $\sim A$; $\varnothing$

11 See figure on page 457. Other correct diagrams are possible.

13 (a) False: $C - M = \{t\}$; $M - C = \{e,m\}$

(b) False: There are elements in each which are not in the other.

(c) True: $C - M = \{t\}$; t is an element of C.

(d) False: $M \cap (C - M) = \varnothing$ and not M

15 It is a true statement since $\varnothing = \varnothing$.

17 (a) $P = \{d,e,f\}$; (b) $P = \{a,d,e,f\}$; (c) $P = \{a,c,d\}$; (d) $P = \{d,e,f\}$; (e) $P = \{a,b,c\}$

19 $Q \cap P$ 21 $(P \cap Q) \cup (P \cap R)$ 23 $P \cap (Q - R)$

25 $\sim(\sim P)$ 27 $Q \cap \sim P$ or $Q - P$ 29 $\sim\varnothing$

31 (a) $P \cup R$ or $P \cup (Q \cup R)$ or $(P \cup Q) \cup R$

(b) $P \cap R$

(c) $P - Q$

(d) $Q \cup (P \cap R)$ or $(Q \cup P) \cap (Q \cup R)$

(e) $P \cup Q$

(f) $R - P$ or $(Q \cup R) - P$

(g) $R - Q$

(h) $Q \cup (R - P)$

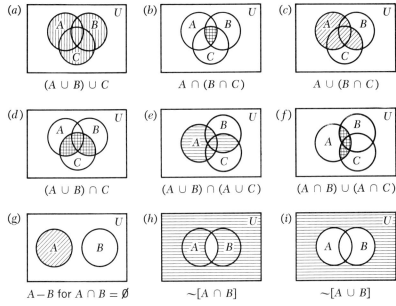

(a) $(A \cup B) \cup C$

(b) $A \cap (B \cap C)$

(c) $A \cup (B \cap C)$

(d) $(A \cup B) \cap C$

(e) $(A \cup B) \cap (A \cup C)$

(f) $(A \cap B) \cup (A \cap C)$

(g) $A - B$ for $A \cap B = \emptyset$

(h) $\sim[A \cap B]$

(i) $\sim[A \cup B]$

EXERCISE 11,
SEC. 2-3

    (i) $P - R$

    (j) $Q - R$ or $Q - (P \cap R)$ or $(P - R) \cap Q$

    (k) $(P - Q) - R$

    (l) $(P - Q) \cap R$

33 (a) $\{a,e,t,n\}$

    (b) $\{a,c,d,e,i,l,m,n,o,r,t,u,y\}$

    (c) $\{b,c,d,f,q,h,i,j,k,o,p,q,s,u,v,w,x,z\}$

    (d) $\{b,f,g,h,j,k,l,m,p,q,r,s,v,w,x,y,z\}$

    (e) $\{b,f,g,h,j,k,p,q,s,v,w,x,z\}$

    (f) $\{l,m,r,y\}$

    (g) $\{d,u,c,i,o\}$

    (h) $\{c,d,i,o,u\}$

    (i) $\{l,m,r,y\}$

    (j) $\{b,c,d,f,g,h,i,j,k,l,m,o,p,q,r,s,u,v,w,x,y,z\}$

35 (a) (See figure on page 458.)

      $(P - Q) \cup (Q - P) = (P \cup Q) - (P \cap Q)$ True

      $(P - Q) \cup (Q - P) = \sim(P \cap Q)$ False

    (b) $(P - Q) \cup (Q - P)$ is a set whose elements are all those in P but not in Q and those in Q but not in P.

      $(P \cup Q) - (P \cap Q)$ is a set whose elements are all those elements of $P \cup Q$ after those common to P and Q have been removed.

      $\sim(P \cap Q)$ is a set whose elements are those of the universal set after those common to P and Q have been removed.

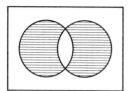 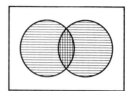 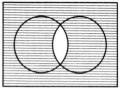

EXERCISE 35,
SEC. 2-3

$(P - Q) \cup (Q - P) \equiv$ $P \cup Q \equiv, \boxplus; P \cap Q \boxplus$ $\sim (P \cap Q) \equiv$

Sec. 2-4, page 50

1 (See figure.) $Y = \{\emptyset, A, B, Q\}$ where $A = \{1\}$ and $B = \{2\}$.

∪	∅	A	B	Q
∅	∅	A	B	Q
A	A	A	Q	Q
B	B	Q	B	Q
Q	Q	Q	Q	Q

∩	∅	A	B	Q
∅	∅	∅	∅	∅
A	∅	A	∅	A
B	∅	∅	B	B
Q	∅	A	B	Q

EXERCISE 1,
SEC. 2-4

3 The elements of S may be represented as follows:
$\emptyset$, $A = \{0\}$, $B = \{1\}$, $C = \{2\}$, $D = \{3\}$, $E = \{0,1\}$, $F = \{0,2\}$, $G = \{0,3\}$,
$H = \{1,2\}$, $I = \{1,3\}$, $J = \{2,3\}$, $K = \{0,1,2\}$, $L = \{0,1,3\}$, $M = \{0,2,3\}$,
$N = \{1,2,3\}$, R. (See diagrams.)

∪	∅	A	B	C	D	E	F	G	H	I	J	K	L	M	N	R
∅	∅	A	B	C	D	E	F	G	H	I	J	K	L	M	N	R
A	A	A	E	F	G	E	F	G	K	L	M	K	L	M	R	R
B	B	E	B	H	I	E	K	L	H	I	N	K	L	R	N	R
C	C	F	H	C	J	K	F	M	H	N	J	K	R	M	N	R
D	D	G	I	J	D	L	M	G	N	I	J	R	L	M	N	R
E	E	E	E	K	L	E	K	L	K	L	R	K	L	R	R	R
F	F	F	K	F	M	K	F	M	K	R	M	K	R	M	R	R
G	G	G	L	M	G	L	M	G	R	L	M	R	L	M	R	R
H	H	K	H	H	N	K	K	R	H	N	N	K	R	R	N	R
I	I	L	I	N	I	L	R	L	N	I	N	R	L	R	N	R
J	J	M	N	J	J	R	M	M	N	N	J	R	R	M	N	R
K	K	K	K	K	R	K	K	R	K	R	R	K	R	R	R	R
L	L	L	L	R	L	L	R	L	R	L	R	R	L	R	R	R
M	M	M	R	M	M	R	M	M	R	R	M	R	R	M	R	R
N	N	R	N	N	N	R	R	R	N	N	N	R	R	R	N	R
R	R	R	R	R	R	R	R	R	R	R	R	R	R	R	R	R

EXERCISE 3,
SEC. 2-4

∩	∅	A	B	C	D	E	F	G	H	I	J	K	L	M	N	R
∅	∅	∅	∅	∅	∅	∅	∅	∅	∅	∅	∅	∅	∅	∅	∅	∅
A	∅	A	∅	∅	∅	A	A	A	∅	∅	∅	A	A	A	∅	A
B	∅	∅	B	∅	∅	B	∅	∅	B	B	∅	B	B	∅	B	B
C	∅	∅	∅	C	∅	∅	C	∅	C	∅	C	C	∅	C	C	C
D	∅	∅	∅	∅	D	∅	∅	D	∅	D	D	∅	D	D	D	D
E	∅	A	B	∅	∅	E	A	A	B	B	∅	E	E	A	B	E
F	∅	A	∅	C	∅	A	F	A	C	∅	C	F	A	F	C	F
G	∅	A	∅	∅	D	A	A	G	∅	D	D	A	G	G	D	G
H	∅	∅	B	C	∅	B	C	∅	H	B	C	H	B	C	H	H
I	∅	∅	B	∅	D	B	∅	D	B	I	D	B	I	D	I	I
J	∅	∅	∅	C	D	∅	C	D	C	D	J	C	D	J	J	J
K	∅	A	B	C	∅	E	F	A	H	B	C	K	E	F	H	K
L	∅	A	B	∅	D	E	A	G	B	I	D	E	L	G	I	L
M	∅	A	∅	C	D	A	F	G	C	D	J	F	G	M	J	M
N	∅	∅	B	C	D	B	C	D	H	I	J	H	I	J	N	N
R	∅	A	B	C	D	E	F	G	H	I	J	K	L	M	N	R

EXERCISE 3, SEC. 2-4

Sec. 2-6, page 57

1 $A \times B = \{(1,a), (1,b), (2,a), (2,b), (3,a), (3,b)\}$
    $B \times A = \{(a,1), (a,2), (a,3), (b,1), (b,2), (b,3)\}$

3 6 in each

5 $A \times A = \{(1,1), (1,2), (1,3), (2,1), (2,2), (2,3), (3,1), (3,2), (3,3)\}$
    $B \times B = (a,a), (a,b), (b,a), (b,b)$

7 (See figure.)

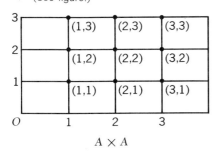

$A \times A$

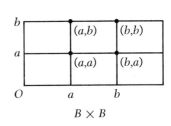

$B \times B$

EXERCISE 7, SEC. 2-6

9 $\{(2,1), (3,1), (3,2)\}$

11 (a,b) is the ordered pair whose first element is a and whose second element is b. $\{a,b\}$ is the set whose elements are a and b. $\{(a,b)\}$ is the set whose one element is the ordered pair (a,b).

13 {(Nashville, Tennessee), (Atlanta, Georgia), (Sacramento, California), (Austin, Texas), (Columbia, Missouri), (Lincoln, Nebraska)}

15 *Relation*   *Domain*   *Range*

Relation	Domain	Range
Is equal to	{1,2,3}	{1,2,3}
Is greater than	{2,3}	{1,2}
Is the capital of	C	S
Is the name of a team in the	T	L

17 *Relation*   *Domain*   *Range*

	Domain	Range
(*a*)	{1,2,3,4}	{1,2,3,4}
(*b*)	{1,2,3}	{2,3,4}
(*c*)	{2,3,4}	{1,2,3}
(*d*)	{1,2,3,4}	{1,2,3,4}
(*e*)	{1,2,3,4}	{1,2,3,4}
(*f*)	{1,2,3,4}	{1,2,3,4}

Sec. 2-9, page 63

1 When to each element of set P there corresponds one and only one element of set Q and to each element of Q there corresponds one and only one element of P.

3 When a one-to-one correspondence exists between pupils and chairs.

5 It is the property the set has in common with every set with which it can be put into one-to-one correspondence.

11 (*a*) 0; (*b*) 1; (*c*) 0; (*d*) 7; (*e*) 60

13 {Linda}; {Linda, Sharon}; {Linda, Sharon, Diana}; {Linda, Sharon, Diana, Marly}; {Linda, Sharon, Diana, Marly, Cathy}; {Linda, Sharon, Diana, Marly, Cathy, Carol}

15 (*a*) ⊥I△; (*b*) □△0I; (*c*) ⊥0⊥⊥

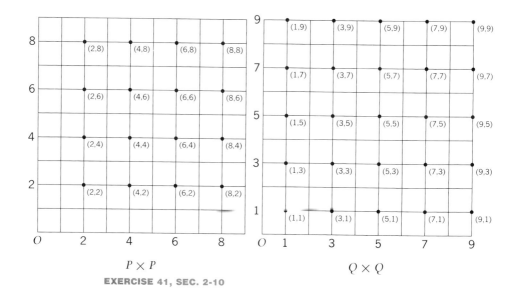

$P \times P$ $Q \times Q$

EXERCISE 41, SEC. 2-10

Sec. 2-10, page 65

27 (*a*) {1,3,5,7,9,11,13,15}; (*b*) {2,4,6}; (*c*) {1,2,3,4,5,6,8,10,12,14}; (*d*) ∅;
   (*e*) {2,4,6,7,8,9,10,11,12,13,14}; (*f*) *R*; (*g*) ∅; (*h*) *U*; (*i*) {8,10,12,14}; (*j*) *E*;
   (*k*) {1,3,5}; (*l*) *R*; (*m*) ∅; (*n*) *S*; (*o*) *R*

29 (*a*) True; (*b*) false; (*c*) true; (*d*) true; (*e*) false; (*f*) true; (*g*) true; (*h*) false;
   (*i*) true; (*j*) true; (*k*) true; (*l*) false; (*m*) false; (*n*) false; (*o*) true; (*p*) true;
   (*q*) true; (*r*) true; (*s*) false; (*t*) false; (*u*) false; (*v*) true; (*w*) false; (*x*) true;
   (*y*) true

31 (*a*) *True.* If *A* and *B* have the same elements, then they have the same cardinal
   number. (*b*) *False.* *A* and *B* may have different elements yet have the same
   cardinal number.

35 They are all equivalent.

37 Relative complementation is closed, but is neither associative nor commutative.

39 {(2,3), (2,5), (2,7), (2,9), (4,5), (4,7), (4,9), (6,7), (6,9), (8,9)}

41 (See figure on page 460.)

45 (*a*) **1000;** (*b*) ⊥□; (*c*) ⊥00△; (*d*) □△⊥; (*e*) □I□⊥

47 (*a*) ⊥I; (*b*) □□; (*c*) II□

Sec. 3-6, page 82

In these answers these symbols are used: A = Associative; Cl = Closure;
Co = Commutative; PV = Place value

1	Cl	3	Cl	5	Co	7	A
9	PV, A, Co	11	A, Co	13	A, Co	15	A, Co

Sec. 3-8, page 85

3 Since we use a decimal numeral system, "carrying" consists of regrouping in
   terms of powers of ten.

5 (*a*) 2,588; (*b*) 1,139; (*c*) 2,611; (*d*) 2,765; (*e*) 1,658; (*f*) 2,199; (*g*) 2,428;
   (*h*) 3,296; (*i*) 2,699; (*j*) 5,289; (*k*) 35,002; (*l*) 7,990; (*m*) 304,641;
   (*n*) 178,824; (*o*) 19,908,288; (*p*) 27,235,565; (*q*) 136,409,105

9 The table is symmetric with the diagonal from the upper left corner to the lower
   right corner.

11 Associative

13 (See figures on page 462.)

17 (*a*) **137;** (*b*) **1605;** (*c*) **20404;** (*d*) **15375;** (*e*) **20135;** (*f*) **10565**

19 (*a*) **2100;** (*b*) **20015;** (*c*) **110300;** (*d*) **122312**

21 (See figures on pages 462 and 463.)

23 (*a*) **1158;** (*b*) **24t40;** (*c*) **15388;** (*d*) **19691**

Sec. 3-12, page 95

In these answers these symbols are used: A = Associative; Cl = Closure;
Co = Commutative; D = Distributive; PV = Place value

1	Cl	3	Cl	5	Co	7	D

9 (20 + 6) is substituted for its equal 26. D and PV

11 3 × 2 is substituted for *b*; A and Co

Five

+	1	2	3	4
1	2	3	4	10
2	3	4	10	11
3	4	10	11	12
4	10	11	12	13

(*a*)

EXERCISE 13, SEC. 3-8

Eight

+	1	2	3	4	5	6	7
1	2	3	4	5	6	7	10
2	3	4	5	6	7	10	11
3	4	5	6	7	10	11	12
4	5	6	7	10	11	12	13
5	6	7	10	11	12	13	14
6	7	10	11	12	13	14	15
7	10	11	12	13	14	15	16

(*b*)

Seven

+	1	2	3	4	5	6
1	2	3	4	5	6	10
2	3	4	5	6	10	11
3	4	5	6	10	11	12
4	5	6	10	11	12	13
5	6	10	11	12	13	14
6	10	11	12	13	14	15

(*a*)

Eleven

+	1	2	3	4	5	6	7	8	9	t
1	2	3	4	5	6	7	8	9	t	10
2	3	4	5	6	7	8	9	t	10	11
3	4	5	6	7	8	9	t	10	11	12
4	5	6	7	8	9	t	10	11	12	13
5	6	7	8	9	t	10	11	12	13	14
6	7	8	9	t	10	11	12	13	14	15
7	0	9	l	10	11	12	13	14	15	16
8	9	t	10	11	12	13	14	15	16	17
9	t	10	11	12	13	14	15	16	17	18
t	10	11	12	13	14	15	16	17	18	19

EXERCISE 21,
SEC. 3-8

(*b*)

Twelve

+	1	2	3	4	5	6	7	8	9	t	e
1	2	3	4	5	6	7	8	9	t	e	10
2	3	4	5	6	7	8	9	t	e	10	11
3	4	5	6	7	8	9	t	e	10	11	12
4	5	6	7	8	9	t	e	10	11	12	13
5	6	7	8	9	t	e	10	11	12	13	14
6	7	8	9	t	e	10	11	12	13	14	15
7	8	9	t	e	10	11	12	13	14	15	16
8	9	t	e	10	11	12	13	14	15	16	17
9	t	e	10	11	12	13	14	15	16	17	18
t	e	10	11	12	13	14	15	16	17	18	19
e	10	11	12	13	14	15	16	17	18	19	1t

**EXERCISE 21,
SEC. 3-8**

(c)

Sec. 3-13, page 96

1	1,036	3	2,169	5	25,185	7	64,169
9	203,721	11	437,072	13	49,584	15	147,614
17	1,371,633	19	6,028,672	21	2,644,462	23	28,371,165

Sec. 3-14, page 98

1 There is no ambiguity about what the product is.
3 Since we use a decimal numeral system, "carrying" consists of regrouping in terms of powers of ten.
7 The sum of each column is 2,476.
9 (a) 180; (b) 300; (c) 60; (d) 286; (e) 312; (f) 396; (g) 231; (h) 204
11 Each system is closed and commutative under multiplication
13 (a) I0⊥I; (b) ⊥□I□⊥△; (c) I⊥□□0□△
15 4,275
17 (a) $63 \times 36 = 63 \times (4 \times 9) = 63 \times (9 \times 4) = 2,268$
    (b) $85 \times 28 = 85 \times (4 \times 7) = 85 \times (7 \times 4) = 2,380$
    (c) $272 \times 54 = 272 \times (6 \times 9) = 272 \times (9 \times 6) = 14,688$
19 (a) **2210**; (b) **240322**; (c) **2640724**
21 (a) **15t8**; (b) **t912**; (c) **3417552**; (d) **t7t16742**
23 (a) 1,976; (b) 18,870; (c) 221,058; (d) 17,464; (e) 1,909,848; (f) 868,944

Sec. 3-17, page 109

1 Under addition: The set of all even natural numbers
   Under multiplication: The set of all composite natural numbers
   The set of all odd natural numbers
   The set of all even natural numbers

3 (a) $2 \times 2 \times 5 \times 7$; (b) $2 \times 2 \times 2 \times 23$; (c) $2^2 \times 3^2 \times 5 \times 11$;
(d) $2 \times 3 \times 5^2 \times 7^2$

5 2; 3; 5; 7; 11; 13; 17; 19; 23; 29; 31; 37; 41; 43; 47; 53; 59; 61; 67; 71; 73; 79;
83; 89; 97; 101; 103; 107; 109; 113; 127; 131; 137; 139; 149; 151; 157; 163;
167; 173; 179; 181; 191; 193; 197; 199

7 1; the numbers are relatively prime.

9 It selects all the prime factors of each number.

11 (See figure.)

Numeral	Base					
	two	three	five	seven	ten	twelve
2	n	p	p	p	p	p
10	p	p	p	p	c	c
11	p	c	c	c	p	p
23	n	n	p	p	p	c
32	n	n	p	p	c	c
47	n	n	n	n	p	c
54	n	n	n	c	c	c

EXERCISE 11,
SEC. 3-18

13 g.c.f. 6; l.c.m. 22,464

15 g.c.f. 1; l.c.m. 30,030

17 (a) Yes; (b) yes; (c) yes

Sec. 3-18, page 111

29 (a) 335; (b) 3,253; (c) 27,155

31 (a) **1712**; (b) **22120**; (c) **31553**; (d) **13223**; (e) **11351**; (f) **15075**

33 (a) 7,644; (b) 182,007; (c) 162,212

37 All the primes listed in the answer to Exercise 5 of Sec. 3-17 and, in addition,
these primes: 211; 223; 227; 229; 233; 239; 241; 251; 257; 263; 269; 271; 277;
281; 283; 293.

39 (a) g.c.f. 15; l.c.m. 40,950
(b) g.c.f. 2; l.c.m. 1,900,080
(c) g.c.f. 1; l.c.m. 14,467
(d) g.c.f. 1; l.c.m. 30,030
(e) g.c.f. 1; l.c.m. 6,630
(f) g.c.f. 38; l.c.m. 43,890

Sec. 4-1, page 120

1 (a) 15 since $10 + 15 = 25$
(b) 7 since $9 + 7 = 16$
(c) No natural number exists since $9 \not< 7$
(d) No natural number exists since $11 \not< 8$
(e) 80 since $25 + 80 = 105$
(f) No natural number exists since $302 \not< 302$

3 (a) $q = 9$ since $45 = 5 \times 9$
   (b) $q = 5$ since $70 = 14 \times 5$
   (c) $q = 8$, $r = 2$ since $26 = 3 \times 8 + 2$
   (d) $q = 1$ since $80 = 80 \times 1$
   (e) $q = 23$, $r = 25$ since $715 = 30 \times 23 + 25$
   (f) $q = 25$ since $625 = 25 \times 25$

5 Exercises 1 and 2 show that the natural numbers are not closed under subtraction. Exercises 3 and 4 show that the natural numbers are not closed under division. Since these two operations are closely related to addition and multiplication it is desirable that these restrictions be removed.

Sec. 4-2, page 125

1 (a) True; (b) false; (c) false; (d) true; (e) true; (f) false; (g) true; (h) false;
   (i) true; (j) true; (k) false; (l) true

3 (a) Let 4 be the cardinal number of set P and 7 the cardinal number of set Q
      where $P \cap Q = \varnothing$.
   (b) Since $A \cap \varnothing = \varnothing$ and $0 = n(\varnothing)$, let 6 be the cardinal number of set A.
   (c) Let 2 be the cardinal number of set P and 6 the cardinal number of set Q
      where $P \cap Q = \varnothing$.
   (d) Since $\varnothing \cap \varnothing = \varnothing$ and $0 = n(\varnothing)$.

5 No. The integer i is negative if and only if $i < 0$.

7 Not necessarily, since it is possible that $a < 0$, in which case $^-a > 0$.

9 Yes, since $^-4 = -4$.

11 Not necessarily, since it is possible that $a = 0$.

13 $a = {}^-2$ by Postulate 4-4

15 $i = 0$ by Postulate 4-3

17 $n = 16$ 19 $n = 15$ 21 $n = 100$

23 Each is the additive inverse of the other by Postulate 4-4

25 $n = 7$ by the cancellation law

27 $a = {}^-3$ 29 $a = b$ 31 $a = b$

33 $a = 6$ 35 $a = 8$ 37 $a = {}^-5 = -5$

Sec. 4-4, page 132

1 The sum of 0 and any specific integer is that integer

3 (a) **3**; (b) **3**; (c) **4**; (d) **5**

11 (a) Associative; (b) commutative and associative; (c) commutative, associative, and additive inverse; (d) commutative, associative, Theorem 4-3, and table of basic facts

13 (a) 30; (b) 25; (c) 244; (d) 109

15 (a) 2,141; (b) 3,743; (c) 1,743; (d) 1,548

17 (a) **101**; (b) **101**; (c) **112**; (d) **1101**

19 (a) Commutative, associative, addition facts, and place value; (b) associative, Theorem 4-3, commutative, additive inverse, addition facts, and place value; (c) associative, Theorem 4-2, and addition facts

23 $^-4$ or -4 25 0   27 5   29 2   31 3

33 13   35 $^-11$ or -11 37 4   39 2

Sec. 4-6, page 137

1 Each operation annuls (cancels) the effect of the other. $(a + b) - b = a$; $(a - b) + b = a$.

3 To compare two numbers; to find the difference between two numbers; to find how many are left after some are removed.

5 (a) No: $a + a = 2a$, which is not in the set.
   (b) No: $a - (-a) = a + a = 2a$, which is not in the set.
   (c) No: $a \times (-a) = -a^2$, which is not in the set.

7 The sum of the excess of nines in the difference and the excess in the subtrahend must be the same as the excess in the minuend.

9 (a) 1,423; (b) 1,978; (c) 3,086

13 (a) **4443**; (b) **4306**; (c) **26t6**; (d) **1e1e**; (e) **899**; (f) **5018**

15 (a) **375**; (b) **612**; (c) **667**; (d) **2505**; (e) **56**; (f) **13**

17 (a) **1212**; (b) **632**; (c) **46**; (d) **253**; (e) **42**; (f) **3045**

Sec. 4-8, page 144

1 (a) Commutative; (b) associative; (c) distributive; (d) multiplicative identity

3 (a) 15; (b) 4; (c) -8; (d) 28; (e) 1; (f) -6; (g) 20; (h) -24; (i) 1

5 (a) 20; (b) 50; (c) 24; (d) -20; (e) 14; (f) -12; (g) -12; (h) -27; (i) -100; (j) 48; (k) 60; (l) -8; (m) -18; (n) 80; (o) 6

7 (a) $x = 2$; (b) $x = -5$; (c) $x = 4$; (d) $x = -3$

11 (a) 921,472; (b) 624,404; (c) 5,572,800; (d) 21,149,408; (e) 43,186,392; (f) 916,861,232

13 (a) **1022**; (b) **2001**; (c) **12201**; (d) **1202212**; (e) **11002202**; (f) **21021210**

15 (a) **1737**; (b) **2536**; (c) **2251**; (d) **17103**; (e) **163025**; (f) **323125**; (g) **4554218**; (h) **31650650**; (i) **28853723**

17 (a) 25,784,076; (b) 41,500,702; (c) 29,967,557

19 (a) **t2e3916**; (b) **27e1e116**; (c) **t68e23e4**

Sec. 4-9, page 150

1 (a) There is no numeral **6** in base **five.**
   (b) There are no numerals **e** and **t** in base ten.
   (c) There is no numeral **9** in base **nine.**

3 **62—four; 28—nine**

5 (a) **10**; (b) **100**; (c) **1000**; (d) **10000**; (e) **100000**; (f) **101**; (g) **110**; (h) **111**

7 Annex a zero; to multiply by 10^n annex n zeros.

9 (a) **306t**; (b) **12222**; (c) **40214**

11 (a) 20,590; (b) **31217**; (c) **11001232**

13 (a) **20330**; (b) **938**; (c) **10112**

Sec. 4-10, page 153

1 (a) 1, 2, 3, 6; (b) 2, 0, (c) 1, 2, 3

3 (a) 1, 3, 11, 33; (b) 3, 11; (c) 1, 3, 11

5 (a) 1, 3, 9, 27, 81; (b) 3; (c) 1, 3, 9, 27

7 (a) 1, 7, 49; (b) 7; (c) 1, 7

9 (a) 1, 2, 3, 5, 6, 7, 10, 14, 15, 21, 30, 35, 42, 70, 105, 210
   (b) 2, 3, 5, 7

(c) 1, 2, 3, 5, 6, 7, 10, 14, 15, 21, 30, 35, 42, 70, 105

11 (a) 1, 2, 4, 8, 16, 32, 64; (b) 2; (c) 1, 2, 4, 8, 16, 32

13 (a) 1, p, p^2; (b) p; (c) 1, p

15 (a) 1, p, p^2, p^3, p^4; (b) p; (c) 1, p, p^2, p^3

17 Not prime   19 Prime   21 Not prime   23 Prime

25 311, 313; 347, 349

27 $8 = 3 + 5$

29 $22 = 3 + 19$; $5 + 17$; or $11 + 11$

31 $34 = 3 + 31$; $5 + 29$; $11 + 23$; or $17 + 17$

33 $122 = 13 + 109$; $19 + 103$; $43 + 79$; or $61 + 61$

Sec. 4-12, page 159

1 To find the quotient; to compare two numbers; to find how many small sets of objects there are contained in a larger set; to find how many objects will be in each set when a larger set is separated into several smaller sets with the same cardinal number

3 They are inverse relations.

7 First, find the excess of nines in the quotient and in the divisor. Second, find the excess in the product of these two excesses. Third, find the excess of nines in the remainder. The excess in the sum of the excesses found in the second and third steps should then be the same as the excess in the dividend.

9 If the divisor is subtracted from the dividend the number of times indicated by the quotient, the final remainder should be the same as the remainder obtained in the division.

11 (a) 52; (b) 105; (c) 83

13 (a) $6{,}798 = 206 \cdot 33 + 0$

(b) $9{,}672 = 537 \cdot 18 + 6$

(c) $41{,}935 = 130 \cdot 321 + 205$

15 (a) 35; (b) 231; (c) 537; (d) 102

17 If the greatest common divisor (the last divisor of the algorithm) is 1, the numbers are relatively prime.

19 (a) **154 R 2**; (b) **115 R 22**; (c) **507**; (d) **67**; (e) **213 R 252**; (f) **2042**

21 (a) **167—nine**; (b) **34—nine**

Sec. 4-13, page 163

1 **22503—six; 1t33—twelve**

3 (a) **e01t—twelve**; (b) **1,859—ten; 24414—five**; (c) **51540—seven**;

(d) **200112—four; 2750—nine**

Sec. 4-14, page 163

25 To provide a system which will be closed under subtraction as well as under addition and multiplication.

27 $0 = n(\varnothing)$ and $7 = n(S)$ where $S \cap \varnothing = \varnothing$.

29 i

31 (a) 0; (b) $^-2 = -2$; (c) 5; (d) 3, (e) $^-2 - -2$; (f) 6; (g) 8; (h) 7; (i) 14; (j) 1

33 (a) -29; (b) 4; (c) -1

35 (a) 28,231; (b) 212,270; (c) 153,635

37   (*a*) **3113**; (*b*) **71**; (*c*) **26006**
39   (*a*) D-3, D-4, and D-6 and the fact that $^-i = -i$, when $i > 0$.
41   In order from left to right the integers are -7, -6, -3, -1, 0, 1, 4, 5, 7, 8
43   (*a*) **365267**; (*b*) **−130602**; (*c*) **100101011**; (*d*) **−8266e92**; (*e*) **22133212**;
     (*f*) **14300415**
45   (*a*) **tete**; (*b*) 217; (*c*) **2735**
47   (*a*) 0 since $0 \times 4 = 0$; (*b*) undefined
49   (*a*) Composite; (*b*) prime; (*c*) composite; (*d*) prime; (*e*) prime; (*f*) prime;
     (*g*) composite; (*h*) prime
51   (*a*) g.c.d. 317, l.c.m. 37,089 × 34 or 10,778 × 117
     (*b*) g.c.d. 263, l.c.m. 121,506 × 85 or 22,355 × 462

Sec. 5-1, page 174

1   Abelian group
3   Abelian group
5   Not a group. There exists no inverse for any element of the set other than for
    the identity 1.
7   Not a group. There exists no inverse for any element of the set other than for
    the identity 1.
9   Not a group. There is no identity element.
11 Not a group. There exists no inverse for any element of the set other than for
    the identity 1.
13 Not a group. The set is not closed since d is not in the set.

Sec. 5-3, page 180

1   In Figure 5-8 there occur only the remainders upon division by 5. In the table
    of basic addition facts the actual sums would occur. For example, in base **five**
    3 + 2 = 10 and **4 + 3 = 12**. In the table of modular addition $3 + 2 \equiv 0$ and
    $4 + 3 \equiv 2$. These are numerals in base ten.
3   There are 365 days in a year except for a leap year in which there are 366 days.
    $365 \equiv 1$ (mod 7) and $366 \equiv 2$ (mod 7).
7   Not a group. 0 has no inverse.
9   An abelian group.
11 Not a group. The set does not have closure. Also, 2, 3, and 4 have no inverses.
13 Not a group. 0 has no inverse.
17 In each case the smaller number is the remainder obtained when the larger
    number is divided by 7.

Sec. 5-4, page 186

1   (*a*) Is divisible; (*b*) is not divisible; (*c*) is not divisible
3   (*a*) The number formed by the last four digits is a number divisible by 16.
    (*b*) If T = thousands digit, h = hundreds digit, t = tens digit, and u = ones digit,
         then the test is that $8T + 4h + 10t + u$ be divisible by 16.
5   The number formed by the last two digits is divisible by 25.
7   The number formed by the last four digits is divisible by 625.
9   $10 \equiv 10$ (mod 25), and so both tests for divisibility by 25 become the same test;
    $10 \equiv 10$ (mod 125) and $100 \equiv 100$ (mod 125), and so both tests for 125 become
    the same test.

11 Divisible by 2, 3, 4, 5, 6, 12, 15
13 Divisible by 2, 3, 4, 6, 9, 11, 12, 18, 22
15 Divisible by 2, 11, 22
17 Divisible by 2, 3, 4, 5, 6, 8, 11, 12, 15, 22, 24
19 **2:** Sum of the digits divisible by **2; 4:** same as for **2**
21 **2:** Sum of the digits divisible by **2**
    3: Digit in ones place divisible by **3**
    4: Same as for **2**
    6: Combination of tests for **2** and **3**
    8: Same as for **2**

Sec. 5-5, page 186

19 If we use S_k (read "S sub k") to represent the set of positive integers congruent to zero modulo k, the relations between the sets of Exercise 18 are $S_{10} \subset S_2$; $S_{10} \subset S_5$; $S_9 \subset S_3$; $S_6 \subset S_2$; $S_6 \subset S_3$; $S_4 \subset S_2$.
21 (a) 2; (b) 0; (c) 3; (d) 4; (e) 3; (f) 0; (g) 0; (h) 4; (i) 4; (j) 4; (k) 2; (l) 0; (m) 0; (n) 0; (o) 0; (p) 0; (q) 6; (r) 5; (s) 2; (t) 7
23 No; 0, 2, 4, and 6 do not have inverse elements.
25 Yes
27 If the operation is addition, the set will have no identity element; if the operation is multiplication, the factors of the modulus will have no inverse elements.
29 In each case the sum of the digits must be divisible by the respective number, **2, 3,** or **6.**
31 For **2** and **4** the units digit must be divisible by **2** and **4,** respectively. For 7 the sum of the digits must be divisible by **7.**

Sec. 6-1, page 195

1 (a), (b), (d), (e), (f), (h), (i), (j), (k), (l), (m), (n), (p), (q), (s), (u)

Sec. 6-2, page 204

3 (a) $\dfrac{29}{21}$; (b) $\dfrac{10}{16}$ or $\dfrac{5}{8}$; (c) $\dfrac{109}{66}$; (d) $\dfrac{137}{28}$; (e) $\dfrac{43}{10}$; (f) $\dfrac{135}{36}$; (g) $\dfrac{2}{24}$ or $\dfrac{1}{12}$;
   (h) $\dfrac{-62}{45}$; (j) $\dfrac{-1}{6}$

7 (a) No; (b) Rational numbers are not associative under subtraction

11 (a) $\dfrac{8}{36}$ or $\dfrac{2}{9}$; (b) $\dfrac{7}{15}$; (c) $\dfrac{6}{18}$ or $\dfrac{1}{3}$; (d) $\dfrac{21}{16}$; (e) $\dfrac{90}{45}$ or $\dfrac{2}{1}$ or 2;
   (f) $\dfrac{-21}{6}$ or $\dfrac{-7}{2}$; (g) $\dfrac{126}{-315}$ or $\dfrac{-2}{5}$; (h) $\dfrac{42}{78}$ or $\dfrac{7}{13}$; (i) $\dfrac{-42}{35}$ or $\dfrac{-6}{5}$

33 $S = \left\{\dfrac{0}{1}, \dfrac{0}{2}, \dfrac{0}{3}, \dfrac{0}{4}, \dfrac{0}{5}, \dfrac{0}{6}, \dfrac{0}{7}, \dfrac{0}{8}, \dfrac{0}{9}\right\}$

35 $S = \left\{\dfrac{2}{2}\right\}$

37 $\varnothing$

39 $I = \left\{\dfrac{e}{f} \,\middle|\, e, f \in Z, f > 0, e = f\right\}$

Sec. 6-17, page 237

29 (a) =; (b) <; (c) <; (d) >; (e) =; (f) >; (g) =; (h) >; (i) <

35 In Exercises 32 and 34 one counterexample is sufficient to show that the property does not hold for *all* rational numbers. In Exercises 31 and 33 one example where the property does hold is *not* sufficient to show that it holds for *all* rational numbers.

39 No since there is not an additive identity

41 (a) 93; (b) $-\dfrac{180}{7}$ or $-25\frac{5}{7}$; (c) $\dfrac{84}{5}$ or $16\frac{4}{5}$; (d) $\dfrac{455}{2}$ or $277\frac{1}{2}$

45 0.5%; 5%; 33%; $\dfrac{1}{3}$; 0.339; $\dfrac{2}{5}$; 0.401; 0.428; $\dfrac{3}{7}$; 0.429; 0.8; $\dfrac{9}{11}$; 0.08571; $\dfrac{6}{7}$; 0.857

47 $66\frac{2}{3}$%   49 $35.00   51 $798.00
53 70.8%   55 7.2%   57 251.3 gallons
59 Copper, 307.5 pounds; lead, 15 pounds; zinc, 175.5 pounds

Sec. 7-2, page 250

3 Both are rational numbers.
5 (a) 1.7321; (b) 2.6458; (c) 2.8284; (d) 3.8730; (e) 12.2475
7 Yes
15 3.142
17 1,038 miles per hour

Sec. 7-4, page 257

5 (a) 0.7071; (b) 1
7 (a) $0.\overline{857142}$; (b) $0.875\overline{0}$ or $0.874\overline{9}$; (c) $0.\overline{6}$; (d) $7.5\overline{0}$ or $7.4\overline{9}$

13 (a) $-3 < x < 3$; (b) $-\dfrac{1}{2} \le x \le \dfrac{1}{2}$; (c) $-6 \le x < 2$; (d) $-2 \le x \le 4$

Sec. 7-6, page 263

1 $Z \subset Q \subset R \subset K$, where R = set of all real numbers and K = set of all complex numbers.

3 Any irrational number $\in R \subset K$.

5
6: $N,\ Z,\ Q,\ R,\ K$ $-\dfrac{7}{8}$: $Q,\ R,\ K$

$\sqrt{2}$: $I,\ R,\ K$ -1.5: $Q,\ R,\ K$
-4: $Z,\ Q,\ R,\ K$ $\sqrt[3]{4}$: $I,\ R,\ K$

π: $I,\ R,\ K$ $\dfrac{1}{\pi}$: $I,\ R,\ K$

$2i$: K $3 - 4i$: K

7

	Sum	Product
(a)	$5 + 3i$	$4 + 7i$
(b)	$2 + 0 \cdot i$ or 2	$2 + 0 \cdot i$ or 2
(c)	$2 + 5i$	$-4 + 2i$
(d)	$7 - 2i$	$11 - 7i$

(e) $0 - i$ or $-i$ $12 + 0 \cdot i$ or 12

(f) $6 + 3i$ $8 + 12i$

(g) $2\sqrt{2} + 0 \cdot i$ or $2\sqrt{2}$ $5 + 0 \cdot i$ or 5

(h) $1 + \sqrt{3}i$ $-\dfrac{1}{2} + \dfrac{\sqrt{3}}{2}i$

(i) $1 + 0 \cdot i$ or 1 $1 + 0 \cdot i$ or 1

(j) $\left(\dfrac{\sqrt{3}}{2} + 2\right) - \frac{1}{2}i$ $\left(\dfrac{1}{2} + \sqrt{3}\right) + \left(1 - \dfrac{\sqrt{3}}{2}\right)i$

9   (a) $1 + i$; (b) $7 - 9i$; (c) $(\sqrt{2} - 4) + (1 - \sqrt{3})i$; (d) $-12 + i$

11   (See figure.)

$\times$	1	-1	i	$-i$
1	1	-1	i	$-i$
-1	-1	1	$-i$	i
i	i	$-i$	-1	1
$-i$	$-i$	i	1	-1

EXERCISE 11,
SEC. 7-6

Sec. 7-7, page 264

23   (a) 3.3166; (b) 6.2450; (c) 24.9800

25   (a) 0.3015; (b) $3.3166 \times 0.3015 = 0.999955$

27   (a) $-3, -2, -1, 0, 1, 2, 3$; (b) $-1, 0, 1$; (c) 0; (d) 0; (e) none; (f) none;
   (g) 0; (h) 0; (i) $-2, -1, 0, 1, 2$

29   2.744×10^7 tons or 27,440,000 tons

31   120,000

35   (a) $p = 126$, $r = 0.15$; $b = 840$
   (b) $p = 150$, $b = 225$; $r = 0.66\frac{2}{3}$
   (c) $r = 0.12\frac{1}{2} = 0.125$; $b = 648$; $p = 81$

37   \$300 loss or 2.44% loss

39   (a) $|x| < 2$; (b) $|x| \leq 1$; (c) $|x| \leq \dfrac{5}{8}$; (d) $|x| < \dfrac{5}{4}$

41   (a) 50; (b) $28 - 24i$; (c) 1; (d) $5 - 10i$

43

	Additive inverse	*Multiplicative inverse*
(a)	$-5 - 4i$	$\dfrac{5}{41} - \frac{4}{41}i$
(b)	i	$-i$
(c)	$-3 + 2i$	$-\dfrac{3}{13} - \frac{2}{13}i$
(d)	$1 - i$	$\dfrac{1}{2} + \frac{1}{2}i$

Sec. 8-3, page 283

5 Planes M and N have at least line $\overleftrightarrow{PQ}$ in common.
11 No four points coplanar: 10 lines and 10 planes
    Four points coplanar: 10 lines 7 planes

Sec. 8-4, page 291

1 $\angle ABC$ and $\angle CBA$ are the same angle since they have the same vertex and are formed by the same rays; $\angle ABC$ and $\angle BCA$ are not the same angle since they have different vertices and are formed by different rays.
3 (a) The intersection is the half plane on the P side of l.
   (b) The intersection is the portion of the plane between the two lines l and m.
   (c) The intersection is the portion of the plane between the two lines l and m.
11 35, 145, and 145   13 $m\angle A - m\angle C = 45$

Sec. 8-6, page 300

7 Use a straightedge to see whether it can be made to contact each of the points when placed on the plane.
9 Six lines can be drawn; only four if three points are collinear
11 Fresno, California
13 St. Louis and Denver, 1 hour; Dutch Harbor and Camden, 6 hours; Honolulu and Fresno, 2 hours; Honolulu and Denver, 3 hours

Sec. 8-8, page 312

9 (a) Yes; (b) yes; (c) no

Sec. 8-10, page 315

1 $\angle A \cong \angle D$; $\angle B \cong \angle E$; $\angle C \cong \angle F$; $\overline{AB} \cong \overline{DE}$; $\overline{BC} \cong \overline{EF}$; $\overline{CA} \cong \overline{FD}$.
   Definition 8-24.
9 The lines which join midpoints of opposite sides of the hexagon

Sec. 8-13, page 327

39 (a)

	Points	*Lines*	*Planes*
Pyramids	5	10	5
Prism	6	15	5

(b) *Pyramid:* 16; *ABC, ACB, BAC, ACD, ADC, DAC, ADE, AED, DAE, ABE, AEB, BAE, BCD, CDE, DEB, EBC*
  Prism: 18; *PQR, QRP, RPQ, LMN, MNL, NLM, LPR, PRN, RNL, NLP, QRN, RNM, NMQ, MQR, QML, MLP, LPQ, PQM*

(c) *Pyramid:* 8; $E - AB - C$, $B - AC - D$, $C - AD - E$, $B - AE - D$, $A - BE - D$, $A - BC - E$, $A - CD - B$, $A - DE - C$
  Prism: 9; $P - MQ - R$, $Q - LP - R$, $Q - NR - P$, $M - PQ - R$, $L - PR - Q$, $N - QR - P$, $Q - LM - N$, $P - LN - M$, $R - MN - L$

(d) $\angle PQR$ is a plane angle of the dihedral angle P-MQ-R.
(e) 2; (f) 2; (g) They are skew lines; (h) 3; (i) 3

41 (*a*) 15 lines and 20 planes

(*b*) Lines: $\overleftrightarrow{PQ}$, $\overleftrightarrow{PR}$, $\overleftrightarrow{PS}$, $\overleftrightarrow{PT}$, $\overleftrightarrow{PU}$, $\overleftrightarrow{QR}$, $\overleftrightarrow{QS}$, $\overleftrightarrow{QT}$, $\overleftrightarrow{QU}$, $\overleftrightarrow{RS}$, $\overleftrightarrow{RT}$, $\overleftrightarrow{RU}$, $\overleftrightarrow{ST}$, $\overleftrightarrow{SU}$, $\overleftrightarrow{TU}$

Planes: *PQR, PQS, PQT, PQU, PRS, PRT, PRU, PST, PSU, PTU, QRS,*
QRT, QRU, QST, QSU, QTU, RST, RSU, RTU, STU

43 $m \angle A = 125$, $m \angle B = 35$, $m \angle C = 55$

Sec. 9-2, page 337

1 Approximate numbers are numbers such as those which express the results obtained from estimating, measuring, or rounding. Exact numbers are numbers such as those which express the results of counting or of operating on exact numbers. For example, $\frac{2}{3}$ is an exact number, but the decimal 0.667 is an approximation to $\frac{2}{3}$.

3 Exact. The number is used in the definition which gives meaning to the concept of 1 mile.

5 This number could be either, depending upon whether it is the result of an actual count of the number of volumes in the library or not.

7 3.162 is an approximate value of the positive square root of the exact number 10, while 5 is an exact value of the real cube root of the exact number 125.

9 Both numbers are approximate. 100 yards is a measure of distance, and 9.1 seconds is a measure of time.

11 Approximate, since it is the result of a measurement

13 1970 is exact since it is a counting number; 2,781,829 is approximate; although it is a form of counting number, it must represent an estimate.

15 The 2 and $2.29 are exact numbers since they express the result of counting. The 4 pounds 8 ounces, as a weight, is approximate; however, the 4 pounds must be considered as an exact count of the number of pounds in the weight. The unit is the ounce and the weight is 72 ounces, which is approximate. From one point of view $2.29 must be considered an approximate number, since it expresses the result of an operation involving an exact number, the price per pound, and an approximate number, the weight of the chickens.

17 The 8 cents and 48 cents are exact numbers. The 6 ounces expresses a weight, and so the 6 is an approximate number.

Sec. 9-4, page 340

1 Whole units are counted on the scale; fractional parts are estimated.

3 Exact   5 Direct   7 Indirect   9 Indirect

11 Indirect   13 Indirect   15 Direct   17 Indirect

Sec. 9-7, page 347

3 1 in. = 2.540 cm   5 1 grain = 0.00229 ounce

7 25,000,000,000,000 miles

9 0.0000000000000000000000000000091 gram

11 16 yd 7 in.   13 11 bu 5 qt   15 34 years 8 months 14 days

17 2 g 1 dg 6 cg 3 mg   19 (*a*) 7 l 4 dl 1 cl 6 ml; (*b*) 7.416 l

21 30 bu 3 pk 6 qt 1 pt   23 22 days 3 hr 34 min 30 sec

25 2 gal 1 qt $1\frac{2}{5}$ pt   27 1 g 6 dg 5 cg   29 1 to 2

31 1 to 200,000   33 1.65 g   35 0.000005 to 1

Sec. 9-11, page 359

1 All digits except zero when it is used merely for placing the decimal point

3 4 in. means correct to the nearest inch, or 4 in. $\pm \frac{1}{2}$ in.; $4\frac{0}{8}$ in. means correct to the nearest $\frac{1}{8}$ in., or 4 in. $\pm \frac{1}{16}$ in.

15 ft 4 in. means correct to the nearest inch, or 15 ft 4 in. $\pm \frac{1}{2}$ in.; $15\frac{1}{3}$ ft means correct to the nearest $\frac{1}{3}$ ft, or $15\frac{1}{3}$ ft $\pm \frac{1}{6}$ ft.

5 (*a*) 1 in.; (*b*) 1 pk; (*c*) 1 oz; (*d*) 1 sec; (*e*) 1 pk; (*f*) 1 qt; (*g*) 0.00001 in.; (*h*) 0.001 mile; (*i*) 0.1 sec; (*j*) 1,000 miles; (*k*) 1 dollar; (*l*) 0.1 in.

7 (*a*) 123.6 in. is both more precise and more accurate.

(*b*) 360 sec is more precise; they are of the same degree of accuracy.

(*c*) 3 yd is more precise; they are of the same degree of accuracy.

(*d*) 58 ft is more precise; 572 miles is more accurate.

(*e*) 8<u>0</u> oz is more precise; 195 lb is more accurate.

(*f*) 3 qt 1 pt is more precise; 5 gal 3 qt is more accurate.

9 (*a*) $a + b + c$ where a, b, and c each represent the length of one side

(*b*) $3s$ where s represents the common length of the three sides

(*c*) $2l + 2w$ where l represents the length and w the width of the rectangle

(*d*) $4s$ where s represents the common length of the four sides

(*e*) $5s$ where s represents the common length of the five sides

(*f*) $6s$ where s represents the common length of the six sides

11 Perimeter, 225.6 in.; area, 3,180 sq in.

13 Circumference: (*a*) 16 in.; (*b*) 0.842 ft

Area: (*a*) 2<u>0</u> sq in.; (*b*) 0.0564 sq ft

15 4.58 sq in.

17 (*a*) 0.064 sec

(*b*) 6,<u>0</u>00,000,000,000,000,000,000,000,000 grams

(*c*) 25,000,000,000,000 miles

(*d*) 0.0000000000000000000000003 gram

(*e*) 186,000 miles per sec

19 499 sec or 8 min 19 sec

21 Proton: $1.00758 \times 1.660 \times 10^{-24} = 1.673 \times 10^{-24}$ gram

Neutron: $1.00897 \times 1.660 \times 10^{-24} = 1.675 \times 10^{-24}$ gram

Electron: $(5.5 \times 10^{-4}) \times (1.660 \times 10^{-24}) = 9.1 \times 10^{-28}$ gram

23 71 ft

25 75.4 gal

Sec. 9-14, page 374

1 $AB = 5$; $RS = 2\underline{0}$

3 Length of lake is 125 yd

5 $\sin A = \dfrac{20}{29}$, $\cos A = \dfrac{21}{29}$, $\tan A = \dfrac{20}{21}$, $\cot A = \dfrac{21}{20}$

7 $\sin A = \dfrac{9}{41}$, $\cos A = \dfrac{40}{41}$, $\tan A = \dfrac{9}{40}$, $\cot A = \dfrac{40}{9}$

9 They are congruent.

11 It is congruent with $\angle A$ of Exercise 4.

13 (See figure on page 477.)

15 22°37′

Exercise	4	5	6	7	8	10
sin B	4/5	21/29	140/149	40/41	36/39	140/175
cos B	3/5	20/29	51/149	9/41	15/39	105/175
tan B	4/3	21/20	140/51	40/9	36/15	140/105
cot B	3/4	20/21	51/140	9/40	15/36	105/140

EXERCISE 13, SEC. 9-14

17   (a) Not defined for $m \angle A = 90°$; (b) not defined for $m \angle A = 0°$
19   54°; 39°; 22°; 24°   21   3°50′
23   54°15′   25   41°50′
27   51°52′   29   84°42′

Sec. 9-15, page 379

1   $\cos A = \dfrac{24}{25}$, $\tan A = \dfrac{7}{24}$, $\cot A = \dfrac{24}{7}$

For the remaining exercises answers are given only for those cases in which the identity is not true.

3   0°   5   90°   9   0°
11   0° and 90° in the two cases involving cot A and tan A
15   90°   17   0° and 90°   21   90°
23   90°   25   0° and 90°

Sec. 9-16, page 386

1   Mean, 40; median, 40; no mode
3   Mean, 97; median, 96; mode 96

5		*Ex 1*	*Ex 2*	*Ex 3*
	Variance	24	2.39	30.88
	Standard deviation	4.9	1.55	5.56

7   The scores are all 10.

Sec. 9-17, page 386

23   As an exact number it tells the number of cents the buyer actually paid. In this use it is a counting number. The exact price of one can of evaporated milk would be 19½ cents. In this usage the 20 cents is an approximate number.
25   (a) The first measure is correct to the nearest ¼ ft, or the same precision as that implied by $2\frac{3}{4}$ ft. The second measure is correct to the nearest ⅛ ft, or the same precision as that implied by $2\frac{6}{8}$ ft.
    (b) 8.6 implies 8.6 ± 0.05 in.; 8.60 in. implies 8.60 in. ± 0.005 in.
27   (a) Direct; (b) indirect; (c) indirect; (d) indirect
29   (a) 34 hr 9 min 40 sec or 1 d 10 hr 9 min 40 sec
    (b) 24 gal 1 pt
    (c) 9 yd 1 ft
    (d) 19 m 7 dm 2 cm 2 mm

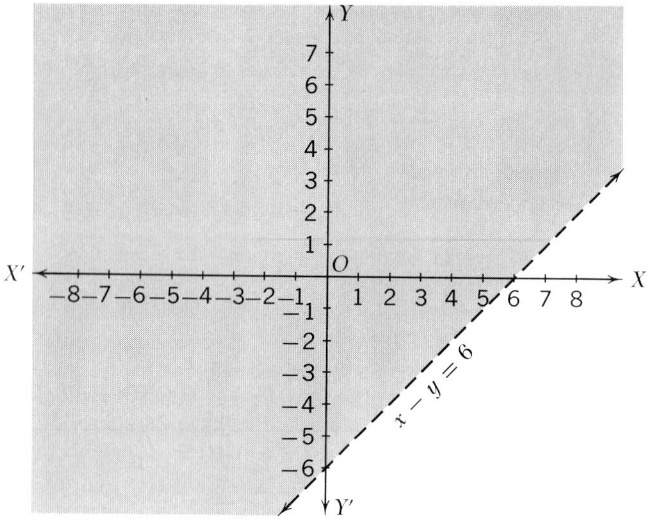

EXERCISE 27, SEC. 10-6

Incomplete graph $x - y < 6$.
(The line is not included.)

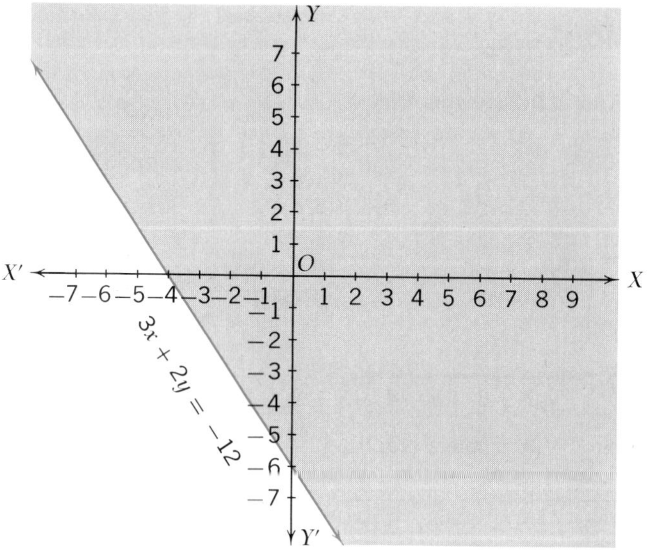

EXERCISE 29, SEC. 10-6

Incomplete graph $3x + 2y \geq -12$.
(The line is included.)

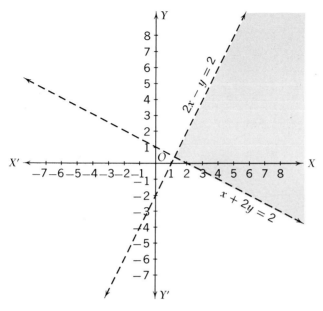

Incomplete graph $x + 2y > 2$, $2x - y > 2$.
(The lines are not included.)

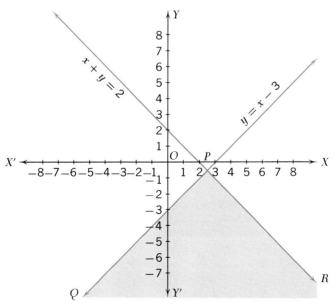

Incomplete graph $x + y \leq 2$, $y \leq x - 3$.
(The rays $\overrightarrow{PQ}$ and $\overrightarrow{PR}$ are included.)

(*c*)

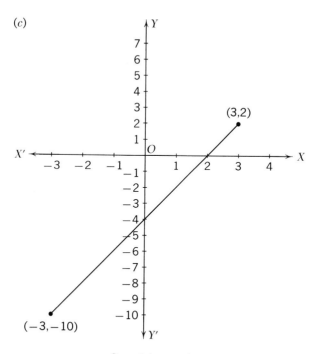

Complete graph.
(Endpoints (3,2), (−3,−10) are included.)

(*d*)

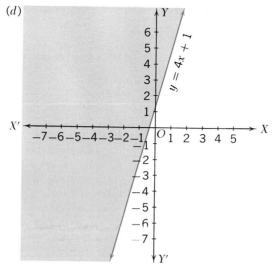

Incomplete graph $y \geq 4x + 1$.
(The line is included.)

39 (a) $x = -1; y = 2;$ (b) inconsistent; (c) $x = 4; y = 0;$ (d) The solution set consists of all ordered pairs satisfying the relation $y = -4x + 9$. (For graphs see the accompanying figures.)

41 (a) $-7 \leq x \leq 7;$ (b) $-1 \leq x \leq 5;$ (c) $-9 < x < -1;$ (d) $0 \leq x \leq 2;$ (e) $-6 < x < 0;$ (f) $4 < x < 14$

43 400 ft   45 80 revolutions per minute

(a)

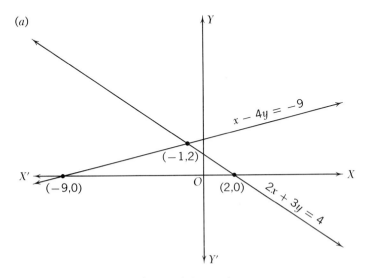

Incomplete graph

(b)

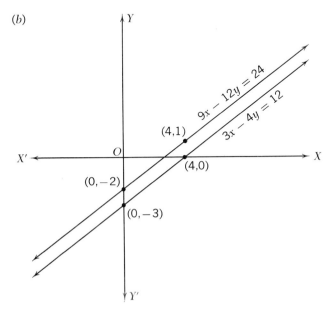

Incomplete graph

(c)

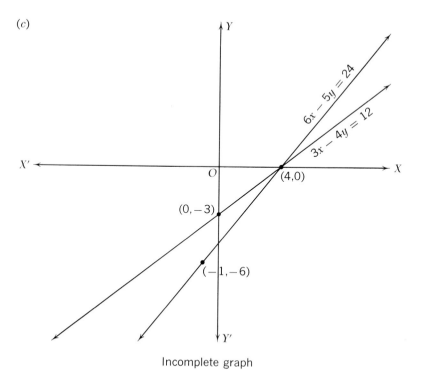

Incomplete graph

(d)

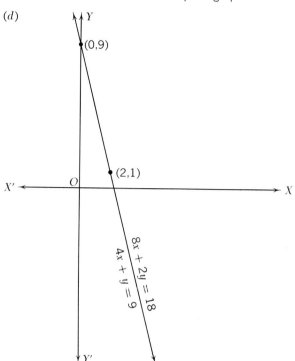

Incomplete graph

INDEX

INDEX